Advanced Free Radical Reactions for Organic Synthesis

Advanced Free Radical Reactions for Organic Synthesis

Hideo Togo

Department of Chemistry
Faculty of Science
Chiba University
Japan

2004

ELSEVIER

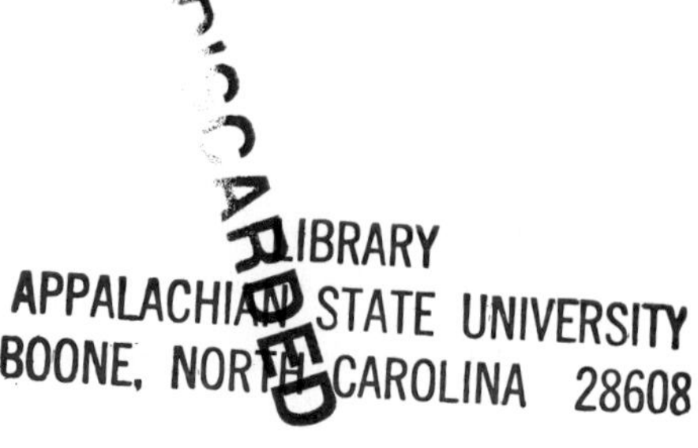

Amsterdam · Boston · Heidelberg · London · New York · Oxford · Paris
San Diego · San Francisco · Singapore · Sydney · Tokyo

ELSEVIER B.V.
Sara Burgerhartstraat 25
P.O.Box 211, 1000 AE
Amsterdam, The Netherlands

ELSEVIER Inc.
525 B Street, Suite 1900
San Diego, CA 92101-4495
USA

ELSEVIER Ltd
The Boulevard, Langford Lane
Kidlington, Oxford OX5 1GB
UK

ELSEVIER Ltd
84 Theobalds Road
London WC1X 8RR
UK

First edition 2004

Library of Congress Cataloging in Publication Data
A catalog record is available from the Library of Congress.

British Library Cataloguing in Publication Data
A catalogue record is available from the British Library.

ISBN: 0-08-044374-5

♾ The paper used in this publication meets the requirements of ANSI/NISO Z39.48-1992 (Permanence of Paper).
Printed in Hungary.

Dedicated to my family and parents,
and the late Professor Derek H. R. Barton

Preface

This book covers the fundamental properties of organic free radicals and their synthetic uses. It consists of twelve chapters, starting from fundamentals and physical properties of organic free radicals, reduction and functional group conversion, cyclization, addition, alkylation onto aromatics, Barton reaction and related reactions, Barton-McCombie reaction, Barton decarboxylation, free radical reaction with metal hydrides, stereo-selective free radical reactions, free radicals in biology, and free radicals for green chemistry. The important factors in these free radical reactions are some radical specific reactions, as mentioned in each chapter. Since the basic study on free radical reactions has been established by Barton, Ingold, Stork, Beckwith, Giese, *etc*., free radical reactions have become an increasingly important and attractive tool in organic synthesis, especially in the last two decades. Recently, in addition to a typical but toxic radical reagent, *i.e.* tributyltin hydride, much less toxic but more effective radical reagents such as tris(trimethylsilyl)silane, 1,1,2,2-tetraphenyldisilane, samarium (II) iodide, indium, *N*-acyloxy-2-thiopyridone, triethylborane, *etc*. have been developed. The author hopes that the free radical reactions will be widely applied to the synthesis of biologically attractive compounds with high chemoselectivity and stereoselectivity, and green chemistry, based on the advantages of free radicals.

Finally, I would like to thank Dr. Adrian Shell and Mr. Derek Coleman in Elsevier Ltd.

Hideo Togo
Aug., 2003, Chiba, Japan

Contents

List of Abbreviations

Chemicals

AIBN	α,α'-azobis(isobutyronitrile)
CAN	cerium(IV) ammonium nitrate
DIBAL	diisobutylaluminium hydride
DMAP	4-(dimethylamino)pyridine
DMSO	dimethyl sulfoxide
HMPA	hexamethylphosphoramide
LAH	lithium aluminum hydride
LDA	lithium diisopropylamide
*m*CPBA	*m*-chloroperoxybenzoic acid
NBS	*N*-bromosuccinimide
NCS	*N*-chlorosuccinimide
$O_2^{-\cdot}$	superoxide anion radical
PCC	pyridinium chlorochromate
PDC	pyridinium dichromate
Py	pyridine
TBAF	tetrabutylammonium fluoride
TEMPO	2,2,6,6-tetramethyl-1-piperidinyloxy free radical

Protecting groups

Ac	acetyl
Ar	aryl
Bn	benzyl
Boc	*tert*-butoxycarbonyl
Bz	benzoyl
Cbz	benzyloxycarbonyl
Et	ethyl
i-Bu	isobutyl
i-Pr	isopropyl
Me	methyl
n-Bu	*n*-butyl
n-Pr	*n*-propyl
Ph	phenyl
s-Bu	*sec*-butyl
t-Bu	*tert*-butyl
TBDPS	*tert*-butyldiphenylsilyl
TBS	*tert*-butyldimethylsilyl
Tf	trifluoromethanesulfonyl
TMS	trimethylsilyl

Tr	triphenylmethyl
Ts	*p*-toluenesulfonyl

Symbols

Δ	refluxing conditions
Hg-hν	irradiation with a mercury lamp
Ea	activation energy
k	rate constant
r.t.	room temperature
W-hν	irradiation with a tungsten lamp

– 1 –

What are Free Radicals?

1.1 GENERAL ASPECTS OF FREE RADICALS

1.1.1 Aspects of free radicals

Generally, molecules bear bonding electron pairs and lone pairs (a non-bonding electron pair or unshared electron pair). Each bonding or non-bonding electron pair has two electrons, which are in opposite spin orientation, $+1/2$ and $-1/2$, in one orbital based on Pauli's exclusion principle, whereas an unpaired electron is a single electron, alone in one orbital. A molecule that has an unpaired electron is called a free radical and is a paramagnetic species.

Three reactive species, a methyl anion, methyl cation, and methyl radical, are shown in Figure 1.1. Ethane is composed of two methyl groups connected by a covalent bond and is a very stable compound. The methyl anion and methyl cation have an ionic bond mainly between carbons and counter ions, respectively, and are not particularly unstable, though there are some rather moisture-sensitive species. However, the methyl radical is an extremely unstable and reactive species, because its octet rule on the carbon is not filled. The carbon atom in the methyl cation adopts sp^2 hybridization and the structure is triangular (120°) and planar. The carbon atom in the methyl anion adopts sp^3 hybridization and the structure is tetrahedral (109.5°). However, the carbon atom in the methyl radical adopts a middle structure between the methyl cation and the methyl anion, and its pyramidal inversion rapidly occurs as shown in Figure 1.1, even at extremely low temperature.

From the above, it is apparent that free radicals are unique and rare species, and are present only under special and limited conditions. However, some of the free radicals are familiar to us in our lives. Thus, molecular oxygen is a typical free radical, a biradical species. Standard and stable molecular oxygen is in triplet state ($^3O_2$), and the two unpaired electrons have the same spin orientation in two orbitals (parallel), respectively, having the same orbital energy, based on Hund's rule. Nitrogen monoxide and nitrogen dioxide are also stable, free radical species. Moreover, the reactive species involved in immunity are oxygen free radicals, such as superoxide anion radical ($O_2^{-\bullet}$) and singlet molecular oxygen ($^1O_2$). So, free radicals are very familiar to us in our lives and are very important chemicals.

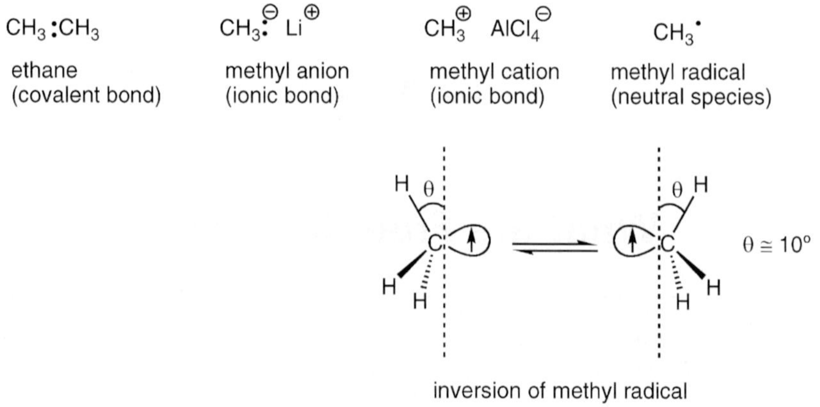

$CH_3{:}CH_3$	$CH_3{:}^{\ominus}\ Li^{\oplus}$	$CH_3^{\oplus}\ AlCl_4^{\ominus}$	$CH_3^{\bullet}$
ethane (covalent bond)	methyl anion (ionic bond)	methyl cation (ionic bond)	methyl radical (neutral species)

$\theta \cong 10°$

inversion of methyl radical

Figure 1.1

Historically, the triphenylmethyl radical (**1**), studied by Gomberg in 1987, is the first organic free radical. The triphenylmethyl radical can be obtained by the reaction of triphenylmethyl halide with metal Ag as shown in eq. 1.1. This radical (**1**) and the dimerized compound (**2**) are in a state of equilibrium. Free radical (**1**) is observed by electron spin resonance (ESR) and its spectrum shows beautiful hyperfine spin couplings. The spin density in each carbon atom can be obtained by the analysis of these hyperfine spin coupling constants as well as information on the structure of the free radical.

$$Ph_3CX + Ag \xrightarrow[(-AgX)]{} \underset{\mathbf{1}}{Ph_3C^{\bullet}} \rightleftharpoons \frac{1}{2}\ \mathbf{2} \qquad (1.1)$$

The structure of dimer (**2**) was characterized by NMR. Thus, one triphenylmethyl radical reacts at the *para*-position of a phenyl group in another triphenylmethyl radical, not the central sp^3 carbon (to form hexaphenylethane), to form dimer (**2**). However, *tris*(*p*-methylphenyl)methyl radical does not dimerize. So, the electronic effect in free radicals is quite large.

Molecular oxygen and nitrogen monoxide are specifically stable free radicals. However, in general radicals are reactive species, and radical coupling reaction, oligomerization, polymerization, etc. occur rapidly, and their control is not so easy. This is one of the main reasons why most organic chemists do not like radical reactions for organic synthesis. However, mild and excellent free radical reactions have recently been established. Here, the fundamentals of organic free radicals, such as the kinds of radicals, reaction styles of radicals, etc. will be introduced.

1.1.2 Types of free radicals

Most organic radicals are quite unstable and very reactive. There are two kinds of radicals, neutral radicals and charged radicals as shown below, i.e. a neutral radical (**3**), a cation radical (**4**) and an anion radical (**5**) (Figure 1.2).

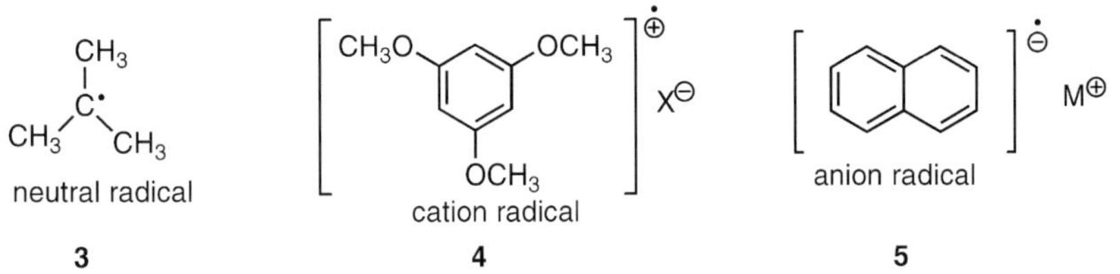

Figure 1.2

Moreover, there are two types of radicals, the σ radicals and the π radicals. An unpaired electron in the σ-radical is in the σ orbital, and an unpaired electron in the π radical is in the π orbital, respectively. Therefore, the radicals (**4**) and (**5**) above are π radicals. *t*-Butyl radical (**3**) is also π radical, since this radical is stabilized by the hyperconjugation. However, the phenyl radical and the vinyl radical are typical σ radicals. Normally, π radicals are stabilized by the hyperconjugation effect or the resonance effect. However, σ radicals are very reactive because there is no such stabilizing effect (Figure 1.3).

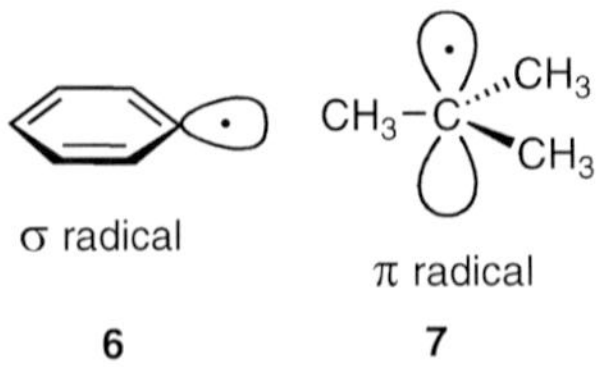

Figure 1.3

This result can be explained by the following fact. The bond dissociation energies of the C–H bond in $(CH_3)_3C$–H (isobutane) and C_6H_5–H (benzene) are ~91 kcal/mol and ~112 kcal/mol, respectively. So, the bond dissociation energy of the C–H bond in benzene is 21 kcal/mol stronger than that in isobutane. This suggests that the phenyl radical is more unstable by about 21 kcal/mol than the *t*-butyl radical, and therefore should be more reactive.

1.1.3 Reaction styles of radicals

In polar reactions, heterolytic (unsymmetrical) bond cleavage (heterolysis) and bond formation occur, while homolytic (symmetrical) bond cleavage (homolysis) and bond formation occur in radical reactions as shown below (Scheme a).

Typical radical reactions are substitution and addition reactions as shown below (Scheme b). A typical substitution reaction is the halogenation of methane with chlorine gas under photolytic conditions, and generally available chlorohydrocarbons are prepared by this method. The chlorination reaction proceeds through a chain pathway via the initiation step, propagation step, and termination step as shown below (Scheme 1.1).

The driving force of this reaction is the heat of the formation, namely, the difference in the bond dissociation energies of the products and the starting materials. Thus, the bond

Polar reaction

$RX \longrightarrow R^{\oplus} + X^{\ominus}$ Heterolytic bond cleavage

$R^{\oplus} + X^{\ominus} \longrightarrow RX$ Heterolytic bond formation

Radical reaction

$RX \longrightarrow R^{\bullet} + X^{\bullet}$ Homolytic bond cleavage

$R^{\bullet} + X^{\bullet} \longrightarrow RX$ Homolytic bond formation

Scheme a

Intermolecular radical substitution reaction (RH $\rightarrow$ RCl)

$RH + Cl^{\bullet} \longrightarrow R^{\bullet} + HCl$

$R^{\bullet} + Cl_2 \longrightarrow RCl + Cl^{\bullet}$

Intermolecular radical addition reaction (C-C bond formation)

$RX + {\bullet}M \longrightarrow R^{\bullet} + MX$

$R^{\bullet} + CH_2{=}CH(CN) \longrightarrow RCH_2\dot{C}HCN$

Scheme b

Cl_2	$\xrightarrow{h\nu}$	$2\,Cl^{\bullet}$	Initiation step
$CH_4 + Cl^{\bullet}$	$\longrightarrow$	$CH_3^{\bullet} + HCl$	Propagation step
$CH_3^{\bullet} + Cl_2$	$\longrightarrow$	$CH_3Cl + Cl^{\bullet}$	
$2\,Cl^{\bullet}$	$\longrightarrow$	Cl_2	Termination step
$2\,CH_3^{\bullet}$	$\longrightarrow$	CH_3CH_3	
$CH_3^{\bullet} + Cl^{\bullet}$	$\longrightarrow$	CH_3Cl	

Scheme 1.1

dissociation energies of Cl–Cl (molecular chlorine) and CH_3–H (methane) are 58 kcal/mol and 104 kcal/mol, respectively, and 162 kcal/mol in total (starting materials), while those of H–Cl (hydrogen chloride) and CH_3–Cl (methyl chloride) are 103 kcal/mol and 84 kcal/mol, respectively, and 187 kcal/mol in total (products). Therefore, the products are in total 25 kcal/mol more stable than the starting materials (exothermal), and this difference is the driving force of the reaction. The formation of methyl chloride in this reaction is a substitution reaction; one hydrogen atom of methane is substituted by one chlorine atom, through a homolytic pathway. Therefore, this type of reaction is called the S_H2 (Substitution Homolytic Bimolecular) reaction and is the fundamental reaction style in radical reactions. This reaction proceeds through a chain pathway, via an initiation step, propagation step, and termination step.

When molecular bromine or molecular iodine is used instead of molecular chlorine in this reaction, the chain reaction does not proceed effectively. The bond dissociation energies of Br–Br and I–I are 46 and 36 kcal/mol in the starting materials, and those of CH_3–Br, CH_3–I, H–Br, and H–I in the products are 70, 56, 88, and 71 kcal/mol, respectively. Thus, the difference in the bond dissociation energies between the starting materials and the products in these reactions tends to be small. Especially, iodination does not proceed at all. Therefore, the considerable difference in bond dissociation energies between the starting materials and the products is the driving force of radical reactions.

1.1.4 Orientation in radical additions

The addition reactions of HBr to isobutene in a polar reaction and in a radical reaction, respectively, are shown below in Scheme 1.2, and opposite orientation is observed.

In the polar reaction, a proton in HBr first adds to the terminal sp^2 carbon in isobutene to produce a stable *tert*-butyl cation (**8**), and then it reacts with the counter bromide anion to form *tert*-butyl bromide. Thus, the proton in HBr adds to the less substituted sp^2 carbon in alkene to produce a more stable carbocation. This is based on the Markovnikov rule. In radical reactions, the hydrogen atom of HBr is abstracted first by the initiator, $PhCO_2^{\cdot}$ (or $Ph^{\cdot}$) derived from $(PhCO_2)_2$, and the formed bromine atom then adds to the terminal sp^2 carbon in isobutene to form the stable β-bromo *tert*-butyl radical (**9**), and then it reacts with HBr to produce *iso*-butyl bromide and a bromine atom. This bromine atom again

Polar reaction $(CH_3)_2C{=}CH_2 \xrightarrow{HBr} (CH_3)_2\overset{\oplus}{C}{-}CH_3 \;\; Br^{\ominus}$ (**8**) $\longrightarrow CH_3{-}C(CH_3)_2{-}Br$

Radical reaction

$(PhCO_2)_2 \longrightarrow 2\,PhCO_2^{\cdot}$

$HBr + PhCO_2^{\cdot} \longrightarrow Br^{\cdot} + PhCO_2H$

} Initiation step

$(CH_3)_2C{=}CH_2 + Br^{\cdot} \longrightarrow (CH_3)_2\dot{C}{-}CH_2Br$ (**9**)

9 $+ HBr \longrightarrow (CH_3)_2CH{-}CH_2Br + Br^{\cdot}$

} Propagation step

$CH_3{-}C(CH_3)(H){-}CH_2^{\oplus}$ **8′** $CH_3{-}C(CH_3)(Br){-}CH_2^{\cdot}$ **9′**

Scheme 1.2

adds to the terminal sp^2 carbon in isobutene, and the chain reaction occurs. So, the *anti*-Markovnikov addition product is obtained in a radical reaction, and, consequently, the opposite addition-orientation products are obtained in a polar reaction and in a radical reaction, respectively. However, it is an important fact that both the polar reaction and the radical reaction do not produce unstable intermediates (**8′**: primary carbocation) and (**9′**: primary carbon-centered radical), respectively; instead, they produce the more stable intermediates (**8**) and (**9**).

Why are intermediates (**8**) and (**9**) more stable than intermediates (**8′**) and (**9′**)? This can be explained by the inductive effect (I effect) and the hyperconjugation effect. The methyl group has an electron donation ability through the σ bond. So, the *tert*-butyl cation and the *tert*-butyl radical can be stabilized by the inductive effect of the methyl group (Figure 1.4). Normally, the inductive effect is increased in the following order:

$CH_3^{\bullet}$ < *prim.*-R$^{\bullet}$ < *sec.*-R$^{\bullet}$ < *tert.*-R$^{\bullet}$ (R = alkyl group)

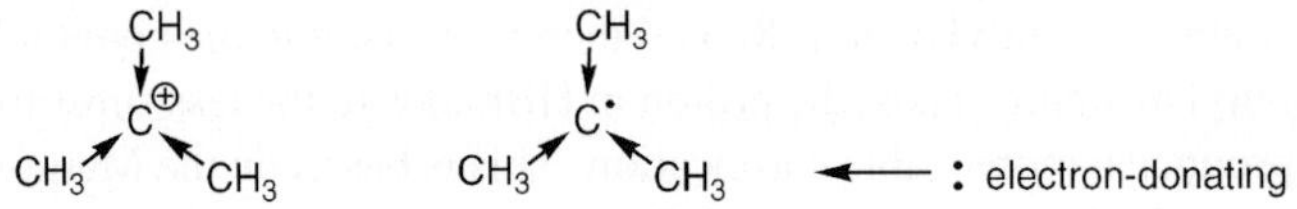

←: electron-donating

Figure 1.4 Inductive effect in *tert*-butyl cation and *tert*-butyl radical

Another effect is the hyperconjugation effect, which comes from the following resonance (Figure 1.5).

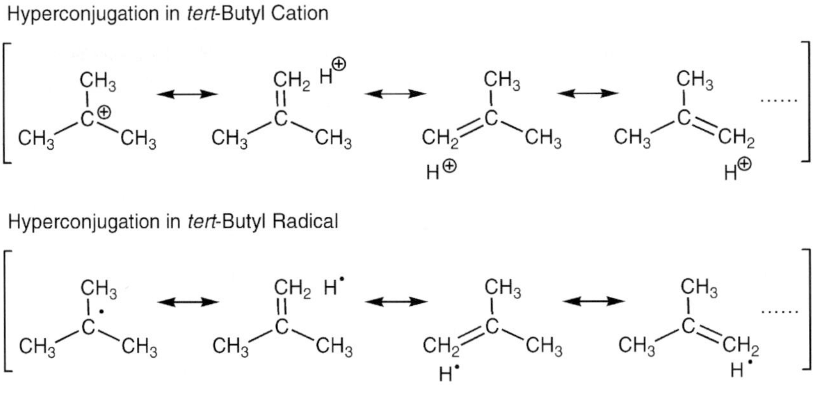

Hyperconjugation in *tert*-Butyl Cation

Hyperconjugation in *tert*-Butyl Radical

Figure 1.5

The inductive effect depends on the electronegativity of atoms and functional groups, and works through the σ bond. Hyperconjugation is like the resonance above (Figure 1.5) and is the orbital interaction between the cation-centered p_π orbital and the C–H σ bond in methyl groups, and the interaction between the radical-centered p_π orbital and the C–H σ bond in methyl groups. Thus, hyperconjugation is the orbital interaction between the central p_π orbital and the C–H σ bond at the β position and is called σ-p_π orbital interaction as shown in Figure 1.6.

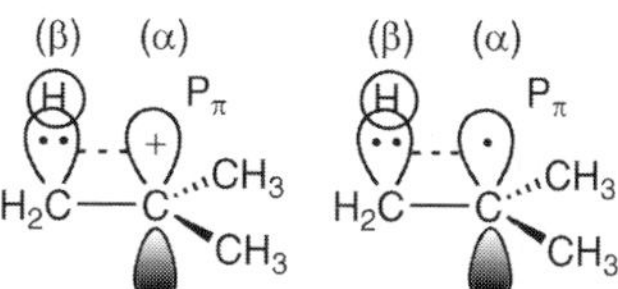

Figure 1.6 σ-p_{π} Orbital interaction in hyperconjugation

1.1.5 Reactivity in radical additions

In polar reactions, there are negatively charged nucleophilic species and positively charged electrophilic species. On the other hand, the radical species are mainly neutral. However, these neutral radical species can be also divided into two types, nucleophilic radical species and electrophilic radical species. These electronic characters come from the spin energy level of the radical species. Thus, electron density of the *tert*-butyl radical is moderately high due to the inductive effect of its three methyl groups, and the spin energy level in singly occupied molecular orbital (SOMO) is high. Therefore, when the *tert*-butyl radical is treated with olefins, it behaves as a nucleophilic radical. So, π-deficient olefins such as acrylonitrile or ethyl acrylate are much more reactive than π-excess olefins such as ethyl vinyl ether, to give the corresponding C–C bond formation products (eqs. a, b in Scheme 1.3). The electron density of the diethyl malonyl radical is rather low due to the resonance effect by two ester groups. Thus, the diethyl malonyl radical is stabilized, and the spin energy level in SOMO is low. Therefore, when the diethyl malonyl radical is treated with olefins, it behaves as an electrophilic radical. So, π-excess olefins are much more reactive than π-deficient olefins in reaction with the diethyl malonyl radical, to give the corresponding C–C bond formation products (eqs. c, d in Scheme 1.3).

a) $(CH_3)_3C^{\bullet} + CH_2{=}CH{-}CN \xrightarrow{\text{rapid}} (CH_3)_3C{-}CH_2\dot{C}H{-}CN$

b) $(CH_3)_3C^{\bullet} + CH_2{=}CH{-}OC_2H_5 \longrightarrow$ no reaction

c) $(H_5C_2O_2C)_2CH^{\bullet} + CH_2{=}CH{-}CN \longrightarrow$ no reaction

d) $(H_5C_2O_2C)_2CH^{\bullet} + CH_2{=}CH{-}OC_2H_5 \xrightarrow{\text{rapid}} (H_5C_2O_2C)_2CH{-}CH_2\dot{C}H{-}OC_2H_5$

Scheme 1.3

1.1.6 Reaction patterns of radicals

There are three types of typical radical reactions, in addition to the addition reactions mentioned above in Scheme 1.3, as follows:

β-Cleavage reaction

The most typical β-cleavage reaction is the decarboxylation of an acyloxyl radical ($RCO_2^{\cdot}$, oxygen-centered radical) to form an alkyl radical and CO_2. These reactions are observed in the Kolbe electrolytic oxidation and the Hunsdiecker reaction, as shown in eq. a of Scheme 1.4. The driving force of this β-cleavage reaction is the formation of stable CO_2 gas, and the formation of a more stable alkyl radical (carbon-centered radical) than the oxygen-centered radical. Alkoxyl radicals, especially *tert*-alkoxyl radicals, induce a β-cleavage reaction to generate the alkyl radicals and stable ketones. For example, the *tert*-butoxyl radical readily gives rise to β-cleavage to give a methyl radical and acetone (eq. b). Generally, the β-cleavage reaction does not occur in alkyl radicals; however, strained carbon-centered radicals, such as cyclopropylmethyl radical and cyclobutylmethyl radical rapidly induce the β-cleavage reaction to give 3-buten-1-yl (eq. c) and 4-penten-1-yl radicals respectively.

a) $RCO_2^{\cdot}$ —β-cleavage→ $R^{\cdot} + CO_2$

b) $(CH_3)_3C{-}O^{\cdot}$ —β-cleavage→ $CH_3^{\cdot} + CH_3C(=O)CH_3$

c) cyclopropyl–$CH_2^{\cdot}$ —β-cleavage→ $^{\cdot}CH_2CH_2CH{=}CH_2$

Scheme 1.4

Cyclization reaction

A typical cyclization reaction example is the cyclization of the 5-hexen-1-yl radical, which cyclizes to give a cyclopentylmethyl radical (primary alkyl radical) and a cyclohexyl radical (secondary alkyl radical), as shown in eq. 1.2. Generally, the radical cyclization proceeds via a kinetically controlled pathway, so the less stable cyclopentylmethyl radical is formed predominantly.

5-hexen-1-yl radical —cyclization→ cyclopentylmethyl radical + cyclohexyl radical (1.2)

Hydrogen atom abstraction via 6 (7)-membered transition state

An oxygen- or nitrogen-centered radical abstracts an inert hydrogen atom at the 5- or 6-position via a 6- (1,5-H shift) or 7-membered transition state (1,6-H shift) to form a carbon-centered radical as shown in eq. 1.3. The driving force of this reaction is the formation of a strong O–H or N–H bond. This is really a radical specific reaction. In an oxygen-centered radical, i.e. an alkoxyl radical, the reaction is called the Barton reaction. In a nitrogen-centered radical, i.e. an aminium radical, the reaction is called the Hofmann–Löffler–Freytag reaction.

Tetrahydrofuran, tetrahydropyran, pyrrolidine, and piperidine skeletons can be constructed by these reactions.

X• H R (n) ⟶ XH • R (n) $n = 1$ 1,5-H shift; $n = 2$ 1,6-H shift (1.3)

X = O, NR

1.1.7 Generation of radicals

Typical generation methods of radicals are mentioned below.

Thermolysis of peroxides or azo compounds

The formation of oxygen- and carbon-centered radicals by the thermolysis of peroxides or azo compounds is well known. Today, these compounds have been also used as radical initiators. For example, treatment of a CCl_4 solution of toluene and *N*-bromosuccinimide (NBS) in the presence of a catalytic amount of benzoyl peroxide in refluxing conditions gives benzyl bromide in good yield as shown in Scheme 1.5. This is called the Wohl–Ziegler reaction.

Refluxing treatment of a mixture of cyclohexyl bromide and Bu_3SnH in the presence of a catalytic amount of 2,2′-azobis (isobutyronitrile) (AIBN) in benzene produces cyclohexane in good yield as shown in Scheme 1.6. The Bu_3SnH/AIBN system is the most popular radical reaction system in organic synthesis.

Decarboxylation of carboxylic acids

The Kolbe and Hunsdiecker reactions are popular, but are now old radical decarboxylation reactions of carboxylic acids. The Barton radical decarboxylation with *N*-acyl ester of *N*-hydroxy-2-thiopyridone is the best and most useful for organic synthesis. The driving force of the Barton radical decarboxylation is the weak N–O bond of the starting Barton ester (**10**) and the formation of highly stable CO_2. Therefore, the generation of carbon-centered radicals and their synthetic use can be carried out readily by heating the solution at 80 °C or irradiating it with a tungsten lamp (W–hν) at room

Scheme 1.5

Scheme 1.6

temperatures as shown in eq. 1.4.

$$CH_3-\overset{O}{\overset{\|}{C}}-O-N(\text{Barton ester } \mathbf{10}) \xrightarrow{W-h\nu} CH_3^{\bullet} + CO_2 + \text{2-pyridylthiyl}^{\bullet} \longrightarrow CH_3-S-\text{2-pyridyl} \quad (1.4)$$

Barton ester **10**

Photochemical reaction of carbonyl groups

Irradiation of ketones or aldehydes with a UV lamp induces electron transition from the highest occupied molecular orbital (HOMO) to the lowest unoccupied molecular orbital (LUMO). Here, the lone pair (n) orbital on the carbonyl oxygen atom to the $\pi^*_{C=O}$ oribital, namely, n$-\pi^*$ electron transition, generates a biradical. The n- and π^*-orbitals are perpendicular, and so n$-\pi^*$ electron transition is not favorable although it is not impossible. After the generation of the biradical, there are two reaction pathways, Norrisch I and Norrisch II, as shown in Figure 1.7.

In the presence of a moderate hydrogen donor such as isopropanol, the oxygen-centered radical of the biradical abstracts a hydrogen atom from the α-position of isopropanol to give pinacole. For example, the benzophenone biradical, generated from the irradiation of benzophenone, abstracts a hydrogen atom from isopropanol to form an α,α-diphenyl-α-hydroxymethyl radical, which is then coupled to give benzopinacol (**12**) (eq. 1.5).

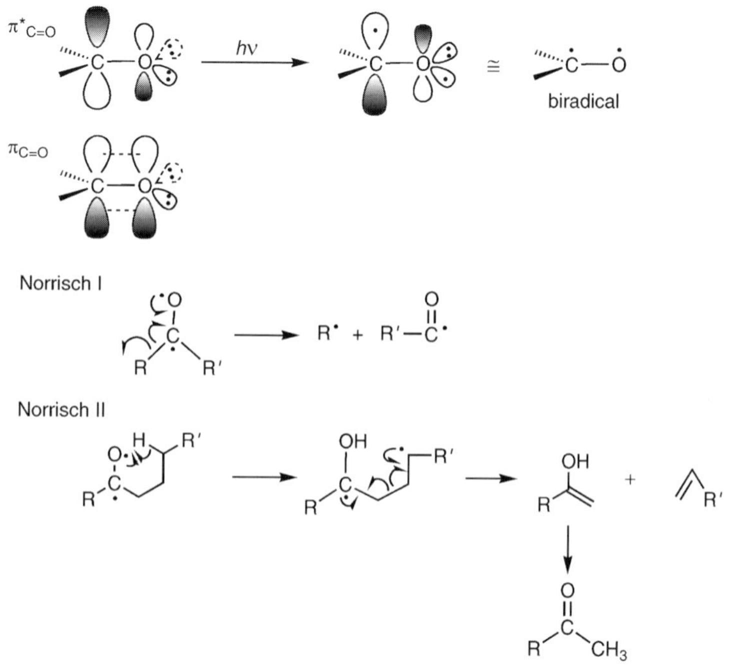

Figure 1.7

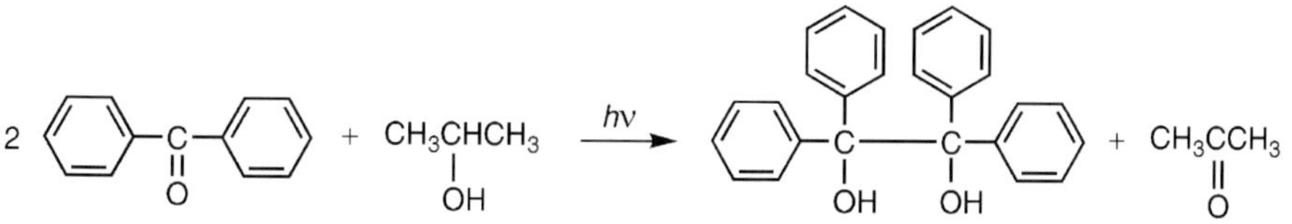

(1.5)

Oxidative conditions

Single-electron oxidants such as Mn^{3+}, Cu^{2+}, and Fe^{3+} abstract one electron from the substrates to produce carbon-centered radicals, as shown in eq. 1.6.

$$\mathbf{13a} \rightleftharpoons \mathbf{13b} \xrightarrow[\text{AcOH}]{\text{Mn(OAc)}_3} CH_3COĊHCOCH_3 + H^{\oplus} \quad (1.6)$$

Fe^{2+} with hydrogen peroxide is called the Fenton system. The first step in this reaction is the electron transfer from Fe^{2+} to hydrogen peroxide to produce extremely reactive $HO^{\cdot}$ (hydroxyl radical) and HO^{-} (hydroxide anion). Once $HO^{\cdot}$ is formed, it rapidly abstracts a hydrogen atom from the substrates to generate carbon-centered radicals.

Reductive conditions

Single-electron reductants such as Fe^{2+}, Cu^{+}, Ti^{3+}, and Sm^{2+} give one electron to the substrates to form carbon-centered radicals, as shown in eq. 1.7.

$$\underset{\mathbf{14}}{CH_3COCH_3} \xrightarrow[\text{THF}]{\text{SmI}_2} (CH_3)_2\dot{C}OSmI_2 \longrightarrow \frac{1}{2}\underset{\mathbf{15}}{CH_3\text{-}C(CH_3)(OH)\text{-}C(CH_3)(OH)\text{-}CH_3} \quad (1.7)$$

These radicals formed are formally neutral and, therefore, the solvent effect is smaller than that in polar reactions. The driving force of these radical reactions is the difference in bond dissociation energy between the starting materials and the products. Therefore, carbonyl, ester, amino, and hydroxy groups, bearing strong bond dissociation energy, are not affected by the radical reactions. This suggests that sugars, nucleosides, and peptides can be used in radical reactions, without the requirement of serious protection of those functional groups.

1.2 FAMILIAR AND CLOSE RADICALS IN OUR LIVES

The closest and most familiar radical is molecular oxygen. Molecular oxygen is a biradical and, therefore, it can be transported to all parts of the body through the binding and dissociation onto the heme part of hemoglobin through breathing. Molecular oxygen is a biradical and each spin orientation is the same (parallel, triplet state) based on Hund's rule, as shown in Figure 1.8 (left), and this molecular oxygen is shown as $^3O_2$. Nitrogen monoxide and nitrogen dioxide are also radicals. Active oxygen radicals related to immunity and cancer induction in living bodies are singlet molecular oxygen ($^1O_2$) and superoxide anion radical ($O_2^{\cdot-}$) as shown in the middle and the right of Figure 1.8. $^1O_2$ is unstable and much more reactive than $^3O_2$, because each spin orientation is opposite. Electronegativity of the oxygen atom is high and so molecular oxygen can be easily reduced to $O_2^{\cdot-}$. It is also a reactive oxygen radical, and a really reactive and important species in immunity reactions.

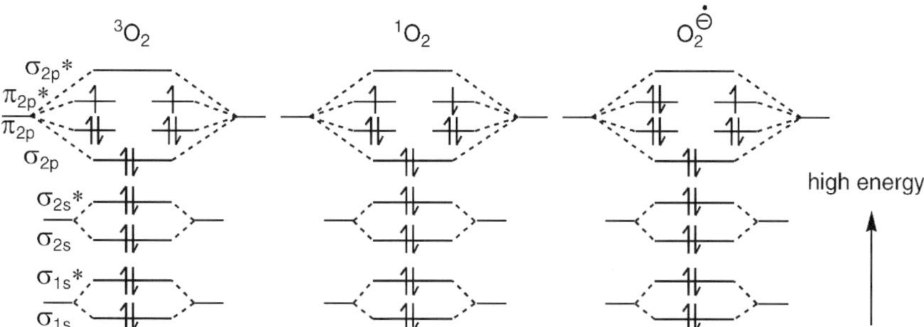

Figure 1.8 Electron configuration of molecular oxygens and related radicals.

$^1O_2$ and $O_2^{\bullet -}$ are important radical species for the maintenance of health in living bodies. However, these radical species induce disease when they are formed in stages where they are not required. For example, when $O_2^{\bullet -}$ is formed in healthy fatty membranes, which consist of unsaturated fatty acids such as arachidonic acid (**16**), it abstracts an allylic hydrogen atom of the unsaturated fatty acids and oxidizes it to a hydroxy group and, finally, the functional ability of the fatty membrane is lost as shown in Scheme 1.7. $O_2^{\bullet -}$ also abstracts a hydrogen atom from peptides, DNA, and RNA, giving rise to their C–C, C–O, and C–N bond cleavages. This is one major cause of inflammation, ageing, cancer, etc. [1, 2].

arachidonic acid **16** → ($O_2^{\bullet\ominus}$) **17** + $HOO^\bullet$ → ... → ($Fe^{2\oplus}$ ($Cu^{\oplus}$), ($-HO^{\ominus}$)) → ... (**16** → **17**)

16 $= CH_3(CH_2)_4(CH{=}CHCH_2)_4CH_2CH_2CO_2H$

Scheme 1.7

How can we keep our health against these reactive oxygen radicals? Fortunately, vitamin C (hydrophilic), vitamin E (hydrophobic), flavonoids, and other polyphenols can function as anti-oxidants. These anti-oxidants are phenol derivatives. Phenol is a good hydrogen donor to trap the radical species and inhibits radical chain reactions. The formed phenoxyl radical is actually stabilized by the resonance effect as shown in eq. 1.8. Thus, phenol and polyphenol derivatives are excellent hydrogen donors to inhibit the radical reactions and, therefore, they are called radical inhibitors.

PhOH → ($O_2^{\bullet\ominus}$, ($-HOO^{\ominus}$)) → [phenoxyl radical resonance structures] (1.8)

For example, when $O_2^{\cdot-}$ is formed in the hydrophilic stage, vitamin C (**18**, L-ascorbic acid; present in hydrophilic stage) assists the hydrogen atoms to form dehydroascorbic acid (**19**) via monodehydroascorbic acid, and hydrogen peroxide (eq. 1.9).

vitamin C **18** $\xrightarrow[(-HOO^{\ominus})]{O_2^{\cdot\ominus}}$ $\xrightarrow[(-HOO^{\ominus})]{O_2^{\cdot\ominus}}$ **19** (1.9)

Moreover, when $O_2^{\cdot-}$ is formed in the hydrophobic stage, vitamin E (**20**, tocopherol) creates a hydrogen atom. The hydrogen peroxide formed is decomposed to water and molecular oxygen catalyzed by catalase enzyme (protein containing Fe-complex), and the oxidized vitamin E radical is reduced to vitamin E again by vitamin C (eq. 1.10)

vitamin E **20** $\underset{\text{L-ascorbic acid}}{\overset{O_2^{\cdot\ominus}}{\rightleftharpoons}}$ **21** + H_2O_2 (1.10)

$R_1, R_2 = H, CH_3$

Concretely, these anti-oxidants prevent higher unsaturated fatty acids such as linolic acid $[CH_3(CH_2)_4CH{=}CHCH_2CH = CH(CH_2)_7COOH]$ and arachidonic acid $[CH_3(CH_2)_4(CH{=}CHCH_2)_4(CH_2)_2COOH]$, which constitutes the cell membrane, from oxidation by active oxygen radicals. Thus, vitamin E and vitamin C protect living bodies against oxidation by active oxygen radicals. Oxidized vitamin E in living bodies is regenerated by reduction with vitamin C. However, oxidized vitamin C cannot be regenerated, and so vitamin C must be supplied constantly in living bodies.

Typical flavonol, anthocyanidine (anthocyanin is a sugar-binding anthocyanidine), catechin, uric acid, and tannin are shown in Figure 1.9. All these compounds bear phenolic hydroxy groups which can function as anti-oxidants [3, 4]. Green tea contains high levels of tannin and catechin, and red wine contains a high level of anthocyanidine.

Based on these results, 2,6-di-*tert*-butyl-4-methylphenol (BHT) and 3-*tert*-butyl-4-methoxyphenol (BHA), bearing a phenolic hydroxy group, have been used in recent times as anti-oxidants in many kinds of foods.

Finally, the reduced active oxygen radicals formed from the reactions of $^3O_2$ or $O_2^{\cdot-}$ with vitamin E or vitamin C in living bodies become O_2^{2-} (H_2O_2), which can be further reduced by catalase (to H_2O and molecular oxygen) or glutathione. $O_2^{\cdot-}$ is also reduced to O_2^{2-} by SOD (enzyme: protein containing Cu and Zn complex) (Figure 1.10). However, there is no enzyme that can destroy the most reactive $HO^{\cdot}$. So, once $HO^{\cdot}$ is formed in living bodies, it destroys any kind of DNA and proteins. One typical radiation disease comes from this radical, which is formed from the irradiation of H_2O in a body (the weight percentage of water in a living body is about 70–80%).

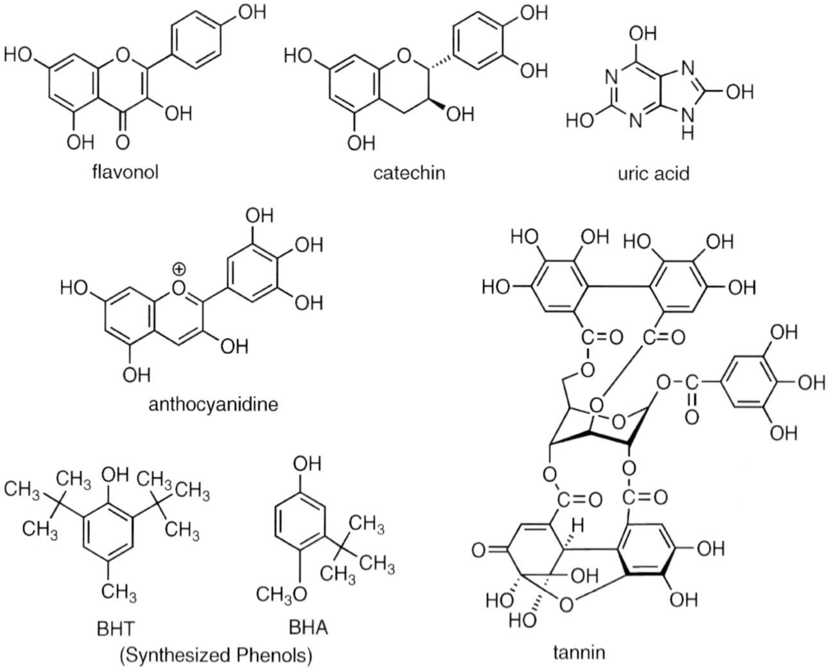

flavonol catechin uric acid

anthocyanidine

BHT BHA
(Synthesized Phenols)

tannin

Figure 1.9 Natural polyphenols and synthesized phenols.

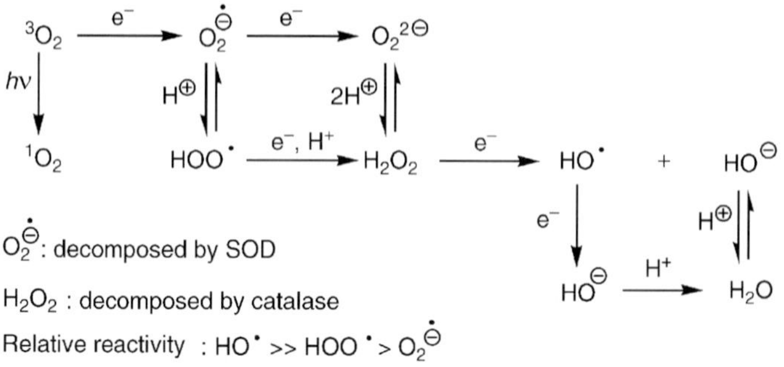

$O_2^{\bullet -}$: decomposed by SOD

H_2O_2 : decomposed by catalase

Relative reactivity : $HO^{\bullet} \gg HOO^{\bullet} > O_2^{\bullet -}$

Figure 1.10

1.3 STABLE FREE RADICALS

Commercially available stable free radicals are shown in Figure 1.11.

Recently reported stable free radicals are shown in Figure 1.12. Most of these stable free radicals are oxygen- or nitrogen-centered radicals, like molecular oxygen, nitrogen monoxide, and nitrogen dioxide, where the oxygen and nitrogen atoms have high electronegativity. Moreover, these free radicals bear quite a large resonance effect and steric effect for high stabilization. Generally, stable radicals are stabilized by thermodynamic control (this is by resonance effect), not kinetic control (this is by steric

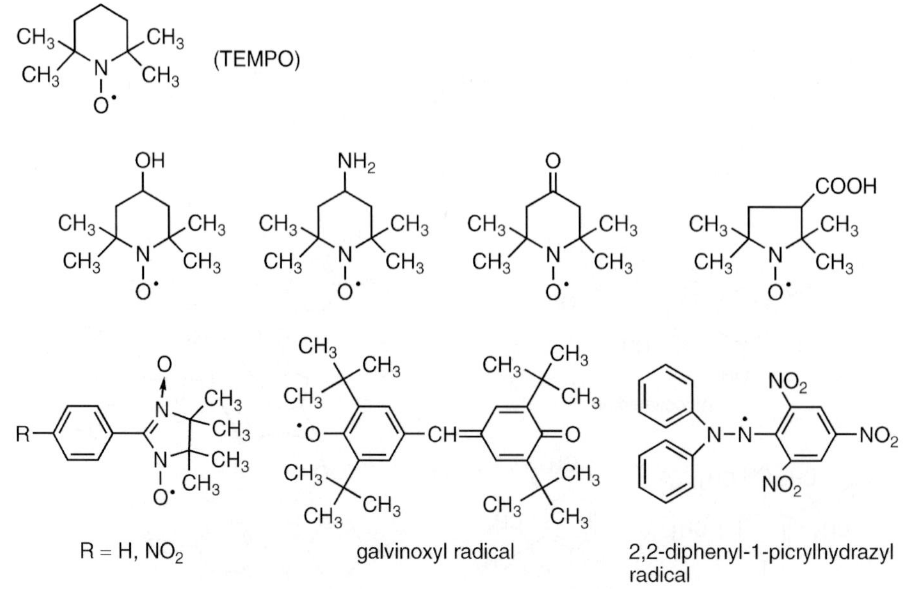

(TEMPO)

$R = H, NO_2$

galvinoxyl radical

2,2-diphenyl-1-picrylhydrazyl radical

Figure 1.11 Commercially available, stable free radicals.

effect). Since general radicals are extremely reactive, it is not possible to stabilize radicals only by steric effect. Thus, all the radicals in Figures 1.11 and 1.12 are stabilized by thermodynamic control. These radicals are important in ESR study for analysis of the spin density and conformation of the radicals. However, from the viewpoint of synthetic organic chemistry, these stable free radicals are not interesting and not attractive, since these free radicals are too stable and essentially they do not react with

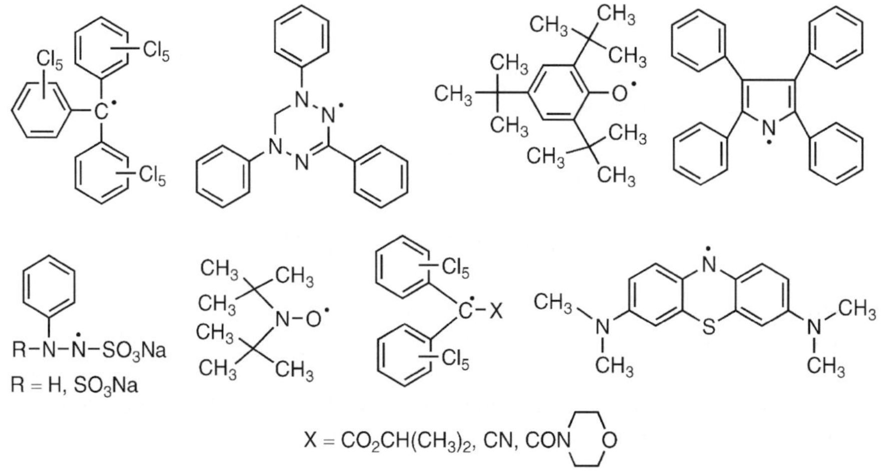

$R = H, SO_3Na$

$X = CO_2CH(CH_3)_2$, CN, CON(morpholino)

Figure 1.12 Recently reported stable free radicals.

organic molecules directly. There is only one synthetic use of these stable radicals, which is to trap reactive radical species formed during the reactions, as a radical scavenger.

Free radicals are directly observed by ESR, where the wavelength is in the microwave range. Generally, wavelength λ for ESR is ~ 3.2 cm. The principle is analogous to that of NMR. Thus, the electron has a magnetic moment (spin) resulting from the rotation of a charged particle about an axis. So, there are two spin states ($+1/2$: α spin and $-1/2$: β spin) corresponding to the two orientations in space (Scheme c).

$\Delta E = h\nu = g\beta H_0$

$h = 6.63 \times 10^{-27}$ erg · s (Plank's constant)

$\beta = 9.27 \times 10^{-21}$ erg · G^{-1} (Bohr magneton)

H_0 = G (gauss), external magnetic field

g = g-factor

$+\frac{1}{2}$ α spin (*anti*-parallel)

energy

$\pm\frac{1}{2}$ ΔE Zeeman splitting

$-\frac{1}{2}$ β spin (parallel)

Scheme c

In the absence of an external magnetic field, the electron spin is oriented randomly, with α and β spin having the same energy. However, when an external magnetic field H_0 is applied to the free electrons, Zeeman splitting occurs and the energy of α and β spin becomes different. β Spin has parallel orientation of the magnetic moment of the electron with respect to the field, and α spin has anti-parallel orientation of the magnetic moment of the electron with respect to the field. The population of the two spins are given by Boltzmann's distribution. Though exposure to an external magnetic field, transition from β to α spin by the absorption of energy ΔE occurs. This transition corresponds to the ESR spectrum. In ESR, there are three parameters, i.e. g-factor, hyperfine coupling constant a, and line-width, and the first two parameters are the most important. The g-factor corresponds to the electronic environment of radicals, i.e. it corresponds to the chemical shift in NMR. Normally, the g value is in the range of 2, especially for π radicals. For hyperfine coupling, when the electron is close to an atom with a non-zero nuclear such as ^1H or ^{13}C, interaction between the electron and the nucleus occurs, and hyperfine coupling is observed. For example, quartet (strength: 1:3:3:1) hyperfine coupling in the ESR spectrum of $CH_3^{\cdot}$ is observed, and the coupling constant a is 23G. Coupling constant a is related to the spin density ρ_c as follows (McConnell equation):

$$a = A\rho_c \quad A: \textit{proportional constant}, \ \rho_c: \textit{spin density on carbon}.$$

By the measurement of ESR, information on the physical character of radicals and the spin density of radicals can be obtained [5].

Recent reports on the g factor and the coupling constants a of moderately stable radicals are shown below. A triphenylmethyl radical, which is generated by the reaction of triphenylmethyl halide with Ag, does not form a head-to-head dimer, hexaphenylethane, as mentioned previously (eq. 1.1). However, R–$C_{60}^{\cdot}$ (**22**) couples form a head-to-head dimer, R–C_{60}–C_{60}–R [6, 7]. Here, with an increase of both bulkiness and electronegativity of the R group, R–$C_{60}^{\cdot}$ becomes a more stable radical (eq. 1.11). This radical is a π radical, so the *g* value is 2.00.

C_{60} —(RBr, hν or RI, $(Bu_3Sn)_2$, hν)→ R – $C_{60}^{\cdot}$ (**22**) ⇌ $\frac{1}{2}$ R – C_{60}– C_{60} – R 50 ~ 60%

R	*g* value
i-Pr	2.00223
CCl_3	2.00341
CBr_3	2.00910
C_3F_7	2.00289

= R – C_{60}– C_{60} – R

(1.11)

The following α ester radical (**23**) is just stabilized by the resonance effect of one ester group. This effect is not as strong, so the α ester radical (**23**) can be observed using ESR only at < − 30 °C, and it couples to a dimer soon at room temperature [8].

t-Bu–C(R_1)(H)–C(R_2)(CO_2R^3)–Br —($(Bu_3Sn)_2$, *hν*)→ *t*-Bu–C(R_1)(H)–$\dot{C}$(R^2)(CO_2R^3) (**23**)

R^1	R^2	R^3	H_α	H_β	$H_{\beta'}$	H_δ	$H_{\delta'}$(G)	*g*
$CO_2C_2H_5$	H	C_2H_5	20.4	7.6	–	1.54	0.36	2.0035
CH_3	CH_3	CH_3	–	4.8	21.9	–	–	2.0034

t-Bu, H, CO_2CH_3, CH_3O_2C, H

θ (θ = 20° is the most stable angle)

(1.12)

Today, many stable radicals are known, as shown in Figures 1.11 and 1.12. However, most of them are nitroxyl radicals like NO or NO_2. Standard generation methods of nitroxyl radicals are as follows. One is the oxidation of amines or hydroxyamines by PbO_2, or by less toxic oxidants such as oxone, $Cu(OAc)_2$, *m*CPBA (eqs. 1.13 and 1.14). Another one is the reaction of nitro compounds with Grignard reagents (eq. 1.15) [9–14].

Ph_2NH —(Oxone(4 eq.), acetone, CH_2Cl_2, Na_2HPO_4, $Bu_4N^{\oplus}$ $HSO_4^{\ominus}$ buffer)→ $Ph_2NO^{\cdot}$ **24** 83% a_N = 10.7G (1.13)

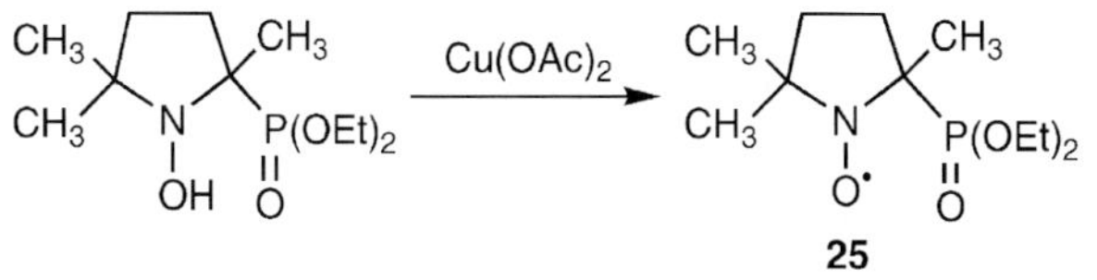

$Cu(OAc)_2$ → **25** *g* value 2.0059; a_N = 13.8 G; a_P = 50.0 G (1.14)

t-BuMgBr, ether → **26** (1.15)

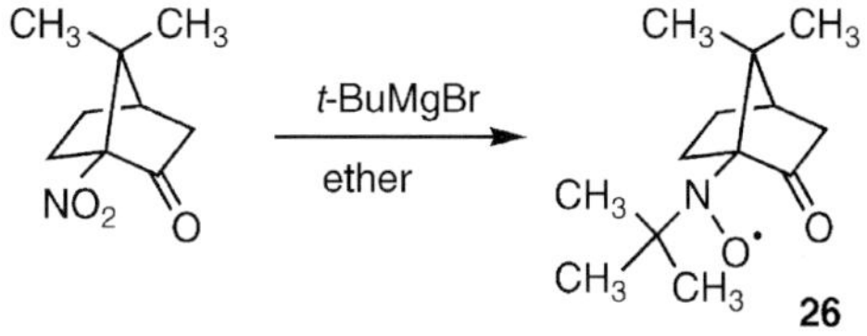

Reaction of nitro compounds such as nitro-*tert*-butane with Bu_3SnH or $(Me_3Si)_3SiH$ produces a nitroxyl radical (**27**). This is just an addition product of $Bu_3Sn^{\cdot}$ or $(Me_3Si)_3Si^{\cdot}$ onto the nitro group [15, 16]. This radical is also a π radical (eq. 1.16).

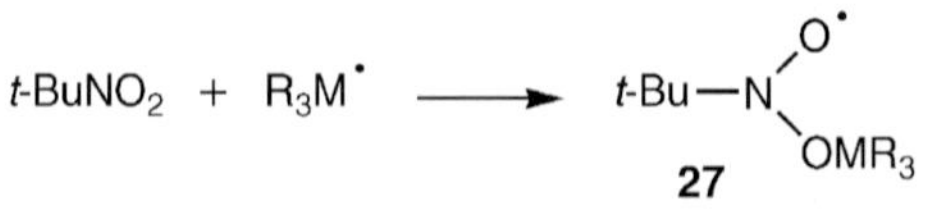

$t\text{-}BuNO_2 + R_3M^{\bullet} \longrightarrow t\text{-}Bu{-}N(O^{\bullet})(OMR_3)$ **27** (1.16)

R_3M	a_N(G)	*g* value
Bu_3Sn	27.9(triplet)	2.0050
$(Me_3Si)_3Si$	29.0(triplet)	2.0053

Generally, nitrogen-centered radicals are very reactive. However, the following sulfenamidyl radical (**28**) bearing a condensed polyaromatic group is stable for a long time (eq. 1.17), due to the resonance effect by *p*-nitrobenzenesulfenyl and condensed polyaromatic groups [17, 18].

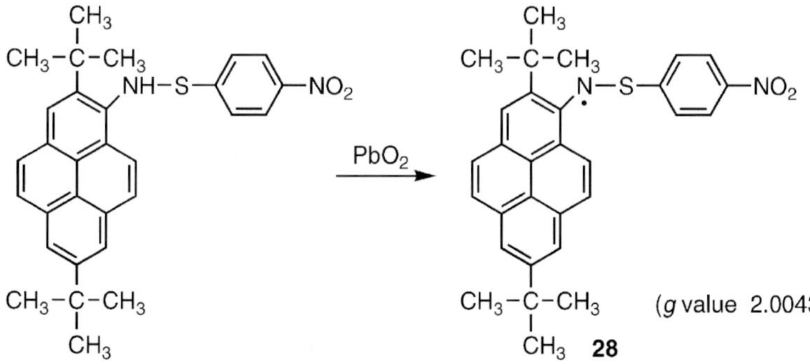

PbO_2 → **28** (*g* value 2.0043) (1.17)

1.4 PHYSICAL AND CHEMICAL CHARACTERISTICS OF FREE RADICALS

1.4.1 Orbital interactions between radicals and olefins

A free radical has an unpaired electron that has the highest energy among all bonding and non-bonding electrons in a molecule. The orbital having this unpaired electron is called SOMO. In the reactions of a free radical with another molecule, SOMO in a free radical interacts with HOMO or LUMO in another molecule, and its reactivity depends on the energy level of SOMO. Namely, an electron-rich free radical having high potential energy, behaves as a nucleophile and interacts with LUMO in another molecule. An electron-poor free radical having low potential energy, behaves as an electrophile and interacts with HOMO in another molecule. This orbital interaction between SOMO–LUMO or SOMO–HOMO is the initial step for the chemical reactions, and the reactions proceed smoothly when the energy difference is small. Two examples for the interactions of $(CH_3)_3C^{\cdot}$ with olefin and $(C_2H_5O_2C)_2CH^{\cdot}$ with olefin are shown in Figure 1.13. $(CH_3)_3C^{\cdot}$ is an electron-rich radical because of the electron-donating effect of three methyl groups through the inductive effect, and its SOMO has high potential energy and nucleophilic character. Therefore, it smoothly interacts with electron-deficient olefins such as phenyl vinyl sulfone, because of the small energy difference in the SOMO–LUMO interaction. $(C_2H_5O_2C)_2CH^{\cdot}$ is an electron-deficient radical because of the electron-withdrawing effect of two ester groups through the resonance effect, and its SOMO has low potential energy and electrophilic character. Therefore, it smoothly interacts with electron-rich olefins such as ethyl vinyl ether because of the small energy difference in the SOMO–HOMO interaction.

Generally, as the potential energy level of SOMO increases (becomes a more reactive radical), free radicals have nucleophilic character, while as the potential energy level of SOMO decreases (becomes a stable radical), free radicals have electrophilic character. Thus, when effective radical reactions are required, small energy difference in SOMO–HOMO or SOMO–LUMO interactions is necessary. For example, the relative reactivities of radical addition reactions of a nucleophilic cyclohexyl radical to alkenes,

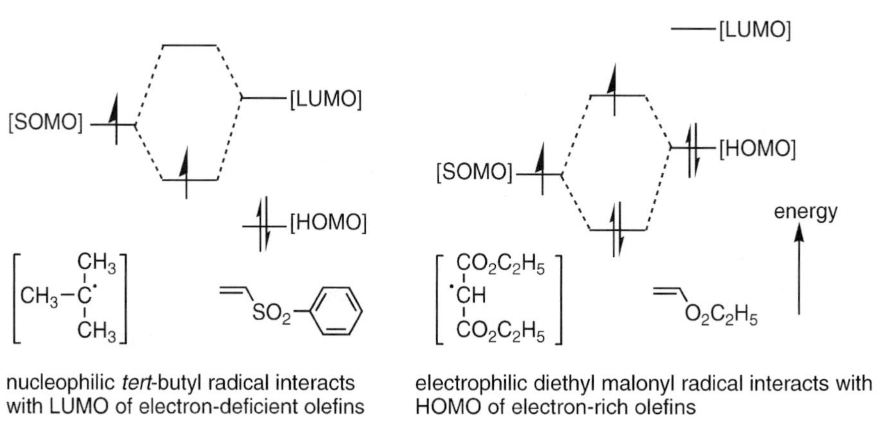

nucleophilic *tert*-butyl radical interacts with LUMO of electron-deficient olefins

electrophilic diethyl malonyl radical interacts with HOMO of electron-rich olefins

Figure 1.13 Interaction between carbon-centered radicals and olefins.

Addition reaction of nucleophilic cyclohexyl radical to alkenes

Relative reactivity :

C_4H_9-*n* : C_6H_5 : CO_2CH_3 : CHO

1 : 84 : 3000 : 8500

Addition reaction of electrophilic diethyl malonyl radical to alkenes

$C_2H_5O_2C$ $C_2H_5O_2C$ + C_6H_5 R → $C_2H_5O_2C$ $C_2H_5O_2C$ C_6H_5 R

Relative reactivity :

C_6H_5 N O : C_6H_5 CH_3 : C_6H_5 $CO_2C_2H_5$

23.0 : 3.5 : 1.0

Figure 1.14

and of an electrophilic malonyl radical to alkenes are shown in Figure 1.14. Here, the former reaction proceeds through the SOMO–LUMO interaction, and the latter reaction proceeds through the SOMO–HOMO interaction. In the former reaction, an electron-withdrawing group in alkenes increases the SOMO–LUMO interaction, while an electron-donating group in alkenes increases the SOMO–HOMO interaction in the latter reaction.

1.4.2 Baldwin's rule

One typical radical reaction is cyclization. This cyclization has been used as an indirect proof for radical reactions and a strategic method for the construction of 5- and 6-membered cyclic compounds. The experienced rule for the cyclization is Baldwin's rule [19]. There are two cyclization modes, i.e. *exo* and *endo*; moreover, there are three types of hybridization in a carbon atom, sp^3 (tetrahedral: *tet*), sp^2 (trigonal; *trig*), and sp (digonal; *dig*). Baldwin's rule is the cyclization rule based on the experimentally obtained cyclization results. The cyclization mode and kinds of hybridization in an intramolecular radical acceptor are shown in Figure 1.15.

Thus, it is called '*exo*', when the cyclization occurs on the inside of the unsaturated carbon–carbon bond, and it is called '*endo*', when the cyclization occurs on the outside of the unsaturated carbon–carbon bond. Moreover, it is '*tet*' (tetrahedral; 109.5°), when the carbon–carbon bond at the reaction site is sp^3 hybridization; it is '*trig*' (trigonal, 120°), when the unsaturated carbon–carbon bond at the reaction site is sp^2 hybridization; and it is '*dig*' (digonal, 180°), when the unsaturated carbon–carbon bond at the reaction site is sp hybridization. For example, there are two types of cyclization manner in 5-hexen-1-yl radical, *exo-trig* and *endo-trig*, based on the above classification. Since a 5-membered cyclopentylmethyl radical is formed through '*exo-trig*' cyclization, it is finally

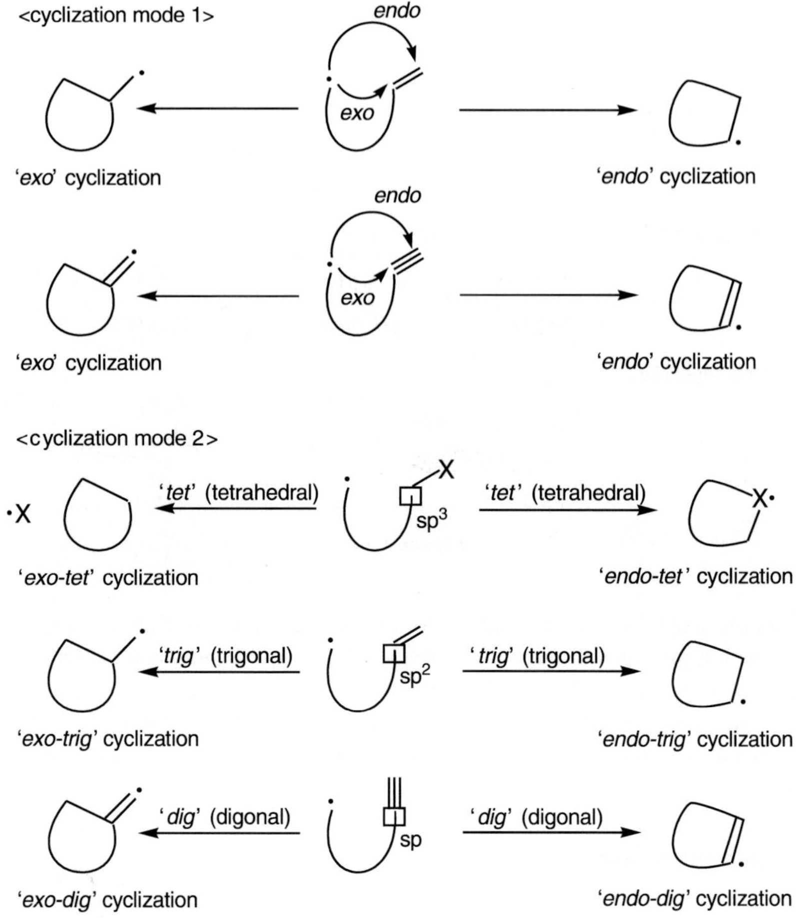

Figure 1.15 Cyclization mode.

classified as 5-*exo-trig* cyclization manner. And it is classified as 6-*endo-trig* cyclization manner, that 6-membered cyclohexyl radical is formed through '*endo-trig*' cyclization. Generally, 5-*exo-trig* cyclization is the main pathway, and this is the Baldwin's rule. The cyclopentylmethyl radical is a primary alkyl radical and the cyclohexyl radical is a secondary alkyl radical. Thus, the formation of the cyclopentylmethyl radical suggests that the cyclization of the 5-hexen-1-yl radical proceeds through a kinetically controlled pathway. Most radical cyclizations occur through the kinetically controlled pathway, since the radicals are generally extremely unstable and reactive. In view of the orbital interaction theory, the most preferable approach angle of a carbon-centered radical onto the unsaturated carbon–carbon bonds depends on the kind of hybridization of the unsaturated carbon–carbon bond. Thus, the most preferable approach angle is $\alpha = 109°$ (perpendicular direction to the plane), when the unsaturated carbon–carbon bond is sp^2 hybridization, and the most preferable approach angle is $\alpha = 120°$, when the unsaturated carbon–carbon bond is sp hybridization as shown in Figure 1.16.

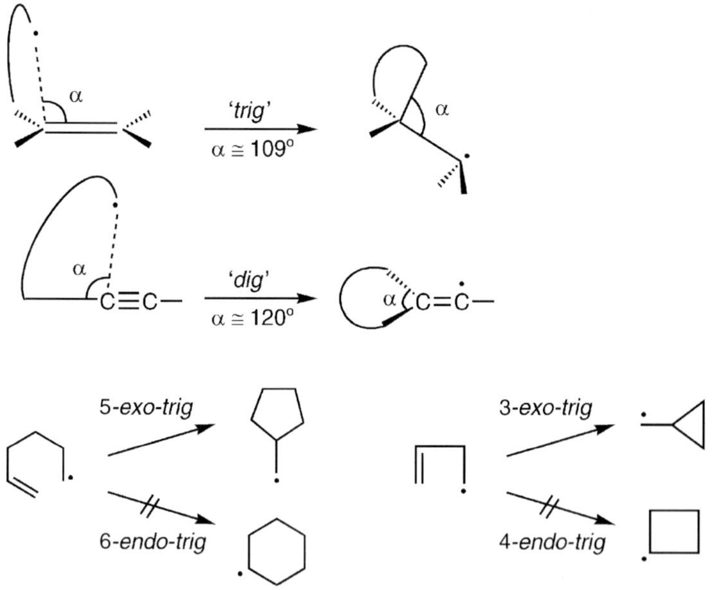

Figure 1.16

From the cyclization of 3-buten-1-yl radical, the cyclopropylmethyl radical through '3-*exo-trig*' manner is generated due to the preferable approach angle, not through '4-*endo-trig*' cyclization. Practically, when the reaction of 5-hexenyl-1-bromide with a Bu_3SnH/AIBN system was carried out in benzene refluxing conditions, a mixture of methylcyclopentane and cyclohexane was obtained in a ratio of 98:2. The transition states in 5-*exo-trig* and 6-*endo-trig* cyclization are shown in eq. 1.18.

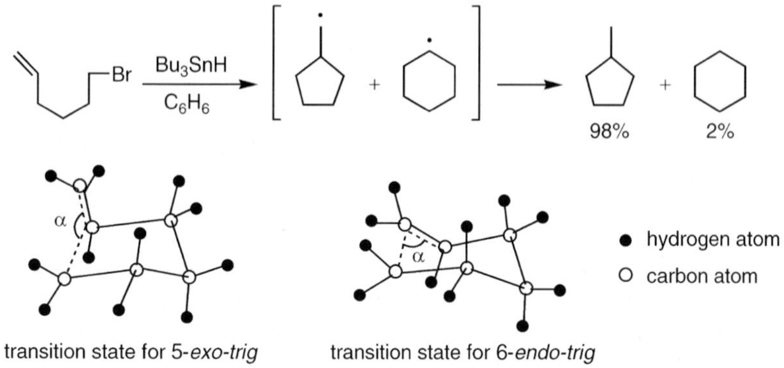

(1.18)

When the two transition states are compared, the radical approach angle in the transition state of 5-*exo-trig* manner is closer to $\alpha = 109°$ than that in 6-*endo-trig* manner.

The introduction of heteroatoms to a radical side chain may change the regioselectivity for cyclization. The change in regioselectivity for cyclization comes from the change in bond length and bond angle of the heteroatoms. In any case, the most preferable approach angle of a carbon-centered radical onto the carbon–carbon double bond is always

Table 1.1

Regioselectivity for ring closure of radicals **29**

29 → 5-*exo-trig* + 6-*endo-trig*

X	*exo* : *endo*
CH_2	40 : 60
O	98 : 2
NTs	100 : 0

$\alpha = 109°$. For example, the ratio of *exo/endo* cyclization of radical (**29**) is shown in Table 1.1, which indicates the dramatic change in the *exo/endo* ratio for X = CH_2, O, and NTs.

1.4.3 Rate constants in radical reactions

Ring-closure

Rate constants for the ring-closure of sp^3 carbon-centered radicals are shown in Table 1.2 [20–24]. The *exo-trig* mode is preferable in the 5-hexen-1-yl radical and the 6-hepten-1-yl radical, respectively. However, the *endo-trig* mode is preferable in the ring-closure of the 7-octen-1-yl radical and the 5-methyl-5-hexen-1-yl radical, respectively. The former reaction indicates that the 8-membered-transition state is more favorable than the 7-membered transition state, and the latter reaction indicates that the methyl group at the 5-position retards the formation of the 5-membered transition state. The introduction of heteroatoms such as oxygen or silicone atom changes the rate constant for ring-closure and cyclization mode. Thus, the rate constants for the ring-closure and the cyclization mode depends on the ring-size of transition state, substituent, and heteroatom in substrates.

The rate constants for the oxygen-centered radical and nitrogen-centered radical (aminyl radical and aminium radical) are also shown in Figure 1.17.

Thus, an oxygen-centered radical such as 4-penten-1-oxyl radical undergoes an extremely rapid 5-*exo-trig* ring-closure, $\sim 10^8\ s^{-1}$, to give 2-methyltetrahydrofuran. Ring-closure of the highly electrophilic 4-pentenyl-1-aminium cation radical is also faster than that of 5-hexen-1-yl radical and neutral 4-pentenyl-1-aminy radical, respectively.

Rate constants for the ring-closure by the sp^2 carbon-centered radical are shown in Table 1.3. The rate constants are increased more than those of sp^3 carbon-centered radicals, because the sp^2 carbon-centered radical is much more reactive than sp^3 carbon-centered radicals. This high reactivity of the sp^2 carbon-centered radical is reflected by the stronger bond-dissociation energy of the sp^2(carbon)–sp^3(carbon) bond than that of the sp^3(carbon)–sp^3(carbon) bond.

Ring-closure reaction of perfluoroalkenyl radicals, which are strong electrophilic sp^3 carbon-centered radicals, is also extremely fast [25–27]. For example, ring-closure of 1,1,2,2-tetrafluoro- and 1,1,2,2,3,3-hexafluoro-5-hexen-1-yl radicals is much enhanced relative to the parent radicals, as shown in Table 1.4.

Table 1.2

Rate constants for radical ring closure (s^{-1}, 25 °C) (sp^3 carbon-centered radicals)

	k_{exo}	k_{endo}
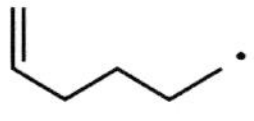	2.3×10^5	4.1×10^3
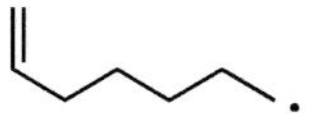	5.4×10^3	7.5×10^2
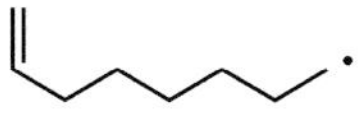	< 0.7	1.2×10^2
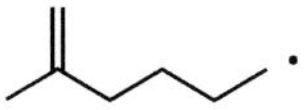	5.3×10^3	9.0×10^3
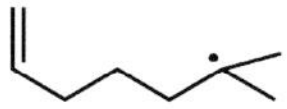	3.5×10^5	6.0×10^3
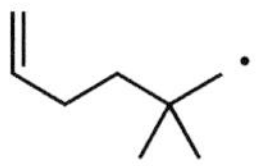	3.6×10^6	1×10^5
	5.1×10^6	1×10^5
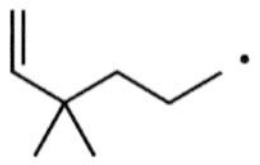	3.2×10^6	1×10^5
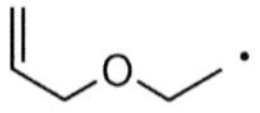 O	8.5×10^6	1×10^5
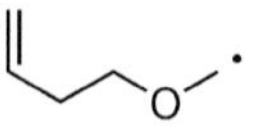 O	5×10^4	–
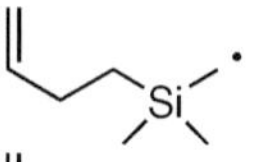 Si	8.7×10^2	1.8×10^3
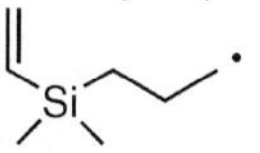 Si	7.4×10^4	5.0×10^3
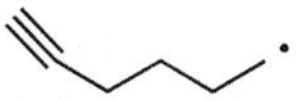	2.8×10^4	6×10^2
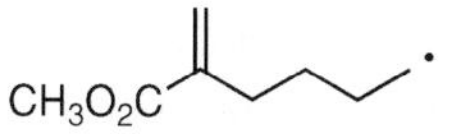 CH_3O_2C	2.9×10^5	2.2×10^6
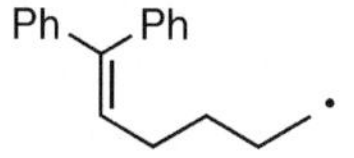 Ph Ph	5×10^7	–
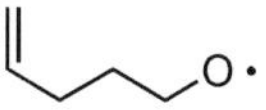 O•	4×10^8	8×10^6 (30 °C)

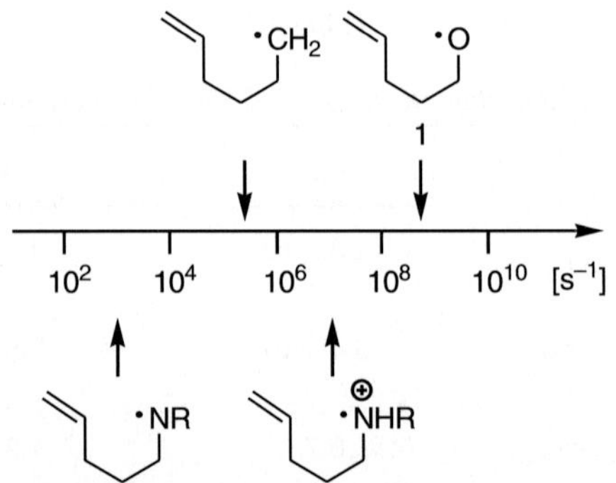

Figure 1.17 Ranges of rate constants for 5-exo-trig cyclization of 5-Hexen-1-yl, 4-Pentenyl-1-aminyl, 4-Pentenyl-1-aminium, and 4-Penten-1-oxyl radicals.

Table 1.3

Rate constants for ring closure (s^{-1}, 25 °C) (sp^2 carbon-centered radicals)

	k_{exo}	k_{endo}
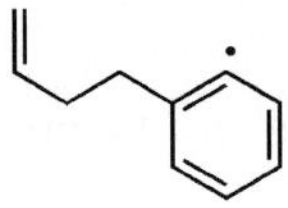	3.1×10^8	6×10^6
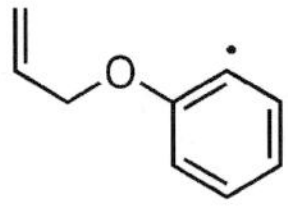	5.3×10^9	5×10^7
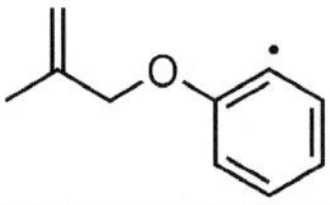	1.7×10^9	3.6×10^7

In nature, many kinds of medium- and large-sized ring lactones and ketones are known. These compounds can be also prepared by the radical ring-closure method. However, the rate constants for ring-closure to medium- and large-sized rings are decreased to $\sim 10^4$ s^{-1}, and most of these ring-closures proceed via the *endo-trig* mode as shown in Table 1.5. This reason can be explained as follows. There is not as much energy difference between the transition states of *exo-trig* and *endo-trig* modes because of the large ring and, therefore, the formation of a secondary alkyl radical through the *endo-trig* mode is preferable to the formation of a primary alkyl radical through the *exo-trig* mode. The introduction of oxygen atoms increases the rate constants for the ring-closure about 10–30 times as shown in Table 1.5 [28].

Now we examine the reactivity of ring-closure for an unsaturated carbon–oxygen and an unsaturated carbon–nitrogen double bond, instead of an unsaturated carbon–carbon double bond. Generally, the ring-closure for these unsaturated carbon–heteroatom double bonds proceed extremely rapidly; however, the reverse ring-opening reaction via β-cleavage also proceeds rapidly as shown in Table 1.6.

Table 1.4

Rate constants for ring closure of fluoroalkenyl radicals (s^{-1}, 25 °C)

	k_{exo}	k_{endo}
O, C F2, ĊF2	3.8×10^7	–
C F2, C F2, ĊF2	4.4×10^7	5.2×10^6
C F2, C F2, C F2, ĊF2	2.0×10^7	–

Table 1.5

Rate constants for ring closure to medium-sized and large-sized rings

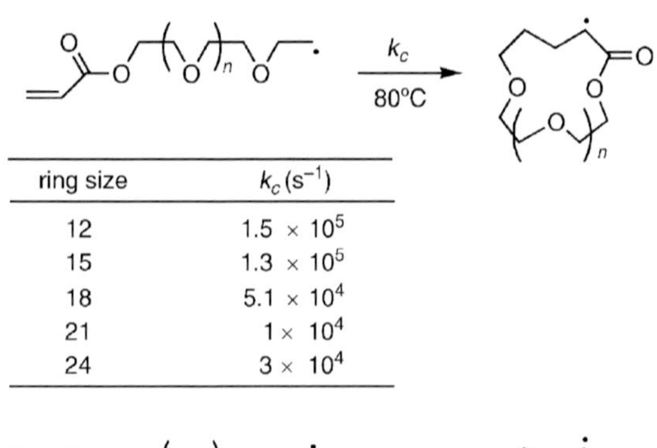

ring size	$k_c\,(s^{-1})$
12	1.5×10^5
15	1.3×10^5
18	5.1×10^4
21	1×10^4
24	3×10^4

O, O, n — k_c, 80°C — O, O, n

ring size	$k_c\,(s^{-1})$
11	2.8×10^3
12	4.8×10^3
15	1.2×10^4
16	1.5×10^4
20	2.1×10^4

These results suggest that the cyclization products to carbonyl and imino groups cannot be obtained so easily. One practical method is to trap the cyclized oxygen- or nitrogen-centered radicals formed through the ring-closure, by oxygen- or nitrogen-favored atoms such as a silyl group [29–40]. Alk-5-enoyl radicals, acyl radicals, cyclize in *exo* and *endo*

Table 1.6

Rate constants for ring closure to carbonyl and imino groups

 4-formylbutyl radical ⇌ cyclopentyloxyl radical (k_1, k_{-1})	$k_1 = 8.7 \times 10^5\ s^{-1}$ (80 °C) $k_{-1} = 4.7 \times 10^8\ s^{-1}$ (80 °C)
 5-formylpentyl radical ⇌ cyclohexyloxyl radical (k_1, k_{-1})	$k_1 = 1.0 \times 10^6\ s^{-1}$ (80 °C) $k_{-1} = 1.1 \times 10^7\ s^{-1}$ (80 °C)
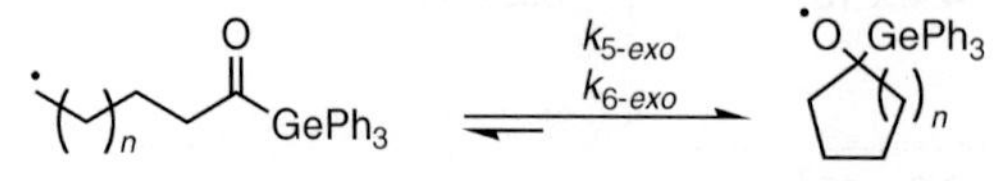 acyl-$GePh_3$ radical ⇌ cyclic O• $GePh_3$ ($k_{5\text{-}exo}$, $k_{6\text{-}exo}$)	$n = 1$ $k_{5\text{-}exo} = 6.4 \times 10^6\ s^{-1}$ (80 °C) $n = 2$ $k_{6\text{-}exo} = 1.3 \times 10^6\ s^{-1}$ (80 °C)
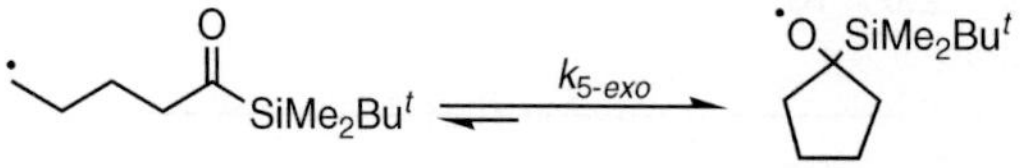 acyl-$SiMe_2Bu^t$ radical ⇌ cyclic O• $SiMe_2Bu^t$ ($k_{5\text{-}exo}$)	$k_{5\text{-}exo} = 5.0 \times 10^6\ s^{-1}$ (80 °C)
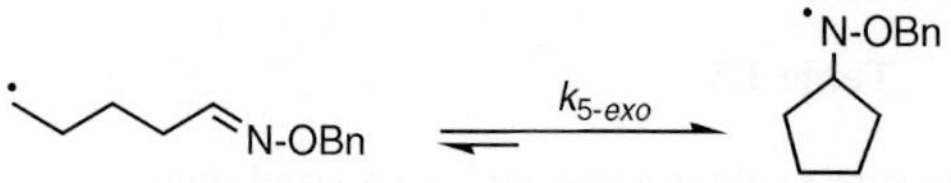 N-OBn imine radical ⇌ cyclic •N-OBn ($k_{5\text{-}exo}$)	$k_{5\text{-}exo} = 4.2 \times 10^7\ s^{-1}$ (80 °C)
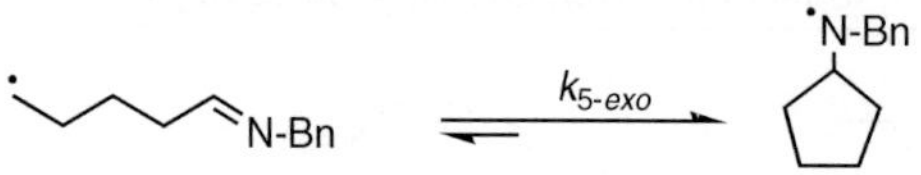 N-Bn imine radical ⇌ cyclic •N-Bn ($k_{5\text{-}exo}$)	$k_{5\text{-}exo} = 6 \times 10^6\ s^{-1}$ (80 °C)
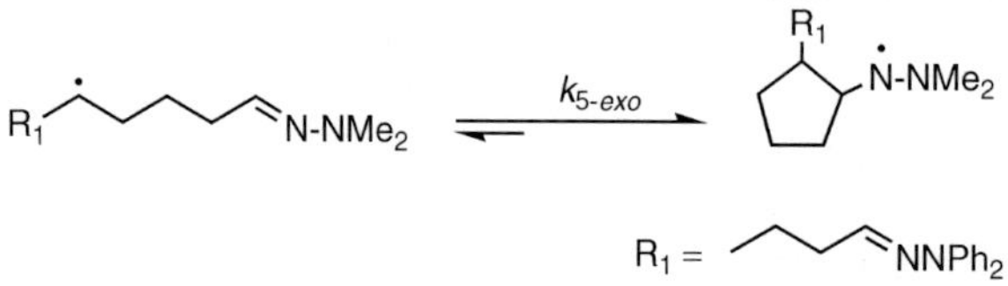 R_1-substituted N-NMe_2 hydrazone radical ⇌ cyclic •N-NMe_2 ($k_{5\text{-}exo}$); R_1 = CH2CH2CH=NNPh2	$k_{5\text{-}exo} = 3.5 \times 10^7\ s^{-1}$ (*cis*) $= 1.8 \times 10^7\ s^{-1}$ (*trans*) (80 °C)

Table 1.7

Rate constants for ring closure of alk-5-enoyl radicals

 alk-5-enoyl radical → 2-alkyl-5-methylcyclopentanone radical	$1.6 \times 10^5\ s^{-1}$
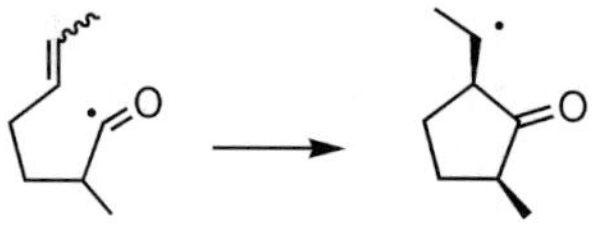 alk-5-enoyl radical → 2-alkyl-5-methylcyclopentanone radical (diastereomer)	$1.3 \times 10^5\ s^{-1}$
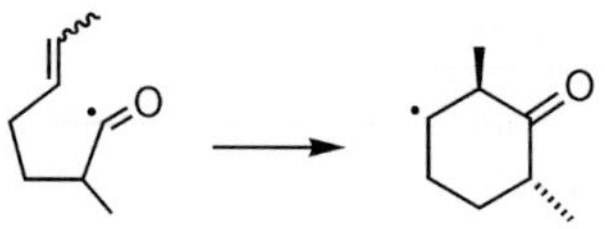 alk-5-enoyl radical → 2,6-dimethylcyclohexanone radical	$2.0 \times 10^4\ s^{-1}$
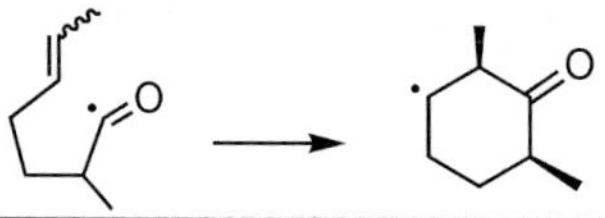 alk-5-enoyl radical → 2,6-dimethylcyclohexanone radical (diastereomer)	$5.2 \times 10^3\ s^{-1}$

Table 1.8

Rate constants for ring-opening of cyclopropylmethyl radicals

Radical		Product	Rate constant
	→		$k = 1.3 \times 10^{8}\ s^{-1}$ (25 °C) activation energy $E_a = 5.94$ kcal/mol
	→		$k = 1.6 \times 10^{8}\ s^{-1}$ (40 °C)
Ph, Ph	→	Ph, Ph	$k = 3.6 \times 10^{8}\ s^{-1}$ (40 °C)
Ph	→	Ph	$k = 3 \times 10^{11}\ s^{-1}$ (25 °C)
EtO_2C	→	EtO_2C	$k = 8 \times 10^{10}\ s^{-1}$ (25 °C)
(F_2)	→	(F_2)	$k = 3.4 \times 10^{11}\ s^{-1}$ (99 °C)

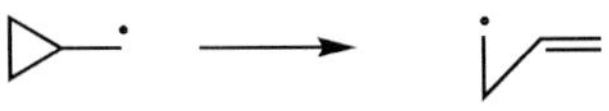

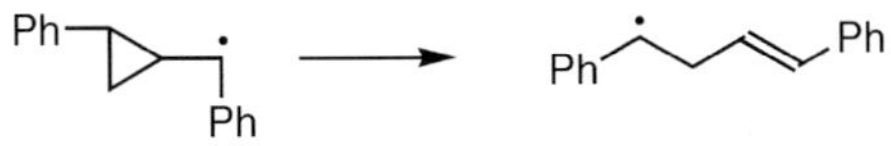

modes to give the corresponding cyclic ketone radicals as shown in Table 1.7, and 1,5- and 1,6-ring closure occurs via a lower energy ‘chairlike’ transition state [29–40].

Ring-opening

5-Membered and 6-membered cyclic compounds are thermodynamically stable; therefore, they do not give rise to ring-opening reactions. However, 3-membered and 4-membered cyclic compounds are thermodynamically unstable due to the ring strain. The rate constants for ring-opening reaction of cyclopropylmethyl radicals and cyclobutylmethyl radicals via β-cleavage are shown in Tables 1.8 and 1.9. Ring-opening of cyclopropylmethyl radicals, especially, is extremely rapid and is nearly at the diffusion control rate [41–48]. The introduction of a phenyl or an ester group for stabilization of the formed radical induces a faster ring-opening reaction than the parent one, and the rate constants are in the $10^{10} \sim 10^{11}\ s^{-1}$ order. The rate constant for the ring opening of (2,2-difluorocyclopropyl)methyl radical is also extremely rapid, and is about 500 times larger than that of the parent unfluorinated radical, and is about 5 times smaller than that of the *trans*-(2-phenylcyclopropyl)methyl radical.

The rate constant for ring-opening of the cyclobutylmethyl radical is reduced to $5 \times 10^{3}\ s^{-1}$, and again, the introduction of a phenyl group accelerates the ring-opening to $10^{6}\ s^{-1}$ order [49–51]. When rate constants for ring-openings of the cyclobutylmethyl radical and the cyclobutylmethyl lithium are compared, we can see their extremely big difference, Δk is $\sim 10^{7}$. Thus, very slow ring-opening reaction occurs in the cyclobutylmethyl anion. Moreover, ring-opening of cyclobutylmethyl magnesium bromide proceeds very slowly even at 90°. Thus, there is a considerable difference in the rate constants of ring-opening between radical and polar reactions. Therefore, the

Table 1.9

Rate constants for ring-opening of cyclobutylmethyl radicals and anions

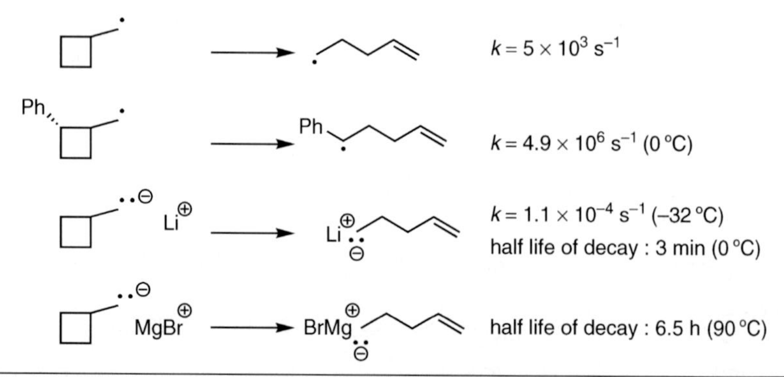

ring-opening reactions of cyclopropylmethyl and cyclobutylmethyl radicals can be used for proof of a radical reaction.

Reduction

The reduction of organic halides has been well used for organic synthesis. The rate constants for the reduction of alkyl, aryl, and vinyl radicals are shown in Table 1.10.

Generally, rate constants for hydrogen atom abstraction from Bu_3SnH by $R^{\cdot}$ are $\sim 10^6\ M^{-1}\ s^{-1}$ for the alkyl radical, and $\sim 10^8\ M^{-1}\ s^{-1}$ for aryl and vinyl radicals [52–55]. Thiols and selenols are also good hydrogen donors and the rate constants for the reaction of alkyl radicals with them are in the range of $10^7 \sim 10^9\ M^{-1}\ s^{-1}$. However, Bu_3SnH and $(Me_3Si)_3SiH$ are more effective hydrogen donors than thiols. The hydrogen-

Table 1.10

Rate constants for reduction of R· with Bu_3SnH

$$R^{\bullet} + Bu_3SnH \xrightarrow[27\,°C]{k} RH + Bu_3Sn^{\bullet}$$

sp³ carbon-centered radical		sp² carbon-centered radical	
$R^{\bullet}$	$k\ (M^{-1}\ s^{-1})$	$R^{\bullet}$	$k\ (M^{-1}\ s^{-1})$
$CH_3^{\bullet}$	1×10^7	$C_6H_5^{\bullet}$	5.9×10^8
$CH_3CH_2^{\bullet}$	2.3×10^6	$(CH_3)_2C{=}CH^{\bullet}$	3.5×10^8
$CH_3CH_2CH_2CH_2^{\bullet}$	2.4×10^6		
$(CH_3)_2CH^{\bullet}$	2.1×10^6		
$(CH_3)_3C^{\bullet}$	1.8×10^6		
▷•	8.5×10^7		

donating ability is decreased as follows, $Bu_3SnH > (Me_3Si)_3SiH > Et_3SiH \sim PhSH$. When a perfluoroalkyl radical is used instead of an alkyl radical, the rate constant for the hydrogen atom abstraction from the hydrogen donor is increased about 10^2–10^3 times as shown in Table 1.11 [56]. This is reflected by the strong bond energy of R_f–H (R_f: perfluoroalkyl) as compared with R–H. Recently, it was reported that the addition of water increases the reduction rate, about a few times [57].

Table 1.11

Rate constants for reactions of electrophilic n-$C_7F_{15}^{\cdot}$ and nucleophilic n-$C_7H_{15}^{\cdot}$ with various hydrogen donors ($M^{-1}\,s^{-1}$, 30 °C)

	Bu_3SnH	$(Me_3Si)_3SiH$	Et_3SiH	PhSH
n-$C_7F_{15}^{\bullet}$	2×10^8	5.1×10^7	7.5×10^5	2.8×10^5
n-$C_7H_{15}^{\bullet}$	2.7×10^6	4.6×10^5	8.5×10^2	1.5×10^8

Vitamin E and vitamin C are also good hydrogen atom donors in living bodies. The rate constants for the reaction of an alkyl radical and an alkoxyl radical with vitamin E are 1.7×10^6 and $3.8 \times 10^9\ M^{-1}\,s^{-1}$, respectively [58, 59]. The rate constants of hydrogen atom abstraction from R–H such as cyclopentane, 1,4-cyclohexadiene, tetrahydrofuran, Bu_3SnH by *tert*-BuO$^{\cdot}$ are shown in Table 1.12.

Table 1.12

Rate constants for hydrogen abstraction by *tert*-butoxyl radical

$$(CH_3)_3C{-}O^{\bullet} + R{-}H \xrightarrow{k} (CH_3)_3C{-}OH + R^{\bullet}$$

R–H	k [$M^{-1} \cdot s^{-1}$]	T [°C]
cyclopentane (C–H)	8.6×10^5	27
cyclohexene (allylic C–H)	5.8×10^6	27
1,4-cyclohexadiene (C–H)	6.8×10^7	50
tetrahydrofuran (α C–H)	8.3×10^6	27
Et_3SiH	5.7×10^6	22
Bu_3SnH	5.0×10^8	30

Reactions with Heteroatoms

Reduction of organic halides and chalcogenides with Bu_3SnH has been used frequently in organic synthesis. Rate constants for the reaction of organic halides and chalcogenides with $Bu_3Sn^{\cdot}$ are shown in Table 1.13.

Table 1.13

Rate constants for reactions of organic halides and chalcogenides with $Bu_3Sn^{\cdot}$

R–X	k ($M^{-1} \cdot s^{-1}$)		R–X	k ($M^{-1} \cdot s^{-1}$)	
n-$BuOCH_2$—SPh	1×10^3	25 °C	n-$C_{10}H_{21}$—Cl	7×10^3	25 °C
n-$BuOCH_2$—SePh	3×10^7	25 °C	n-C_8H_{17}—Br	3×10^7	25 °C
EtO_2CCH_2—SPh	2×10^5	25 °C	$C_6H_5CH_2$—Cl	2×10^6	25 °C
EtO_2CCH_2—SePh	1×10^8	25 °C	C_6F_5—Br	1×10^8	80 °C
EtO_2CCH_2—Cl	1×10^6	25 °C	$CH_3OC_6H_4$—Br	1×10^6	80 °C
n-$BuOCH_2$—Cl	1×10^5	25 °C	$CH_3OC_6H_4$—I	8.8×10^6	80 °C
			4-tert-butylcyclohexenyl–Br	3.4×10^6	80 °C

As can be seen in Table 1.13, organic iodides, bromides, and selenides show high reactivity, over $10^6\ s^{-1}$, and can be adequately used for organic synthesis [60–63]. The reactivity is roughly divided into the following groups:

$\sim 10^9\ M^{-1}\ s^{-1}$: alkyl iodides
10^8–$10^7\ M^{-1}\ s^{-1}$: alkyl bromides, aryl iodides
10^6–$10^5\ M^{-1}\ s^{-1}$: alkyl phenyl selenides, aryl bromides, vinyl bromides, α-chloro esters, α-thiophenyl esters
10^4–$10^2\ M^{-1}\ s^{-1}$: alkyl chlorides, alkyl phenyl sulfides, α-chloro and α-thiophenyl ethers.

On the other hand, reactivity of $Et_3Si^{\cdot}$ to organic halides and chalcogenides is much higher than that of $Bu_3Sn^{\cdot}$, as shown in Table 1.14. However, chemoselectivity of $Et_3Si^{\cdot}$ is generally poor because of its high reactivity. Moreover, the hydrogen-donating ability of Et_3SiH to the alkyl radical formed is poor. Therefore, the radical chain length is quite short, and overall the reduction of organic halides with Et_3SiH does not work so well [64, 65]. $(Me_3Si)_3SiH$ is a good hydrogen-donating agent and much less toxic than Bu_3SnH. The reason for this comes from the fact that the bond dissociation energies of Si–H in $(Me_3Si)_3SiH$ and Sn–H in Bu_3SnH are 79 and 74 kcal/mol, respectively, and the reactivity of the former reagent is closed to that of the latter reagent. So, recently Bu_3SnH has been substituted by less toxic $(Me_3Si)_3SiH$. The rate constants for the reactions of organic halides, chalcogenides, and xanthates with $(Me_3Si)_3S^{\cdot}$ are shown in Table 1.15.

Table 1.14

Rate constants for reactions of organic halides with $Et_3Si^{\cdot}$

R–X	k ($M^{-1} \cdot s^{-1}$, 27 °C)	R–X	k ($M^{-1} \cdot s^{-1}$, 27 °C)
$(CH_3)_2CH$—I	1.4×10^{10}	CCl_3—Cl	4.6×10^{9}
CH_3CH_2—I	4.3×10^{9}	$CH_2{=}CHCH_2$—Cl	2.4×10^{7}
CH_3—I	8.1×10^{9}	$C_6H_5CH_2$—Cl	2.0×10^{7}
C_6H_5—I	1.5×10^{9}	$(CH_3)_3C$—Cl	2.5×10^{6}
$CH_2{=}CHCH_2$—Br	1.5×10^{9}	$CH_3(CH_2)_4$—Cl	3.1×10^{5}
$C_6H_5CH_2$—Br	2.4×10^{9}	C_6H_5—Cl	6.9×10^{5}
$(CH_3)_3C$—Br	1.1×10^{9}		
$CH_3(CH_2)_4$—Br	5.4×10^{8}		
C_6H_5—Br	1.1×10^{8}		

Table 1.15

Rate constants for reactions of heteroatoms with $(Me_3Si)_3Si^{\cdot}$

	k ($M^{-1} \cdot s^{-1}$, 27 °C)		k ($M^{-1} \cdot s^{-1}$, 27 °C)
Cyclohexyl–$\overset{\oplus}{N}{\equiv}\overset{\ominus}{C}$:	4.7×10^{7}	$(CH_3)_3C$—Br	1.2×10^{8}
		$CH_3CH_2CH(CH_3)$—Br	4.6×10^{7}
Cyclohexyl–O–C(=S)–SCH_3	1.1×10^{9}	$CH_3CH_2CH_2CH_2$—Br	2.0×10^{7}
$(CH_3)_3C$—NO_2	1.2×10^{7}		
n-$C_{10}H_{21}$—SePh	9.6×10^{7}		
n-$C_{10}H_{21}$—SPh	5×10^{6}		

Addition Reactions

The addition of alkyl radicals to alkenes is important for C–C bond formation. A *tert*-butyl radical, a typical nucleophilic radical, reacts with acrylonitrile taking a rate constant of 2.4×10^6 M^{-1} s^{-1} (27 °C), through a SOMO–LUMO interaction. However, it reacts with 1-methylcyclohexene, an electron-rich alkene, taking a rate constant of 7.4×10^2 M^{-1} s^{-1} (21 °C). On the other hand, the diethyl malonyl radical, a typical electrophilic radical, shows the opposite reactivity [66–71]. Similarly, the rate constant for the reaction of nucleophilic $C_2H_5^{\cdot}$ and cyclohexene is 2×10^2 M^{-1} s^{-1}, while that of electrophilic $C_3F_7^{\cdot}$ with cyclohexene is 6.2×10^5 M^{-1} s^{-1}.

An acyl radical is also nucleophilic. For example, the rate constant of $(CH_3)_3CCO^{\cdot}$ (*tert*-butylcarbonyl radical, pivaloyl radical) with acrylonitrile is 4.8×10^5 M^{-1} s^{-1} (25 °C), and so its addition reaction proceeds effectively [72].

Table 1.16

Rate constants for reaction of aminium radicals with alkenes (25 °C)

	$CH_2=C(OCH_3)CH_3$	cyclohexene	$CH_2=C(CN)CH_3$
$(C_2H_5)_2NH^{\cdot+}$	1.6×10^8	$< 1 \times 10^6$	–
piperidinium radical ($C_5H_{10}NH^{\cdot+}$)	3.8×10^8	1.2×10^7	$< 1 \times 10^6$

Aminium radical ($R_3N^{\cdot +}$) is electrophilic as shown in Table 1.16 [73].

Others

Nucleophilic radical, $R^{\cdot}$ and activated alkyl iodides, R′I, which have electron-withdrawing groups, react smoothly through a SOMO–LUMO (σ^*) interaction to form RI and stable $R'^{\cdot}$, as shown in Table 1.17. Here, the formed $R'^{\cdot}$ is stabilized through the resonance effect by an ester or a cyano group [74].

Radical decarboxylation of carboxyl radicals ($RCO_2^{\cdot}$), which are generally formed through the Hunsdiecker reaction or Barton decarboxylation reaction, is a β-cleavage reaction. The rate constant of decarboxylation in $RCO_2^{\cdot}$ (aliphatic group) is quite fast and is $\sim 10^9\ s^{-1}$, while that in $ArCO_2^{\cdot}$ (aromatic group) is $\sim 10^5\ s^{-1}$. Therefore, the

Table 1.17

Rate constants for reaction of $R^{\cdot}$ with R′ I (50 °C)

$$C_8H_{17}^{\cdot} + I{-}C(CH_3)(CO_2C_2H_5){-}CO_2C_2H_5 \xrightarrow[C_6H_6]{k_1} C_8H_{17}I + {}^{\cdot}C(CH_3)(CO_2C_2H_5){-}CO_2C_2H_5 \qquad k_1 = 1.8 \times 10^9\ M^{-1} \cdot s^{-1}$$

$$C_8H_{17}^{\cdot} + I{-}CH_2CN \xrightarrow{k_2} C_8H_{17}I + {}^{\cdot}CH_2CN \qquad k_2 = 1.7 \times 10^9\ M^{-1} \cdot s^{-1}$$

$$C_8H_{17}^{\cdot} + I{-}C(CH_3)_2{-}CO_2C_2H_5 \xrightarrow{k_3} C_8H_{17}I + {}^{\cdot}C(CH_3)_2{-}CO_2C_2H_5 \qquad k_3 = 6 \times 10^8\ M^{-1} \cdot s^{-1}$$

$$C_8H_{17}^{\cdot} + I{-}CH_2CO_2C_2H_5 \xrightarrow{k_4} C_8H_{17}I + {}^{\cdot}CH_2CO_2C_2H_5 \qquad k_4 = 2.6 \times 10^7\ M^{-1} \cdot s^{-1}$$

$$C_8H_{17}^{\cdot} + Br{-}CH_2CO_2C_2H_5 \xrightarrow{k_5} C_8H_{17}Br + {}^{\cdot}CH_2CO_2C_2H_5 \qquad k_5 = 7.0 \times 10^4\ M^{-1} \cdot s^{-1}$$

Table 1.18

Rate constants for decarbonylation of acyl radicals (s^{-1}, 23 °C)

$CH_3(CH_2)_{10}-C^{\bullet}(=O) \xrightarrow{k_1} CH_3(CH_2)_{10}^{\bullet} + CO$	$k_1 = 2.1 \times 10^2\ s^{-1}$
$(CH_3CH_2)(CH_3CH_2CH_2CH_2)CH-C^{\bullet}(=O) \xrightarrow{k_2} (CH_3CH_2)(CH_3CH_2CH_2CH_2)CH^{\bullet} + CO$	$k_2 = 1.4 \times 10^4\ s^{-1}$
1-methylcyclohexyl-$C^{\bullet}(=O)$ (CH_3) $\xrightarrow{k_3}$ 1-methylcyclohexyl radical (CH_3) $+ CO$	$k_3 = 5.2 \times 10^4\ s^{-1}$

Hunsdiecker reaction does not work so well in aromatic carboxylic acids [75, 76]. The rate constants for decarbonylation of acyl radicals are lowered as shown in Table 1.18.

Finally, a trapping study of carbon-centered radicals by TEMPO (2,2,6,6-tetramethyl-1-piperidinyloxyl radical) is often used as one form of proof for the formation of carbon-centered radicals. The rate constants for the coupling of carbon-centered radicals and TEMPO are shown in Table 1.19. Activation energy of the radical coupling reaction is nearly zero and, therefore, this coupling reaction is extremely rapid [77–79].

Alkyl radicals ($R^{\cdot}$) react with molecular oxygen with reaction rate constant, $\sim 10^9\ M^{-1}\ s^{-1}$ to give $ROO^{\cdot}$.

Table 1.19

Rate constants for reaction of $R^{\cdot}$ with TEMPO (25 °C)

$C_6H_5-CH_2^{\bullet} + {}^{\bullet}O-N$(TEMPO) $\longrightarrow C_6H_5-CH_2-O-N$(TEMPO)	$k = 4.9 \times 10^8\ M^{-1} \cdot s^{-1}$
$CH_3(CH_2)_7-CH_2^{\bullet} + {}^{\bullet}O-N$(TEMPO) $\longrightarrow CH_3(CH_2)_7-CH_2-O-N$(TEMPO)	$k = 1.2 \times 10^9\ M^{-1} \cdot s^{-1}$
$(CH_3)_3C^{\bullet} + {}^{\bullet}O-N$(TEMPO) $\longrightarrow (CH_3)_3C-O-N$(TEMPO)	$k = 7.6 \times 10^8\ M^{-1} \cdot s^{-1}$

REFERENCES

1. K Kikukawa and H Sakurai, *Free Radicals and Drugs (Japanese)*, Hirokawa, 1991.
2. S Igarashi, *Biology of Vitamins (Japanese)* Shokabo, 1988.
3. KU Ingold, *J. Am. Chem. Soc.*, 1992, **114**, 4589.
4. A Watanabe, N Noguchi, M Takahashi and E Niki, *Chem. Lett.*, 1999, 613.
5. H Ohya and J Yamauchi, *Electron Spin Resonance*, Kodansha, 1997.
6. JR Morton, KF Preston, PJ Frusic, SA Hill and E Wasserman, *J. Am. Chem. Soc.*, 1992, **114**, 5454.
7. M Yoshida, F Sultana, N Uchiyama, T Yamada and M Iyoda, *Tetrahedron Lett.*, 1999, **40**, 735.
8. B Giese, W Damm, F Wetterich and HG Zeitz, *Tetrahedron Lett.*, 1992, **33**, 1863.
9. ME Brik, *Tetrahedron Lett.*, 1995, **36**, 5519.
10. A Mercier, Y Berchadsky, Badrudin, S Pietri and P Tordo, *Tetrahedron Lett.*, 1991, **32**, 2125.
11. FL Moigne, A Mercier and P Tordo, *Tetrahedron Lett.*, 1991, **32**, 3841.
12. P Braslau, H Kuhn, LC Burrill, II, K Lanham and CJ Stenland, *Tetrahedron Lett.*, 1996, **37**, 7933.
13. JM Tronchet, *Tetrahedron Lett.*, 1991, **32**, 4129.
14. K Awaga, T Inabe, U Nagashima, T Nakamura, M Matsumoto, Y Kawabata and Y Maruyama, *Chemistry Lett.*, 1991, 1777.
15. A Kamimura and N Ono, *Bull. Chem. Soc. Jpn.*, 1988, **61**, 3629.
16. M Ballestri, C Chatgillialoglu, M Lucarini and GF Pedulli, *J. Org. Chem.*, 1992, **57**, 948.
17. L Grossi and PC Montevecchi, *Tetrahedron Lett.*, 1991, **32**, 5621.
18. Y Miura, E Yamano, A Tanaka and Y Ogo, *Chemistry Lett.*, 1992, 1831.
19. JE Baldwin, *J. Chem. Soc. Chem. Commun.*, 1976, 734.
20. ALJ Beckwith and CH Schiesser, *Tetrahedron*, 1985, **41**, 3925.
21. JH Horner, FN Martinez, M Newcomb, S Hadida and DP Curran, *Tetrahedron Lett.*, 1997, **38**, 2783.
22. M Newcomb, MA Filipkowski and CC Johnson, *Tetrahedron Lett.*, 1995, **36**, 3643.
23. EW Della, C Kostakis and PA Smith, *Organic Lett.*, 1999, **1**, 363.
24. J Hartung, *Eur. J. Org. Chem.*, 2001, 619.
25. WR Dolbier, Jr., A Li, BE Smart and ZY Yang, *J. Org. Chem.*, 1998, **63**, 5687.
26. A Li, AB Shtarev, BE Smart, ZY Yang, J Lusztyk, KU Ingold, A Bravo and WR Dolbier, Jr., *J. Org. Chem.*, 1999, **64**, 5993.
27. JM Baker and WR Dolbier, Jr., *J. Org. Chem.*, 2001, **66**, 2662.
28. A Philippon, MD Castaing, ALJ Beckwith and B Maillard, *J. Org. Chem.*, 1998, **63**, 6814.
29. ALJ Beckwith and BP Hay, *J. Am. Chem. Soc.*, 1989, **111**, 230.
30. ALJ Beckwith and BP Hay, *J. Am. Chem. Soc.*, 1989, **111**, 2674.
31. ALJ Beckwith and KD Raner, *J. Org. Chem.*, 1992, **57**, 4954.
32. ALJ Beckwith, *Tetrahedron*, 1981, **37**, 3073.
33. J Lusztyk, B Maillard, S Deycard, DA Lindsay and KU Ingold, *J. Org. Chem.*, 1987, **52**, 3509.
34. DP Curran, J Xu and E Lazzarini, *J. Am. Chem. Soc.*, 1995, **117**, 6603.
35. DP Curran, JY Xu and E Lazzarini, *J. Chem. Soc., Perkin Trans. 1*, 1995, 3049.
36. DP Curran and M Palovick, *Synlett*, 1992, 631.
37. C Chatgilialoglu, C Ferreri, M Lucarini, A Venturini and AA Zavitsas, *Chem. Eur. J.*, 1997, **3**, 376.
38. WT Jiaang, HC Lin, KH Tang, LB Chang and YM Tsai, *J. Org. Chem.*, 1999, **64**, 618.
39. P Tauh and AG Fallis, *J. Org. Chem.*, 1999, **64**, 6960.
40. P Tauh and AG Fallis, *J. Org. Chem.*, 1999, **64**, 6960.
41. B Maillard, D Forrest and KU Ingold, *J. Am. Chem. Soc.*, 1976, **98**, 7024.
42. M Newcomb and AG Glenn, *J. Am. Chem. Soc.*, 1989, **111**, 275.
43. M Newcomb, AG Glenn and MB Manek, *J. Org. Chem.*, 1989, **54**, 4603.
44. M Newcomb, AG Glenn and WG Williams, *J. Org. Chem.*, 1989, **54**, 2675.
45. R Hollis, L Hughes, VB Bowry and KU Ingold, *J. Org. Chem.*, 1992, **57**, 4284.
46. M Newcomb and SY Choi, *Tetrahedron Lett.*, 1993, **34**, 6363.

47. JH Horner, N Tanaka and M Newcomb, *J. Am. Chem. Soc.*, 1998, **120**, 10379.
48. F Tian and WR Dolbier, Jr., *Org. Lett.*, 2000, **2**, 835.
49. M Newcomb, JH Horner and CJ Emanuel, *J. Am. Chem. Soc.*, 1997, **119**, 7147.
50. JC Walton, *J. Chem. Soc., Perkin Trans. 2*, 1989, 173.
51. EW Della and DK Taylor, *J. Org. Chem.*, 1995, **60**, 297.
52. LJ Johnston, J Lusztyk, DDM Wayner, AN Abeyarckreyma, ALJ Bechwith, JC Scaiano and KU Ingold, *J. Am. Chem. Soc.*, 1985, **107**, 4594.
53. C Chatgilialoglu, KU Ingold and JC Scaiano, *J. Am. Chem. Soc.*, 1981, **103**, 7739.
54. JW Wilt, J Lusztyk, M Peeran and KU Ingold, *J. Am. Chem. Soc.*, 1988, **110**, 281.
55. M Newcomb, SY Choi and JH Horner, *J. Org. Chem.*, 1999, **64**, 1225.
56. XX Rong, HQ Pan and WR Dolbier, Jr., *J. Am. Chem. Soc.*, 1994, **116**, 4521.
57. C Tronche, FN Martinez, JH Horner, M Newcomb, M Senn and B Giese, *Tetrahedron Lett.*, 1996, **37**, 5845.
58. H Fischer, in New Series, Part D **Vol. 13**, Springer, Berlin, 1984.
59. C Evans, JC Scaiano and KU Ingold, *J. Am. Chem. Soc.*, 1992, **114**, 4589.
60. JW Wilt, FG Belmonte and PA Zieske, *J. Am. Chem. Soc.*, 1983, **105**, 5665.
61. ALJ Beckwith and PE Pigou, *Aust. J. Chem.*, 1986, **39**, 77.
62. D Crich, JT Hwang, S Gastaldi, F Recupero and DJ Wink, *J. Org. Chem.*, 1999, **64**, 2877.
63. DP Curran, CP Jasperse and MJ Totleben, *J. Org. Chem.*, 1991, **56**, 7170.
64. C Chatgilialoglu, KU Ingold and JC Scaiano, *J. Am. Chem. Soc.*, 1982, **104**, 5123.
65. C Chatgilialoglu, C Ferreri and M Lucarini, *J. Org. Chem.*, 1993, **58**, 249.
66. B Giese, *Angew. Chem. Int. Ed.*, 1983, **22**, 753.
67. K Riemenschneider, E Drechsel-Grau and P Boldt, *Tetrahedron Lett.*, 1979, 185.
68. B Giese and W Mehl, *Tetrahedron Lett.*, 1991, **32**, 4275.
69. NA Porter, WX Wu and AT Mcphail, *Tetrahedron Lett.*, 1991, **32**, 707.
70. DP Curran, H Qi, NA Porter, Q Su and WX Wu, *Tetrahedron Lett.*, 1993, **34**, 4489.
71. DV Avila, KU Ingold, J Lusztyk, WR Dolbier, Jr., HQ Pan and M Muir, *J. Am. Chem. Soc.*, 1994, **116**, 99.
72. AG Neville, CE Brown, DM Rayner, J Lusztyk and KU Ingold, *J. Am. Chem. Soc.*, 1991, **113**, 1869.
73. BD Wagner, G Ruel and J Lusztyk, *J. Am. Chem. Soc.*, 1996, **118**, 13.
74. DP Curran, E Bosch, J Kaplan and M Newcomb, *J. Org. Chem.*, 1989, **54**, 1826.
75. T Tateno, H Sakuragi and K Tokumaru, *Chemistry Lett.*, 1992, 1883.
76. C Chatgilialoglu, C Ferreri, M Lucarini, P Pedrielli and GF Pedulli, *Organometallics*, 1995, **14**, 2672.
77. J Chateauneuf, J Lusztyk and KU Ingold, *J. Org. Chem.*, 1988, **53**, 1629.
78. VW Bowry and KU Ingold, *J. Am. Chem. Soc.*, 1992, **114**, 4992.
79. ALJ Beckwith, VW Bowry and KU Ingold, *J. Am. Chem. Soc.*, 1992, **114**, 4983.

– 2 –

Functional Group Conversion

Generally, radicals are very reactive species, so coupling reactions, abstraction of a hydrogen atom from the solvents or reagents, and reactions with molecular oxygen in solution occur rapidly. In particular, since activation energy of the coupling reaction of radicals is nearly zero, the reaction rate for the coupling reaction of carbon-centered radicals is extremely fast, and almost reaches diffusion control rate. Thus, an increase of the concentration of carbon-centered radicals induces the coupling reaction with ease. The pinacol coupling reaction of aldehydes or ketones has, therefore, been well utilized. The rate constants for the coupling reaction of carbon-centered radicals are shown in Table 2.1.

Table 2.1

Rate constants for radical coupling reactions

$$2\,R^{\bullet} \xrightarrow{k} R{-}R$$

R	k ($M^{-1}\cdot s^{-1}$) 25 °C
CH_3	4.5×10^9
c-C_6H_{11}	1.4×10^9
t-Bu	1.1×10^9
$PhCH_2$	1.0×10^9

2.1 RADICAL COUPLING REACTIONS

One typical radical reaction is a coupling reaction. Oxidative decarboxylation coupling reaction of carboxylic acids by electrolysis (Kolbe electrolysis), intramolecular coupling reaction of diesters with Na (acyloin condensation), formation of pinacols from ketones or aldehydes with Na or Mg are well known classical methods [1, 2]. Recently, oxidative decarboxylation coupling reaction of arylacetic acids with HgF_2 via arylmethyl radicals has been reported [3]. Today, SmI_2 is well used as a single electron transfer (SET) reagent for organic synthesis. For example, treatment of aldehydes or ketones (**1**) with SmI_2 rapidly forms the corresponding pinacols (**2**) in good yields via ketyl radicals, as shown in eq. 2.1 [4–6]. The color of SmI_2 in THF solution is deep blue or greenish-blue, and that of Sm^{3+} is colorless. Thus, the disappearance of color in the solution indicates the end of the reaction.

$$2\,\underset{\mathbf{1}}{R\text{-}C(=O)\text{-}R'} \xrightarrow[\text{THF, r.t., 1 min}]{SmI_2} 2\left[R\text{-}\dot{C}(OSmI_2)\text{-}R'\right] \longrightarrow \underset{\mathbf{2}}{R\text{-}C(R')(OH)\text{-}C(R')(OH)\text{-}R} \quad (2.1)$$

Treatment of acyl halides (**3**) with SmI_2 provides 1,2-diketones (**4**) via the coupling of acyl radicals, which are sp^2 carbon-centered radicals (eq. 2.2). Generally, aromatic acid halides are more reactive than aliphatic acid halides.

$$2\,\underset{\mathbf{3}}{R\text{-}C(=O)\text{-}Cl} \xrightarrow[\text{THF, r.t., 10 min}]{SmI_2} 2\left[R\text{-}\dot{C}=O\right] \longrightarrow \underset{\mathbf{4}}{R\text{-}C(=O)\text{-}C(=O)\text{-}R} \quad (2.2)$$

Reactive halide (**5**) bearing electron-withdrawing groups, such as ester, cyano, or sulfonyl groups, at the α-position react with Fe/CuBr to form α,β-diester (**6**) in good yield, through the coupling of the α-ester radical formed via SET as shown in eq. 2.3 [7].

$$2\,\underset{\mathbf{5}}{CH_3(CH_2)_3\text{-}C(Br)(Cl)\text{-}CO_2CH_3} \xrightarrow[\text{DMF, r.t., 24 h}]{\text{Fe, CuBr}} 2\left[CH_3(CH_2)_3\text{-}\dot{C}(Cl)(CO_2CH_3)\right] \longrightarrow \underset{\mathbf{6}\ \ 90\%}{CH_3(CH_2)_3\text{-}C(Cl)(CO_2CH_3)\text{-}C(CO_2CH_3)(Cl)\text{-}(CH_2)_3CH_3} \quad (2.3)$$

Experimental procedure 1 (eq. 2.3).

To a Schlenk bottle, Fe powder (1 mmol), CuBr (1 mmol), DMF or DMSO (1 ml), and α-bromoester (1 mmol) were added, and the mixture was stirred for 24 h at room temperature under atmospheric pressure. After the reaction, HCl (5%, 5 ml) was added to the reaction mixture. The aqueous solution was extracted twice by dichloromethane (2 × 2 ml). The organic layer was dried over Na_2CO_3 and filtered. After removal of the solvent, the residue was distilled at 0.01 mmHg to give diester in 90% yield [7].

The Ce/I_2 system also works for the reductive coupling of ketones or aldehydes to pinacols. However, $Mn(OAc)_3$, $Fe(NO_3)_3$, and $Mn(pic)_3$ (pic = 2-pyridinecarboxylate) are one-electron oxidants. These reagents abstract one electron from cyclopropanols to generate cyclopropanoxyl radicals, which finally produce 1,6-diketones through the β-cleavage of cyclopropanoxyl radicals and the subsequent coupling of the formed β-keto radicals [8, 9].

Bis(tributyl)dithin, $(Bu_3Sn)_2$ is not a reductive reagent like Bu_3SnH. Thus, treatment of α-halo or α-phenylseleno ester (**7**) with $(Bu_3Sn)_2$ under irradiation with a mercury lamp, produces the coupling diester (**8**) as shown in eq. 2.4. Initially, $Bu_3Sn^{\cdot}$ is formed via Sn–Sn homolytic bond cleavage under the irradiation, and it reacts with the α-phenylseleno ester to give the α-ester radical. Since $(Bu_3Sn)_2$ is not a hydrogen donor, the coupling reaction of α-ester radical readily happens.

t-Bu, 2 CH3, CO2C2H5, SePh, **7** —(Bu3Sn)2, Hg-hν, C6H6→ 2 [CH3, t-Bu, CO2C2H5 radical] → CH3 CH3, t-Bu, Bu-t, C2H5O2C, CO2C2H5, **8** (2.4)

The preparation of C_{60} dimers (**10**) by the reaction of R_fI (perfluoroalkyl iodide), C_{60} (**9**), and $(Bu_3Sn)_2$ is set out in eq. 2.5 [10–12] as an interesting study in the use of $(Bu_3Sn)_2$.

2 C60 (**9**) + 2 RfI —(Bu3Sn)2 hν, C6H4Cl2, 5 h→ (meso) **10a** + (racemate) **10b**; [Rf•] → [Rf–C60•]; Rf = CF3CF2CF2 (2.5)

Experimental procedure 2 (eq. 2.5).

A 1,2-dichlorobenzene (5 ml) solution of C_{60} (0.05 mmol), R_fI (0.25 mmol), $(Bu_3Sn)_2$ (0.25 mmol) was irradiated with a halogen lamp (70 W) for 5 to 8 h under a nitrogen atmosphere. After the reaction, the solvent was removed, and the residue was chromatographed on silica gel to give C_{60} dimer in 50% yield [12].

Treatment of 2,5-dimethylfuran (**11**), c-$C_6F_{13}I$, and sodium dithionite ($Na_2S_2O_4$) as SET agent gives the dimerization products (**12**) via the coupling of the adduct radical formed by the addition of $c\text{-}C_6F_{13}^{\cdot}$ to 2,6-dimethylfuran (eq. 2.6) [13].

$$\mathbf{11} \xrightarrow[\substack{Na_2CO_3 \\ CH_3CN/H_2O = 2/1 \\ 0\sim20\,^\circ C}]{\substack{C_6F_{13}I \\ Na_2S_2O_4}} \mathbf{12a}\ 33\% + \mathbf{12b}\ 12\% \tag{2.6}$$

The use of the above methods does not generally result in the coupling reaction of aromatic compounds, ArX, because of the strong bond of $C(sp^2)–X$ in ArX. However, the coupling reaction of a cation radical formed from the single-electron oxidation of aromatics readily occurs. For example, 4-methylquinoline coupled to give bis[2-(4-methylquinolyl)] in 90% yield, by electrolysis [14–19]. Direct irradiation (300 nm) of carbonyl compound (**13**) in dimethylaniline without a solvent gives rise to ethanolamine (**14**) as the major product as shown in eq. 2.7 [20].

$$PhCHO\ (\mathbf{13}) \xrightarrow[PhNMe_2]{h\nu} \left[Ph\dot{C}(OH)H + PhN(Me)\dot{C}H_2 \right] \longrightarrow PhN(Me)CH_2CH(OH)Ph\ \ \mathbf{14}\ 79\% \tag{2.7}$$

2.2 RADICAL REDUCTION

Radical reduction of alkyl and aryl halides is a fundamental and important reaction in organic synthesis, and has been extensively used. Thus, treatment of alkyl halides, selenides, or xanthates with Bu_3SnH, Ph_3SnH, $(Me_3Si)_3SiH$, and $Ph_4Si_2H_2$ in the presence of AIBN gives the corresponding reduction products in good yields [21–36] A typical advantage of the radical reduction is that the reaction can be used for the substrates bearing ester, carbonyl, carbamate, and hydroxy groups under neutral conditions. Generally, alkyl iodides, bromides, selenides, and xanthates are used. The rate constant for the reaction of $Bu_3Sn^{\cdot}$ with organic halides and chalcogenides is over $10^7\ M^{-1}\ s^{-1}$ and can be adequately used for synthetic application. A rough order for the rate constants of organic halides with $Bu_3Sn^{\cdot}$ is as follows: $\sim 10^9\ M^{-1}\ s^{-1}$ for alkyl iodides; $10^8 \sim 10^7\ M^{-1}\ s^{-1}$ for alkyl bromides and aryl iodides; $10^6 \sim 10^5\ M^{-1}\ s^{-1}$ for alkyl selenides, aryl bromides and vinyl bromide; $10^4 \sim 10^2\ M^{-1}\ s^{-1}$ for alkyl chloride and alkyl sulfides. The rate constant for the reactions of $R^{\cdot}$ (sp^3 carbon-centered radical) with Bu_3SnH to form RH is about $10^6\ M^{-1}\ s^{-1}$ (25 °C). The rate constant of the more reactive sp^2 carbon-centered radical, such as phenyl radical or vinyl radical, is about $10^8\ M^{-1}\ s^{-1}$ (30 °C).

The rate constant of $R^{\cdot}$ with $(Me_3Si)_3SiH$ is about $10^5\ M^{-1}\ s^{-1}$. The rate constants of $(Me_3Si)_3Si^{\cdot}$ with alkyl halides and selenides are as follows: $\sim 10^9\ M^{-1}\ s^{-1}$ for alkyl iodides; $10^8 \sim 10^7\ M^{-1}\ s^{-1}$ for alkyl bromides and alkyl methyl xanthates; $\sim 10^7\ M^{-1}\ s^{-1}$ for alkyl selenides; $\sim 10^6\ M^{-1}\ s^{-1}$ for alkyl sulfides.

Shown below are some examples of alkyl bromides bearing various kinds of functional groups such as ester, amino, and hydroxy groups, with Bu_3SnH, to form the corresponding reduction products in good yields.

AIBN
Bu_3SnH
$C_6H_5CH_3$ Δ

15 → **16** 41% (2.8)

Experimental procedure 3 (eq. 2.8).

To a mixture of sugar bromide (5.84 g) in toluene (100 ml) were added Bu_3SnH (12.34 g, 42.4 mmol) and AIBN (0.2 g, 1.2 mmol) under a nitrogen atmosphere. The mixture was heated at 95 °C for 75 min. After the reaction, the mixture was cooled and was poured into petroleum ether. The mixture obtained was filtered and washed with petroleum ether. The solids were recrystallized from ethanol to give 3′-deoxyadenosine in 41% yield [30].

17a → AIBN, Bu_3SnD; **17b** → AIBN, Bu_3SnD

18a (α) **18b** (β) (2.9)

(from **17a**) $C_6H_5CH_3$,		90°C	86% (α : β = 82 : 18)
(from **17b**) $C_6H_5CH_3$,		90°C	65% (α : β = 83 : 17)
(from **17a**) THF,	hν	−20°C	80% (α : β = 98 : 2)

AIBN
Bu_3SnH(dropwise)
C_6H_6 Δ

17a → **19** 80% (2.10)

Experimental procedure 4 (eq. 2.10).

To a mixture of *tetra-O*-acetyl-α-D-glucopyranosyl bromide (8.22 g, 20 mmol) in benzene (80 ml) was added dropwise, a solution of Bu_3SnH (7.0 g, 24 mmol) and AIBN (0.41 g, 2.5 mmol) in benzene (14 ml) over 10 h. After the reaction, the solvent was removed and the residue was chromatographed on silica gel to give 1,3,4,6-*tetra-O*-acetyl-2-deoxy-α-D-arabinohexapyranose in 80% yield [32].

$$PhCH_2CH_2CH_2OTs \; (\mathbf{20}) \xrightarrow[CH_3OCH_2CH_2OCH_3,\ \Delta]{AIBN,\ NaI,\ Bu_3SnH} PhCH_2CH_2CH_3 \; (\mathbf{21}\ 99\%) \tag{2.11}$$

Experimental procedure 5 (eq. 2.11).

To a mixture of alkyl tosylate (0.83 mmol), NaI (1.33 mmol) and AIBN (cat.) in 1,2-dimethoxyethane (5 ml) was added Bu_3SnH (0.83 mmol), and the mixture was refluxed for 1 h. After the reaction, the solvent was removed, and the residue was chromatographed on silica gel to give the reduction product [33].

$$\mathbf{22} \xrightarrow[EtOH,\ r.t.]{Et_3B,\ Air,\ Ph_4Si_2H_2} \mathbf{23}\ 97\% \tag{2.12}$$

Experimental procedure 6 (eq. 2.12).

To a flask equipped with a cooler were added sugar bromide (0.5 mmol), $Ph_4Si_2H_2$ (1.2 mmol), ethanol (2.5 ml), and lastly, Et_3B (1.2 ml, 1 M THP solution). The mixture was stirred under aerobic conditions. After 4 h, Et_3B (1.2 mmol) was added again and stirred for 12 h. After the reaction, the solvent was removed and the residue was chromatographed on silica gel to give the reduction product [36].

For example, treatment of α-glucosyl bromide (**17a**) and β-glucosyl chloride (**17b**) with Bu_3SnD in the presence of AIBN produces the corresponding reduction products, α-D form (**18a**) and β-D form (**18b**), with the same ratio, via an anomeric radical, where generally the α-form is the major product (eq. 2.9). The same treatment of α-glucosyl bromide (**17a**) with Bu_3SnH and AIBN under highly diluted conditions of Bu_3SnH (the dropwise addition of tributylstannane) generates the corresponding 1,2-acetoxy rearranged product (**19**), 2-deoxy sugar, via 1,2-acetoxy migration of the formed anomeric radical (eq. 2.10). This is a very useful method for the preparation of 2-deoxy sugars. The same 1,2-acetoxy rearrangement also occurs with $(Me_3Si)_3SiH$ and AIBN.

Alkyl tosylate (**20**) can be also reduced with a Bu_3SnH and AIBN system in the presence of KI, via the formation of alkyl iodide *in situ* (eq. 2.11).

$(Me_3Si)_3SiH$ and $Ph_4Si_2H_2$ systems initiated by AIBN or Et_3B also reduce alkyl iodides, alkyl bromides, alkyl xanthates, and alkyl selenides to the corresponding reduction products as shown in eq. 2.12.

Radical deoxygenation of alcohols is important, and the reduction of xanthates prepared from alcohols, with Bu_3SnH in the presence of AIBN is called the Barton–McCombie reaction (eq. 2.13) [37–51]. The driving force for the reaction is the formation of a strong C=O bond from the C=S bond, approximately 10 kcal/mol stronger. This reaction can be used for various types of substrates such as nucleosides and sugars. Though methyl xanthates, prepared from alcohols with carbon disulfide and methyl iodide under basic conditions are very frequently used, other thiocarbonates, as shown in eq. 2.14, can also be employed.

24 → 1) CS_2, KOH 2) CH_3I → **25** → AIBN, Bu_3SnH, *p*-cymene, Δ → **26** 65% (2.13)

Experimental procedure 7 (eq. 2.13).

A mixture of Bu_3SnH (200 mg, 0.69 mmol) in *p*-cymene (3 ml) was added dropwise to a solution of methyl xanthate of hederagenin (50 mg, 0.087 mmol) at 150 °C. After 10 h, CCl_4 was added to the mixture and the mixture was refluxed for 3 h. After the reaction, the solvent was removed. A solution of iodine in ether was added to the mixture. The organic layer was washed with 10% KF aq. solution (5 ml), dried over Na_2SO_4, and filtered. After removal of the solvent, methyl oleanoate was obtained by recrystallization [38].

R–OH → R–O–C(=S)–X (2.14)

X = $-SCH_3$, –O–C_6H_5, –N(imidazolyl), –O–C_6H_4–F, –O–C_6H_4–OCH_3, –NH–C_6H_5

Thiocarbonates derived from *sec*- and *tert*-alcohols are more easily reduced than those of *prim*-alcohols. This is because of the slightly stronger C–O bond dissociation energy of thiocarbonates derived from *prim*-alcohols than that of thiocarbonates derived from *sec*- and *tert*-alcohols. This reaction can be carried out with Et_3B at room temperature, instead of AIBN under refluxing conditions, as shown in eq. 2.15.

Polymer-supported di-*n*-butylstannane (di-*n*-butyltin hydride) (**29**) can also reduce alkyl halides and xanthates in solid phase (eq. 2.16).

S
O–C–SCH$_3$
Et$_3$B, Air
Bu$_3$SnH
C$_6$H$_6$ r.t.
27
28 93%

(2.15)

Experimental procedure 8 (eq. 2.15).

To a mixture of *O*-Cyclododecyl *S*-methyl dithiocarbonate (1.0 mmol), Bu_3SnH (320 mg, 1.1 mmol) in benzene (5 ml) was added Et_3B (1.1 ml, 1.1 mmol, 1 M hexane solution). The mixture was stirred for 20 min, then the solvent was removed and the residue was chromatographed on silica gel to give cyclododecane in 93% yield [40].

---

n
Bu$_2$SnH
polymer reagent
29

S
O–C–OPh
AIBN
polymer reagent
C$_6$H$_5$CH$_3$, Δ
30
31 98%

(2.16)

HO
32
1) PhOCSCl, Py 95%
2) (Me$_3$Si)$_3$SiH, 94%
AIBN
C$_6$H$_5$CH$_3$ Δ
33

(2.17)

Experimental procedure 9 (eq. 2.17).

To a mixture of cholesterol (0.77 g, 2 mmol) in dichloromethane (10 ml) were added pyridine (0.6 ml, 8 mmol) and phenoxythiocarbonyl chloride (0.4 ml, 2.2 mmol). After 2 h, methanol (1 ml) was added, and the mixture was washed with 1 M HCl aq. solution twice, then dried over Na_2SO_4. After filtration and removal of the solvent, the residue was recrystallized from acetone to give phenyl thiocarbonate in 95% yield.

To a mixture of phenyl thiocarbonate (0.75 g, 1.5 mmol) in toluene (20 ml) were added $(Me_3Si)_3SiH$ (0.7 ml, 2.2 mmol) and AIBN (50 mg, 0.5 mmol) under a nitrogen atmosphere at 80 °C. After 2 h, the solvent was removed and the residue was recrystallized from acetone to give cholest-5-ene in 94% yield [43].

AIBN
$(Me_3Si)_3SiH$
$C_6H_5CH_3$ Δ
34 → **35** 87%

(2.18)

$Cl-C(=S)-OPh$
DMAP
CH_3CN
36 → **37**

AIBN
reagent
dioxane
Δ
→ **38**

reagent	yield
$(Me_3Si)_3SiH$	78%
Ph_2SiH_2	74%

(2.19)

AIBN
Bu_3SnD
C_6H_6 Δ
39 → **40** 74%(d-content 100%)

(2.20)

Experimental procedure 10 (eq. 2.20).

A mixture of thiocarbamate (0.2 mmol), Bu_3SnD (0.4 mmol), and AIBN (0.04 mmol) in benzene was refluxed for 4 h under an argon atmosphere. After the reaction, the solvent was removed, and the residue was chromatographed on silica gel (eluent: hexane /ethyl acetate = 7/3) to give 2′-deoxy-2′-*d*-3′,5′-*O*-TIPDS-uridine in 74% yield with 100% d-content [46].

$$\mathbf{41} \xrightarrow[\text{AcOEt } \Delta]{\text{AIBN, } Ph_4Si_2H_2} \mathbf{42}\ 86\% \qquad (2.21)$$

Experimental procedure 11 (eq. 2.21).

A mixture of sugar xanthate (0.2 mmol), $Ph_4Si_2H_2$ (0.22 mmol), and AIBN (0.06 mmol) in ethyl acetate (1.5 ml) was refluxed for 16 h under an argon atmosphere. After the reaction, the solvent was removed and the residue was chromatographed on silica gel to give the reduction product [47].

$(Me_3Si)_3SiH$ can also reduce phenoxythiocarbonate derived from cholesterol (**32**) (eq. 2.17), and 2,4-bis(thiocarbonate) (**34**) derived from 2,4-dideoxy-1,6-anhydro-D-glucose (eq. 2.18), in the presence of AIBN under toluene refluxing conditions to generate the corresponding reduction products (**33**), (**35**). Shown above is the deoxygenation of 3′-OH for *anti*-HIV nucleoside (**38**) with Bu_3SnH (eq. 2.19), and the introduction of a deuterium at the 2′-position of nucleoside with Bu_3SnD to form 2′-deoxy-2′-*d*-3′,5′-*O*-TIPDS-uridine (**40**) (eq. 2.20). $Ph_4Si_2H_2$ initiated by AIBN or Et_3B also reduces xanthates (eq. 2.21).

5,10-Dihydro-5,10-disilanthracene and AIBN can reduce xanthates and thiocarbonates in refluxing cyclohexane solution. Reduction of xanthates with monosilanes such as $PhSiH_3$ and Ph_2SiH_2 initiated by AIBN does not work effectively because of their strong Si–H bond dissociation energies. However, the same reactions using either dibenzoyl peroxide or triethylborane as an initiator do induce the effective reduction of xanthates.

2.3 CONVERSION TO HYDROXY GROUPS AND OTHER FUNCTIONAL GROUPS

2.3.1 Conversion to hydroxy groups

The conversion of halides to alcohols is a typical S_N1 or S_N2 reaction in the polar reaction method, and generally the reactions require basic conditions. However, the conversion of halides to alcohols by the radical reaction method can be carried out under neutral conditions. The treatment of alkyl halides with Bu_3SnH /AIBN in toluene under aerobic conditions (atmosphere) gives the corresponding alcohols, by means of the reaction of the alkyl radical with molecular oxygen, and the subsequent reduction of alkyl hydrogen peroxide (ROOH) with Bu_3SnH (eq. 2.22) [52–57]. When $^{18}O_2$ is used instead of $^{16}O_2$ in atmosphere, alcohol containing ^{18}O is obtained, as shown in eq. 2.23.

RX → ROH (Bu_3SnH, Air, 0 ~ 10°C, $C_6H_5CH_3$); RX → [R•] → (O_2) [ROO•] → [ROOH] → (Bu_3SnH) ROH (2.22)

43 → 44 98% (AIBN, $^{18}O_2$, $Bu_2(t\text{-}Bu)SnCl$(cat), $NaBH_4$, t-BuOH 60°C) (2.23)

Experimental procedure 12 (eq. 2.23).

AIBN (5.6 mg, 0.034 mmol), $NaBH_3CN$ (427 mg, 6.8 mmol), and Bu_3SnCl (46.0 mg, 0.17 mmol) were added to a reactor equipped with a Latex balloon. Next, *tert*-BuOH (13.6 ml) and alkyl iodide (3.4 mmol) were added to the reactor, and finally molecular oxygen $^{18}O_2$ (122 ml, 5.1 mmol) was injected and the mixture was heated for 19 h at 60 °C and kept stirred. Then, after the addition of water, the organic layer was extracted with ether four times and dried. After removal of the solvent, the residue was chromatographed on silica gel to give ^{18}O-alcohol in 98% yield [57].

Tetraphenyldistibane ($Ph_2SbSbPh_2$) reacts with alkyl iodides under an oxygen atmosphere to give alcohols.

Treatment of α-iodo lactone (**45**) with triethylborane under oxygen atmosphere gives the corresponding α-hydroxy lactone (**46**), via α-lactone radical species. This reaction comprises of S_H2 reaction by Et• on the iodine atom of α-iodo lactones, reaction of the formed α-lactone radical with molecular oxygen, and subsequent hydrogen-atom abstraction from the solvent to form alkyl hydroperoxide (ROOH). Finally, by the addition of dimethyl sulfide for the reduction of the peroxide, the corresponding α-hydroxy lactone is obtained (eq. 2.24) [58].

45 → 46 82% (1) Et_3B, O_2, CH_2Cl_2 −50°C; 2) CH_3SCH_3) (2.24)

2.3.2 Conversion to other functional groups

In radical reactions, conversion of alkyl halides to oximes is easy. Thus, irradiation of a mixture of alkyl halide (**47**) with amyl nitrite in the presence of $(Bu_3Sn)_2$ gives the oxime

(**48**), through the reaction of the formed alkyl radical with amyl nitrite, and subsequent tautomerization of the nitroso to the oxime as shown in eq. 2.25 [59].

$$C_6H_5CH_2Br \ (\mathbf{47}) \xrightarrow[\text{W-h}\nu,\ \text{r.t.}]{(Bu_3Sn)_2,\ i\text{-AmONO}} C_6H_5CH{=}N{-}OH \ (\mathbf{48}\ 63\%) \quad (2.25)$$

Removal of an amino group readily happens in radical reactions. For example, the amino group is converted to an isonitrile group (**50**) by the treatment with HCO_2COCH_3 (formylation reagent) and subsequent dehydration of the amide with TsCl/Py, and then the formed isonitrile is treated with Bu_3SnH/AIBN to give the corresponding reduction product (**51**), as shown in eq. 2.26a [60–63]. This is the sole method for the practical reduction of an amino group to hydrogen atom (deamination).

$$R{-}NH_2 \ (\mathbf{49}) \xrightarrow[]{1)\ HCO_2COCH_3 \quad 2)\ Py,\ TsCl} R{-}NC \ (\mathbf{50}) \xrightarrow[C_6H_6,\ \Delta]{AIBN,\ Bu_3SnH} RH \ (\mathbf{51},\ 89\%) \quad (2.26a)$$

R– = (steroidal spiroketal skeleton: H_3C, H, CH_3, CH_3, O, O, CH_3)

$$CH_3(CH_2)_{16}CH_2\overset{\oplus}{N}{\equiv}\overset{\ominus}{C}: \ (\mathbf{52}) \xrightarrow[\text{xylene},\ \Delta]{AIBN,\ Bu_3SnH} CH_3(CH_2)_{16}CH_3 \ (\mathbf{53}\ 81\%) \quad (2.26b)$$

Experimental procedure 13 (eq. 2.26b).

A mixture of 1-isocyanooctadecane (1.1 mmol) and AIBN (0.1 g) in dry xylene (50 ml) was added dropwise to a solution of Bu_3SnH (2.2 mmol) in xylene (50 ml) over 2 h. A solution of AIBN (0.1 g) in xylene (50 ml) was also added dropwise to the reaction mixture at the same time over 5 h. After the reaction, the solvent was removed, and the residue was dissolved in pentane. Iodine solution in pentane was added to the residue until the iodine color remained. After removal of the solvent, the residue was purified by preparative TLC (eluent: pentane) to give octadecane in 81% yield [60].

An azide group is also reduced to an amino group with the formation of molecular nitrogen gas, using Bu_3SnH/AIBN [64]. As shown in eq. 2.27, protection of other amino and hydroxy groups is not required to carry out this reaction.

$$\mathbf{54}\ (5'\text{-}N_3\text{ adenosine; }NH_2,\ HO,\ OH) \xrightarrow[C_6H_6,\ \Delta;\ CH_3CON(CH_3)_2]{AIBN,\ Bu_3SnH} \mathbf{55}\ (5'\text{-}NH_2\text{ adenosine; }NH_2,\ HO,\ OH)\ 92\% \quad (2.27)$$

The nitro group is less reactive than halide groups. However, secondary and tertiary alkyl nitro compounds are reduced to the corresponding reduction products with Bu_3SnH/AIBN under refluxing conditions in benzene [65–71]. In this case, it is known that the reaction proceeds via SET from $Bu_3Sn^{\cdot}$ to the nitro group. In practice, an anion radical of the nitro compound can be seen by ESR, and also the byproduct, Bu_3Sn^+, by Sn-NMR. The formed anion radical of the nitro compound produces an alkyl radical and NO_2^- via β-cleavage. One example is shown in eq. 2.28)

56 → (CH_2O, Ac_2O, Base) → **57** 83% → (AIBN, Bu_3SnH, C_6H_6, Δ) → **58** 73% (2.28)

Acyl halides and selenol esters can be reduced to the corresponding aldehydes with Bu_3SnH or $(Me_3Si)_3SiH$ initiated by AIBN under refluxing conditions in benzene (eq. 2.29) [72, 73]. When this reaction is carried out under refluxing conditions (144 °C) in *o*-xylene, decarbonylation of the acyl radical occurs to give the corresponding hydrocarbon (**60**) and carbon oxide (eq. 2.29). The rate constant for decarbonylation of the acyl radical to an alkyl radical and carbon oxide is $\sim 4 \times 10^4\ s^{-1}$(100 °C) in acyl radicals with *sec*- and *tert*-alkyl groups.

59 → (AIBN, Bu_3SnH, C_6H_6, Δ) → **60** + **61** (2.29)

	60	61
benzene (80°C)	8%	92%
o-xylene (144°C)	67%	30%
mesitylene (164°C)	84%	13%

In contrast to eq. 2.29, eq. 2.30 shows the oxidative conversion of aldehydes (**62**) to amides (**63**) via acyl bromides with NBS/AIBN/R'_2NH under refluxing conditions in CCl_4 [74]. The reaction comprises of the abstraction of the formyl hydrogen atom by the succinimidyl radical, bromine atom abstraction from NBS by the acyl radical, and lastly, the polar reaction of acyl bromide with R'_2NH.

$$\underset{\mathbf{62}}{\text{RCHO}} \xrightarrow[\text{R}'_2\text{NH, CCl}_4,\ \Delta]{\text{NBS, AIBN}} \underset{\mathbf{63}}{\text{RC(=O)}-\text{NR}'_2} \qquad (2.30)$$

$$\left[\text{RC(=O)}^\bullet\right] \xrightarrow{\text{NBS}} \text{RC(=O)}-\text{Br}$$

Irradiation with a mercury lamp of (diacyloxyiodo)benzene (**64**) in the presence of disulfide, generates the corresponding alkyl sulfide (**65**) together with iodobenzene and carbon dioxide (eq. 2.31) [75, 76]. This is useful for the preparation of adamantyl sulfides or selenides, since the preparation of such caged *tert*-alkyl sulfides or selenide is not so easy with polar reaction methods.

$$\underset{\mathbf{64}}{\text{PhI}(\text{O-C(=O)-Ad})_2} \xrightarrow[\text{Hg-h}\nu,\ \text{CH}_2\text{Cl}_2,\ \text{r.t.}]{\text{RSSR}} \underset{\mathbf{65}}{\text{R-S-Ad}} \qquad (2.31)$$

Treatment of diselenide with perfluoroalkyl iodide (**66**) in the presence of sodium hydroxymethanesulfinate gives perfluoroalkyl selenide (**67**) in good yield (eq. 2.32). The perfluoroalkyl radical generated by the reaction of perfluoroalkyl iodide with the SO_2 anion radical, reacts on the selenium atom of the diselenide to give perfluoroalkyl selenide and a selenyl radical which dimerizes to the starting diselenide [75, 76].

$$\underset{\mathbf{66}}{C_2F_5I} + \text{PhSeSePh} \xrightarrow[\text{DMF, H}_2\text{O, r.t.}]{\text{HOCH}_2\text{SO}_2\text{Na}} \underset{\mathbf{67}}{C_2F_5-\text{SePh}} \qquad (2.32)$$

$$\text{HOCH}_2\text{SO}_2\text{Na} \longrightarrow SO_2^{\bullet -}$$

$$C_2F_5I + SO_2^{\bullet\ominus} \longrightarrow C_2F_5^\bullet + I^\ominus + SO_2$$

Suitable alkylbenzene side chains are oxidized at the benzylic position under the action of *m*CPBA, air, and $NaHCO_3$ to generate the corresponding ketones. The oxygen-centered radical formed from *m*CPBA, abstracts the benzylic hydrogen atom of the benzylic substrate (**68**) to form a benzylic radical, and then it reacts with molecular oxygen ultimately to provide benzylic ketone (**69**) as shown in eq. 2.33 [77].

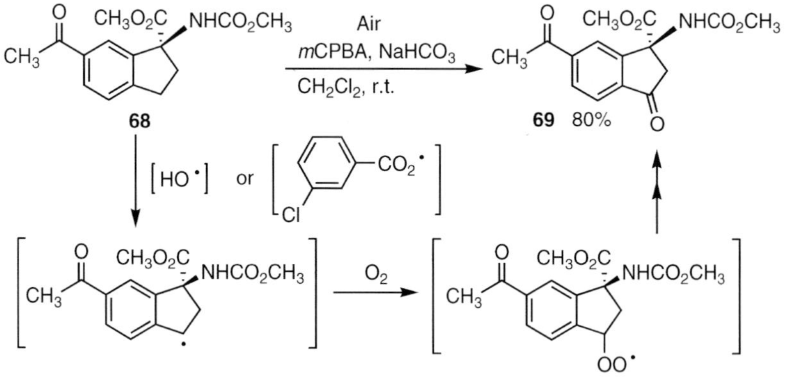

(2.33)

Experimental procedure 14 (eq. 2.33).

To a solution of indane derivative (1.0 mmol) in 10 ml of CH_2Cl_2 were added $NaHCO_3$ (2.0 mmol) and *m*CPBA (2.5 mmol). The mixture was stirred at room temperature under air for 12 to 24 h. When the reaction was completed, the mixture was diluted with 20 ml of CH_2Cl_2 and washed with saturated aqueous $NaHCO_3$ and brine. The organic layer was dried over Na_2SO_4 then the solvent was removed, and the residue was chromatographed on silica gel to give 1-indanone in 80% yield [77].

Secondary β-bromo alcohol (**70**) can be transformed to ketone (**71**) in good yield via the following radical pathway with DBPO [78]. The reaction involves the abstraction of an hydrogen atom at the α-position of the HO group, followed by β-elimination of the bromine atom, and then the tautomerization of the formed enol to the ketone (eq. 2.34).

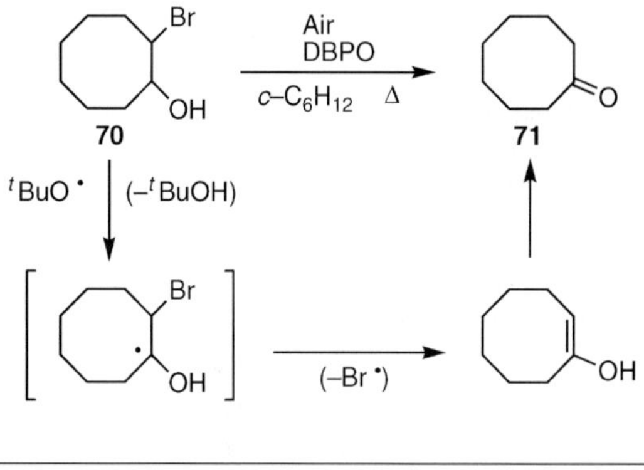

(2.34)

Experimental procedure 15 (eq. 2.34).

Bromohydrin (2 mmol) and di-*tert*-butyl peroxyoxalate (DBPO) (0.3 mmol) were dissolved in 5 ml of cyclohexane. The mixture was purged with argon gas and warmed gradually from 50 °C to reflux for 15 min, and left at refluxing conditions for an additional 10 min. After the reaction, the mixture was diluted with ether, washed with $NaHCO_3$ aqueous solution, and dried over Na_2SO_4. After removal of the solvent, the residue was chromatographed on silica gel to give ketone in 93% yield [78].

Reaction of primary or secondary alcohols (**72**) with $BrCCl_3$ and tetrahydrofuran leads to the generation of 2-tetrahydrofuranyl ethers (**73**) in good yields. The reaction comprises of the abstraction of an α-hydrogen atom of THF by $^{\cdot}CCl_3$, subsequent reaction of the tetrahydrofuranyl radical with $BrCCl_3$, and the polar reaction of the formed α-bromotetrahydrofuran with alcohol to form 2-tetrahydrofuranyl ether, as shown in eq. 2.35 [79].

ROH **72** —(Air, $BrCCl_3$ / THF)→ 2-tetrahydrofuranyl-OR **73**

$\cdot CCl_3$, $CHCl_3$, $BrCCl_3$, ROH, HBr; ROH ≡ CH_3O_2C–C_6H_4–CH_2OH (2.35)

REFERENCES

1. DA White, *Org. Synth.*, 1981, 60.
2. PG Gassman, J Seter and FJ Williams, *J. Am. Chem. Soc.*, 1971, **93**, 1673.
3. MH Habibi and S Farhadi, *Tetrahedron Lett.*, 1999, **40**, 2821.
4. JL Namy, J Souppe and HB Kagan, *Tetrahedron Lett.*, 1983, **24**, 765.
5. P Girard, R Couffignal and HB Kagan, *Tetrahedron Lett.*, 1981, **22**, 3959.
6. J Souppe, JL Namy and HB Kagan, *Tetrahedron Lett.*, 1984, **25**, 2869.
7. N Benircasa, L Forti, F Ghelfi and UM Pagnoni, *Tetrahedron Lett.*, 1995, **36**, 1103.
8. T Imamoto, T Kusumoto, Y Hatanaka and M Yokoyama, *Tetrahedron Lett.*, 1982, **23**, 1353.
9. N Iwasawa, S Hayakawa, K Isobe and K Narasaka, *Chem. Lett.*, 1991, 1193.
10. NA Porter and IJ Rosenstein, *Tetrahedron Lett.*, 1993, **34**, 7865.
11. B Giese, B Ruckert, KS Groninger, R Muhn and HJ Lindner, *Liebigs Ann. Chem.*, 1988, 997.
12. M Yoshida, F Sultana, N Uchiyama, T Yamada and M Iyoda, *Tetrahedron Lett.*, 1999, **40**, 735.
13. S Tews, M Hein and R Miethchen, *Tetrahedron Lett.*, 2002, **43**, 443.
14. Y Kashiwagi, H Ono and T Osa, *Chem. Lett.*, 1993, 257.
15. SN Frank, A Bard and A Ledwith, *J. Electochem. Soc.*, 1968, **90**, 4645.
16. O Hammerich and VD Parker, *Acta. Chem. Scand.*, 1982, **B36**, 519.
17. JM Bobbitt, H Yagi, S Shibuya and JT Stork, *J. Org. Chem.*, 1971, **36**, 3006.
18. S Tews, M Hein and R Miethchen, *Tetrahedron Lett.*, 2002, **43**, 443.
19. SS Kim, YJ Mah and AR Kim, *Tetrahedron Lett.*, 2001, **42**, 8315.
20. SS Kim, YJ Mah and AR Kim, *Tetrahedron Lett.*, 2001, **42**, 8315.

21. ALJ Beckwith and PE Pigou, *Aust. J. Chem.*, 1986, **39**, 77.
22. D Crich, JT Hwang, S Gastaldi, F Recupero and DJ Wink, *J. Org. Chem.*, 1999, **64**, 2877.
23. C Chatgilialoglu, KU Ingold and JC Scaiano, *J. Am. Chem. Soc.*, 1981, **103**, 7739.
24. LJ Johnston, J Lusztyk, DDM Wayner, AN Abeywickreyma, ALJ Bechwith, JC Scaiano and KU Ingold, *J. Am. Chem. Soc.*, 1985, **107**, 4594.
25. JW Wilt, J Lusztyk, M Peeran and KU Ingold, *J. Am. Chem. Soc.*, 1988, **110**, 281.
26. C Chatgilialoglu, J Dickhaut and B Giese, *J. Org. Chem.*, 1991, **56**, 6399.
27. C Chatgilialoglu, D Griller and M Lesage, *J. Org. Chem.*, 1989, **54**, 2492.
28. DPG Hamon and KR Richards, *Aust. J. Chem.*, 1983, **36**, 2243.
29. HG Davies, SM Roberts, BJ Wakefield and JA Winders, *J. Chem. Soc. Chem. Commun.*, 1985, 1166.
30. DG Norman and CB Reese, *Synthesis*, 1983, 304.
31. B Giese and J Dupuis, *Tetrahedron Lett.*, 1984, **25**, 1349.
32. B Giese, KS Gröninger, T Witzel, H-G Korth and R Sustmann, *Angew. Chem. Int. Ed. Eng.*, 1987, **26**, 233.
33. Y Ueno, C Tanaka and M Okawara, *Chem. Lett.*, 1983, 795.
34. C Chatgilialoglu, D Griller and M Lesage, *J. Org. Chem.*, 1988, **53**, 3641.
35. M Ballestri, C Chatgilialoglu, KB Clark, D Griller, B Giese and B Kopping, *J. Org. Chem.*, 1991, **56**, 678.
36. O Yamazaki, H Togo, S Matsubayashi and M Yokoyama, *Tetrahedron*, 1999, **55**, 3735.
37. DHR Barton and SW McCombie, *J. Chem. Soc. Perkin Trans. 1*, 1975, 1574.
38. DHR Barton, WB Motherwell and A Stange, *Synthesis*, 1981, 743.
39. DHR Barton, W Hartwig, RS Motherwell, WB Motherwell and A Stange, *Tetrahedron Lett.*, 1982, **23**, 2019.
40. K Nozaki, K Oshima and K Utimoto, *Tetrahedron Lett.*, 1988, **29**, 6125.
41. M Gerlach, F Jördens, H Kuhn, WP Neumann and M Peterseim, *J. Org. Chem.*, 1991, **56**, 5972.
42. WP Neumann and M Peterseim, *Synlett*, 1992, 801.
43. D Schummer and G Höfle, *Synlett*, 1990, 705.
44. DHR Barton, DO Jang and JC Jaszberenyi, *Tetrahedron Lett.*, 1992, **33**, 6629.
45. S Takamatsu, T Maruyama, S Katayama, N Hirose and K Izawa, *Tetrahedron Lett.*, 2001, **42**, 2321.
46. M Oba and K Nishiyama, *Tetrahedron*, 1994, **50**, 10193.
47. H Togo, S Matsubayashi, O Yamazaki and M Yokoyama, *J. Org. Chem.*, 2000, **65**, 2816.
48. T Gimisis, M Ballestri, C Ferreri and C Chatgilialoglu, *Tetrahedron Lett.*, 1995, **36**, 3897.
49. DHR Barton, DO Jang and JC Jaszberenyi, *Synlett*, 1991, 435.
50. DHR Barton, DO Jang and JC Jaszberenyi, *Tetrahedron Lett.*, 1993, **49**, 2793.
51. DHR Barton, DO Jang and JC Jaszberenyi, *Tetrahedron Lett.*, 1990, **31**, 4681.
52. E Nakamura, T Inubushi, S Aoki and D Machii, *J. Am. Chem. Soc.*, 1991, **113**, 8980.
53. AGM Barrett and LM Melcher, *J. Am. Chem. Soc.*, 1991, **113**, 8177.
54. S Moutel and J Prandi, *Tetrahedron Lett.*, 1994, **35**, 8163.
55. E Nakamura, K Sato and Y Imanishi, *Synlett*, 1995, 525.
56. T Nagashima and DP Curran, *Synlett*, 1997, 330.
57. M Sawamura, Y Kawaguchi and E Nakamura, *Synlett*, 1997, 801.
58. N Kihara, C Ollivier and P Renaud, *Org. Lett.*, 1999, **1**, 1419.
59. RJ Fletcher, M Kizil and JA Murphy, *Tetrahedron Lett.*, 1993, **36**, 323.
60. DHR Barton, G Bringmann, G Lamotte, WB Motherwell, RSH Motherwell and AEA Porter, *J. Chem. Soc. Perkin Trans. 1*, 1980, 2657.
61. DHR Barton, W Hartwig and WB Motherwell, *Chem. Commun.*, 1982, 447.
62. DHR Barton, G Bringmann and WB Motherwell, *Synthesis*, 1980, 68.
63. DHR Barton, G Bringmann and WB Motherwell, *J. Chem. Soc. Perkin Trans. 1*, 1980, 2665.
64. MC Samano and MJ Robins, *Tetrahedron Lett.*, 1991, **32**, 6293.
65. N Ono, H Miyake, R Tamura and A Kaji, *Tetrahedron Lett.*, 1981, **22**, 1705.
66. N Ono, I Hamamoto, H Miyake and A Kaji, *Chem. Lett.*, 1982, 1079.
67. N Ono, H Miyake and A Kaji, *Chem. Commun.*, 1982, 33.
68. N Ono, H Miyake and M Fujii, *Tetrahedron Lett.*, 1983, **24**, 3477.

69. N Ono, A Kamimura, H Miyake, I Hamamoto and A Kaji, *J. Org. Chem.*, 1985, **50**, 3692.
70. J Dupuis, B Giese, J Hartung and M Leising, *J. Am. Chem. Soc.*, 1985, **107**, 4332.
71. N Ono, M Fujii and A Kaji, *Synthesis*, 1987, 532.
72. J Pfenninger, C Heuberger and W Graf, *Helv. Chim. Acta.*, 1980, **63**, 2328.
73. M Ballestri, C Chatgilialoglu, N Cardi and A Sommazzi, *Tetrahedron Lett.*, 1992, **33**, 1787.
74. IE Marko and A Mekhalfia, *Tetrahedron Lett.*, 1990, **31**, 7237.
75. H Togo, T Muraki and M Yokoyama, *Synthesis*, 1995, 155.
76. E Magnier, E Vit and C Wakselman, *Synlett*, 2001, 1260.
77. D Ma, C Xia and H Tian, *Tetrahedron Lett.*, 1999, **40**, 8915.
78. D Dolenc and M Harej, *J. Org. Chem.*, 2002, **67**, 312.
79. JM Barks, BC Gilbert, AF Parsons and B Upeandran, *Tetrahedron Lett.*, 2000, **41**, 6249.

– 3 –

Intramolecular Radical Cyclizations

3.1 CYCLIZATION TO SMALL-SIZED RINGS

3.1.1 Cyclization onto olefinic groups and acetylenic groups by sp^3 carbon-centered radicals

Formation of 5- and 6-membered rings

The formation of 5- and 6-membered rings via a radical pathway is very important in organic synthesis. Although the most typical system is the reaction of organic halides or selenides with Bu_3SnH/AIBN, other less toxic reagent systems have been recently developed. This cyclization reaction is based on Baldwin's rule, as mentioned in Chapter 1, and so, 5-*exo-trig* manner and 6-*exo-trig* manner are the most typical and popular cyclization modes [1–4]. Eq. 3.1 shows the ratio of the reaction products in the reduction of 1-bromo-5-hexene (**1**) with Bu_3SnH and $(Me_3Si)_3SiH$, respectively. Generally, methylcyclopentane (**2a**), formed through the 5-*exo-trig* manner, is the main product, based on a kinetic controlled pathway. Since $(Me_3Si)_3SiH$ has stronger bond dissociation energy of Si–H (79 kcal/mol) than that of Sn–H in Bu_3SnH (74 kcal/mol), the amount of cyclized products with $(Me_3Si)_3SiH$ is increased, in comparison with Bu_3SnH.

1 —(AIBN, 70 °C)→ **2a** + **2b** + **2c** (3.1)

	2a	2b	2c
$(Me_3Si)_3SiH$	93%	2%	4.1%
Bu_3SnH	83%	1.2%	15%

$[Bu_3SnH]$ (M)	Temp(°C)	Direct reduction	5-*exo-trig*	6-*endo-trig*
0.5	40	55.0	23.1	21.9
0.025	40	8.9	47.1	44.0
0.5	100	25.6	27.6	46.8
0.025	100	4.1	37.4	58.5

(3.2)

The results for the cyclization of the 1,1,5-trimethyl-5-hexen-1-yl radical, formed from the reaction of 1,1,5-trimethyl-5-hexenyl-1-bromide (**3**) with Bu_3SnH/AIBN, are shown in eq. 3.2. These results indicate that the ratio of 5-*exo-trig* manner/6-*endo-trig* manner depends on both the temperature and concentration of Bu_3SnH. Thus, as the concentration of Bu_3SnH at 40 °C is decreased (which means the decrease of the concentration of the hydrogen donor), both a decrease of the direct reduction product, 2,6-dimethyl-1-heptene (**4a**), and an increase of the cyclization products (**4b**) and (**4c**) are observed. With the same concentration of Bu_3SnH at 100 °C, a similar tendency is observed. However, the amount of cyclized products is further increased, particularly that of the cyclized product via 6-*endo-trig* manner. Generally, cyclization requires more activation energy than that of the direct reduction, and the formation of thermodynamically controlled products is preferable at high temperature. The (1,2,2-trimethylcyclopentyl)methyl radical formed via 5-*exo-trig* manner is a primary alkyl radical, and the 2,2,6-trimethylcyclohexyl radical formed from 6-*endo-trig* manner is a secondary alkyl radical, respectively. So, at high temperature, the formation of the 2,2,6-trimethylcyclohexyl radical is preferable.

Introduction of heteroatoms to the radical chain changes the regioselectivity of the ring closure from 5-*exo-trig* to 6-*endo-trig* as shown in eq. 3.3, depending on the bond length and the electronic character of the formed carbon-centered radicals ($RSCH_2^{\cdot}$ and $RSO_2CH_2^{\cdot}$). This result indicates that the α-sulfonylmethyl radical proceeds mainly through a thermodynamically controlled pathway.

Bu_3SnH, 80 °C

	5a		**5b**
X = S	1	:	8
X = SO_2	1	:	37

(3.3)

The effect of the side-chained alkyl group on the *cis*/*trans* ratio in the cyclization must be considered. For example, the effects of a methyl group, 2-methyl, 3-methyl, and 4-methyl, in the cyclization of the 5-hexen-1-yl radical is shown in eq. 3.4. The main cyclization product of the 2-methyl-5-hexen-1-yl radical is a *trans* isomer, that of the 3-methyl-5-hexen-1-yl radical is a *cis* isomer, and that of the 4-methyl-5-hexen-1-yl radical is again a *trans* isomer, through the preferable conformation with less steric hindrance (methyl group takes equatorial position) in the transition states.

favor → trans isomer, *trans*/*cis* = 1.8/1.0 (3.4a)

favor → *cis* isomer, *cis*/*trans* = 2.5/1.0 (3.4b)

favor → *trans* isomer, *trans*/*cis* = 4.8/1.0 (3.4c)

The rate constants of these cyclizations via 5-*exo-trig* manner are $2.4 \times 10^5\ s^{-1}$, $7.0 \times 10^5\ s^{-1}$, $7.5 \times 10^5\ s^{-1}$, respectively, and they are close to that of the parent 5-hexen-1-yl radical ($2.4 \times 10^5\ s^{-1}$). Moreover, the introduction of dimethyl groups at the 2,2-, 3,3-, and 4,4-positions of the 5-hexen-1-yl radical, increases the rate constants for cyclization about 10 times and they become approximately $10^6\ s^{-1}$. These cyclization reactions proceed via the SOMO–LUMO interaction. Therefore, the introduction of three fluorine atoms to the olefinic group in the 5-hexen-1-yl radical accelerates the cyclization rate, 2 times. Probably, the substitution of hydrogen atoms by fluorine atoms onto the olefinic group induces the decrease of LUMO energy [5–12].

In any case, 5-*exo-trig* manner via a radical pathway is the most strategic method for the construction of 5-membered cyclic compounds [13–20]. For example, eqs. 3.5a, 3.5b, and 3.5c show the preparation of γ-lactones (**7**) and γ-lactols (**9**), (**12**) from the selenide (**6**) and bromides (**8**), (**11**) with Bu_3SnH/AIBN. The acetal and ketal groups, which are acid-sensitive groups, are not affected during the radical reactions, and, moreover, the starting materials are simply obtained from the reaction of vinyl ethers with carboxylic acids or alcohols in the presence of PhSCl or NBS. 5-*Exo-trig* cyclization onto the allene group also goes on to form antibiotics, the Batryodiplodin model compound (**16**) as shown in eq. 3.5d.

6 —AIBN, Bu_3SnH, C_6H_6, Δ→ **7a** + **7b** 63%(4 : 1) (3.5a)

OBu^i Br, OC$_2$H$_5$ **8** — AIBN, Bu$_3$SnH, C$_6$H$_6$, Δ, 1 h → OBu^i, OC$_2$H$_5$, H **9** 91% — PDC → OBu^i, O, H **10** 70% (3.5b)

Br, OEt **11** — AIBN, Bu$_3$SnH, C$_6$H$_6$, Δ → CH$_3$, H, OEt **12** 81% — Jone's oxidation → CH$_3$, H, O **13** 97% (3.5c)

EtO, CH$_3$, Br, Br **14** — Et$_3$B, O$_2$, Bu$_3$SnH, C$_6$H$_5$CH$_3$ −78 °C → EtO, Br, CH$_3$ **15** — Et$_3$B, O$_2$, (Me$_3$Si)$_3$SiH, −78 °C → EtO, CH$_3$ **16** 62% [HO, CH$_3$, CH$_3$, O] (3.5d)

(–)Batryodiplodin

n-C$_6$H$_{13}$, Br, OEt **17** — AIBN, Bu$_3$SnCl, NaBH$_3$CN, *t*-BuOH Δ → *n*-C$_6$H$_{13}$, OEt **18** 97% (3.6)

TsOH, C$_6$H$_6$ Δ → *n*-C$_6$H$_{13}$, O **19** 89%

Ph, S, SCH$_3$, C, O **20** — Et$_3$B, Bu$_3$SnH, C$_6$H$_6$, −78 °C, 1h → Ph, O, O **21** 78%

H$_2$O (3.7)

[Ph, SSnBu$_3$, C–SCH$_3$, O] → [Ph, SSnBu$_3$, SCH$_3$, O] → Ph, S, O

Eq. 3.6 shows the direct preparation of furan derivative (**19**) by the successive treatment of vinyl ethers with propargyl alcohol/NBS, radical cyclization with Bu_3SnH/AIBN, and lastly, aromatization with *p*-TsOH.

Eq. 3.7 shows the cyclization of the xanthate ester (**20**) formed from homopropargyl alcohol, through 5-*exo-dig* manner of the formed carbon-centered radical. This reaction is not a deoxygenation reaction of alcohol, but the cyclization of the carbon-centered radical formed from the addition of $Bu_3Sn^{\cdot}$ to the thiocarbonyl-sulfur atom, onto the side-chained triple bond to form thionolactone, the hydrolysis of which readily creates lactone (**21**) [21–23].

Experimental procedure 1 (eq. 3.7).

To a toluene (20 ml) solution of dithiocarbonate (1.0 mmol) and Et_3B (1.0 M hexane solution, 1.1 ml, 1.1 mmol) was added dropwise Bu_3SnH (0.5 M, 1.1 mmol) at −78 °C under an argon atmosphere. The mixture was then stirred for 30 min at the same temperature. Next, 1 M HCl solution was added and the reaction mixture was extracted with $CHCl_3$. The organic layer was dried over Na_2SO_4 and filtered. After removal of the solvent, the residue was chromatographed on silica gel to produce α-benzylidene-γ-butyrolactone in 78% yield [21].

As shown in eqs. 3.5 to 3.7, furan, thiophene, and pyrrolidine derivatives can be obtained via 5-*exo-trig* and 5-*exo-dig* by the introduction of heteroatoms such as oxygen, sulfur, and nitrogen atoms onto the 5-hexen-1-yl radical and 5-hexyn-1-yl radical chains [24–43]. These reactions can also be carried out with a polymer-supported solid-phase cyclization method using polymer-supported tin hydride, instead of Bu_3SnH. 1,1,2,2-Tetraphenyldisilane ($Ph_4Si_2H_2$) is a stable and useful radical reagent. Treatment of 2-bromo-3,4,6-tri-*O*-acetyl-1-*O*-allyl glycoside (**22**) with $Ph_4Si_2H_2/Et_3B$ (r.t.) and $Ph_4Si_2H_2$ /AIBN (reflux) gives the corresponding bicyclic sugar (**23a**) alone in both cases, while the same treatment with Bu_3SnH/Et_3B and Bu_3SnH /AIBN provides a mixture of bicyclic sugar (**23a**) and the direct reduction product (**23b**), in both cases [44]. Less toxic $Ph_4Si_2H_2/Et_3B$ and $Ph_4Si_2H_2$/AIBN systems can be used for the preparation of various types of bicyclic sugars.

22 —(initiator, MH, AcOEt)→ **23a** + **23b** (3.8)

MH	initiator	yields (%)	
$Ph_4Si_2H_2$	(r.t.) Et_3B	84	0
	(Δ) AIBN	78	0
Bu_3SnH	(r.t.) Et_3B	37	44
	(Δ) AIBN	65	32

Experimental procedure 2 (eq. 3.8).

A mixture of sugar bromide (0.3 mmol), $Ph_4Si_2H_2$ (0.36 mmol), and AIBN (0.15 mmol) in ethyl acetate (3 ml) was refluxed under an argon atmosphere. After 4 h, K_2CO_3 (0.15 mmol) and water were added to the reaction mixture, and the organic layer was extracted with chloroform. The organic layer was dried over Na_2SO_4 and filtered. After removal of the solvent, the residue was purified by column chromatography on silica gel or preparative TLC (eluent: hexane/ethyl acetate = 1/1 ~ 3/1) to give bicyclic sugar in 78% yield [44].

Experimental procedure 3 (eq. 3.8).

To a flask equipped with a refluxing cooler were added sugar bromide (0.3 mmol), $Ph_4Si_2H_2$ (0.36 mmol), and ethyl acetate (3 ml). Et_3B (0.2 ml, 1 M THF solution) was added to the mixture under aerobic conditions at room temperature. After 1 h, K_2CO_3 (0.15 mmol) and water were added to the mixture, and the organic layer was extracted with chloroform. The organic layer was dried over Na_2SO_4 and filtered. After removal of the solvent, the residue was purified by column chromatography on silica gel or preparative TLC (eluent: hexane/ethyl acetate = 1/1 ~ 3/1) to give bicyclic sugar in 84% yield [44].

Gallium hydride, $HGaCl_2$, also acts as a radical mediator like Bu_3SnH, as shown in eq. 3.9 [45].

HGaCl$_2$, Et$_3$B, THF, 0 °C, 3 h: **24** → **25** 87% (70 / 30) (3.9)

The preparation of alkaloids (cyclic compounds containing nitrogen atoms) is very important, as most alkaloids are biologically active. Eq. 3.10 shows the preparation of alkaloids (**27**), a precursor of gephyrotoxin, via an α-amino radical formed from γ-thiophenoxylactam (**26**). Other alkaloids such as indolizidine and pyrrolizidine can be also prepared similarly via α-amino radicals and α-acylamino radicals.

AIBN, Bu$_3$SnH, C$_6$H$_6$, Δ: **26** → **27a** 51% + **27b** 31% (3.10)

The bond dissociation energy of C–Cl in alkyl chlorides is about 80 kcal/mol and those of C–Br and C–I in alkyl bromides and alkyl iodides are about 68 kcal/mol and 53 kcal/mol, respectively. Therefore, the reactivity of alkyl chlorides toward $Bu_3Sn^{\cdot}$ is rather poor, and is not useful for organic synthesis. However, the chlorine atom in the trichloromethyl or chlorodifluoromethyl group is very reactive and it readily reacts with $Bu_3Sn^{\cdot}$ to form the corresponding electron-deficient α,α-dihalo carbon-centered radical due to its bearing two halogen atoms at the α-position. Thus, the formed carbon-centered radical is electrophilic and cyclizes at the olefinic group via the SOMO–HOMO orbital interaction as shown in eqs. 3.11a and 3.11b [46–49].

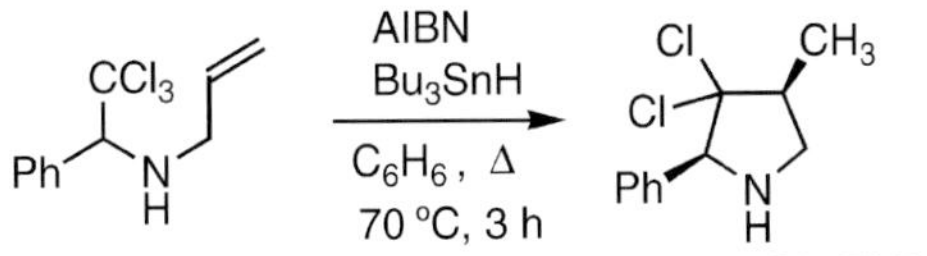

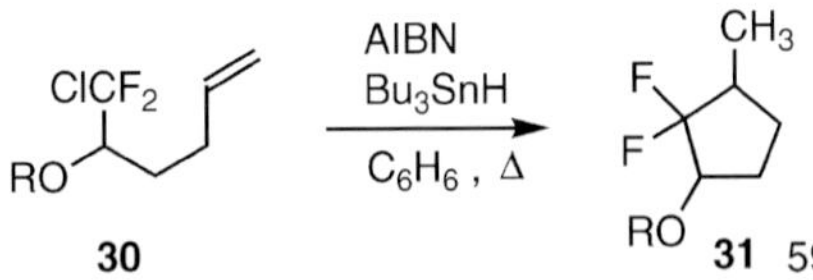

(3.11a)

(3.11b)

(3.11c)

Prolinates (**33**) containing α,α-difluoromethylene group can be also prepared via *N*-protected α-methyl difluoroalaninyl radicals (eq. 3.11c).

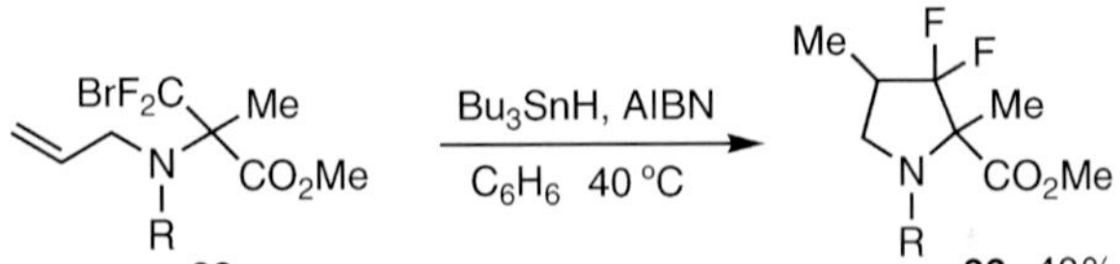

Treatment of epoxide (**34**) with Bu_3SnH/AIBN in the presence of MgI_2 first forms iodohydrin synthon, which rapidly reacts with $Bu_3Sn^{\cdot}$ to form a cyclohexanol derivative (**35**) via 6-*exo-trig* ring closure of β-hydroxyl radical as shown in eq. 3.12 [50–54]. Since the epoxides can be obtained from alkenes with peroxides, this is an indirect radical cyclization method of alkenes.

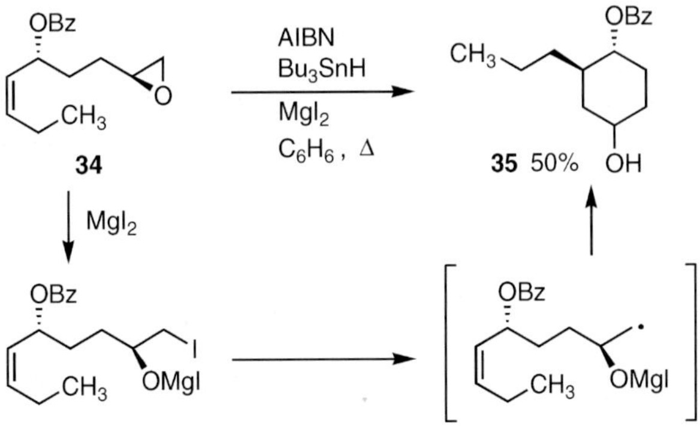

(3.12)

γ-Lactones containing fluorine atoms at the α-position are attractive compounds because of their biological activity. However, α,α,α-bromodifluoroacetate (**36**), in eq. 3.13, does not cyclize directly via 5-*exo-trig* manner, because the formed α-difluoroacetate radical may be stabilized. Instead, only the direct reduction product is formed. Treatment of TMS-acetal (**37**), prepared from the reduction of α,α,α-bromodifluoroacetate (**36**) with DIBAL, with Bu_3SnH/AIBN produces the corresponding lactol (**38**) via 5-*exo-trig* manner, and finally the oxidation with PDC gives α,α-difluoro-γ-lactone (**39**).

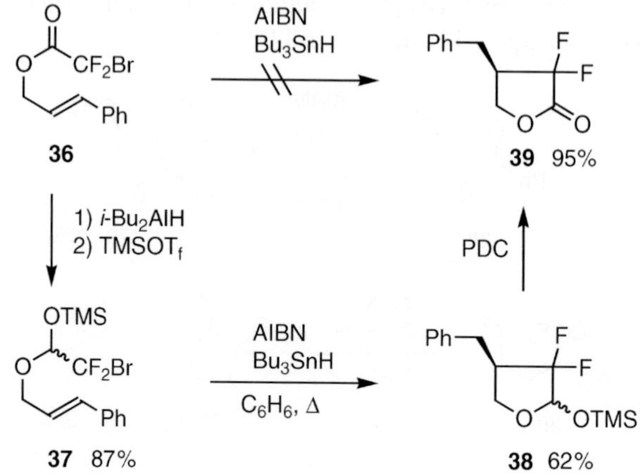

(3.13)

5-*endo-trig* ring-closure is now considered to be a disfavored process, based on Baldwin's rule. However, 5-*endo-trig* ring-closure does occur when the reaction proceeds via thermodynamic control, depending on the stability of the formed carbon-centered radicals. Thus, in eq. 3.14a, the cyclized carbon-centered radical [I] formed via 5-*endo-trig* manner is more stable than the precursor α-acetamide radical, because radical [I] is stabilized by capto-dative effect bearing both an electron-donating nitrogen group and an electron-withdrawing ester group at the α-position. Thus, the cyclization occurs through a thermodynamically controlled pathway [55–60].

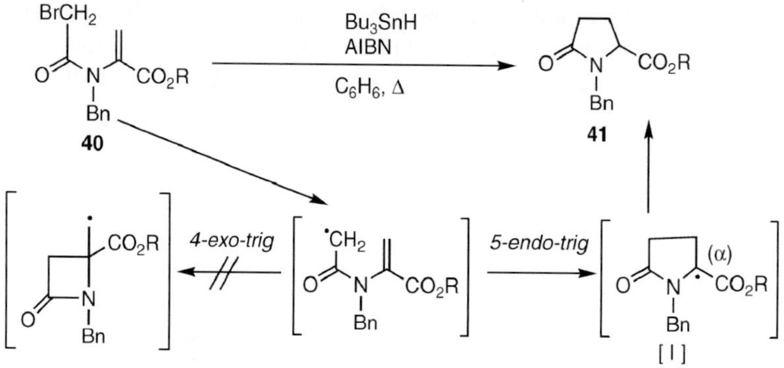

(3.14a)

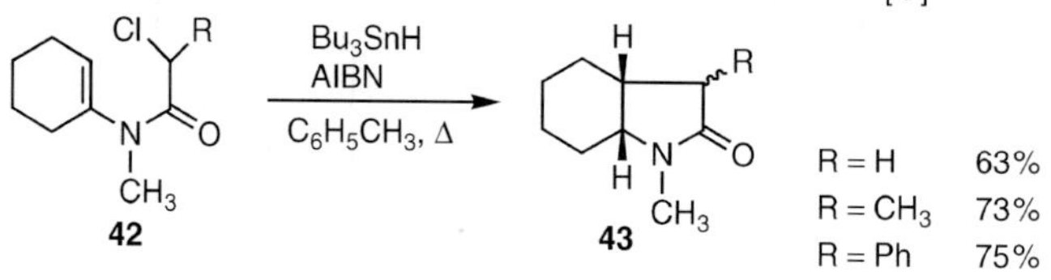

(3.14b)

α-Sulfonamidyl radical generated from α-halosulfonamide (**44**) with Bu_3SnH/AIBN prefers 5-*exo-trig* ring-closure to form 5-membered sultam (**45**) (eq. 3.15a). However, 7-*endo-trig* ring-closure predominates over 6-*exo-trig* ring closure as shown in eq. 3.15b [61–65].

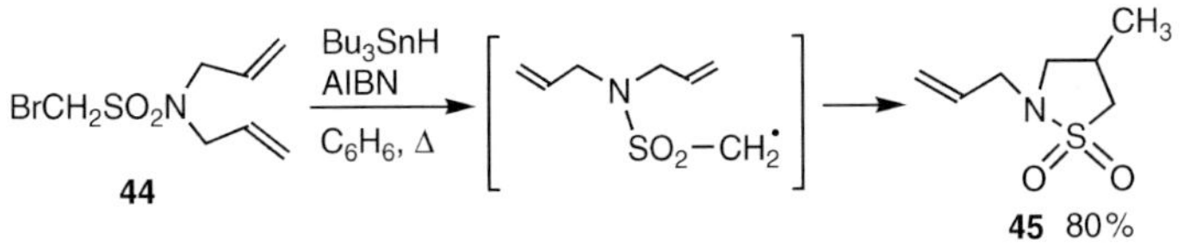

(3.15a)

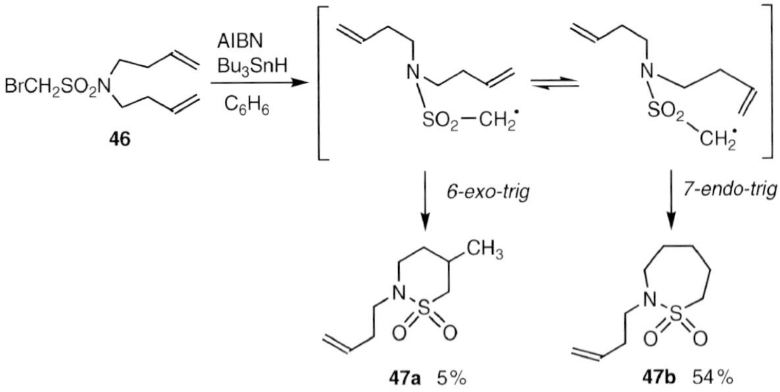

(3.15b)

Since Bu_3SnH is a reducing agent, cyclization onto the stable aromatic ring does not generally happen. However, when the aromatic ring is activated to an electron-deficient ring such as the pyridinium ring, cyclization onto the aromatic ring proceeds with ease. For example, treatment of *N*-iodoalkylpyridinium salts (**48**) with Bu_3SnH/AIBN under refluxing conditions generates the corresponding cyclization products (**49**) (eq. 3.16a). By this method, quinolidine alkaloids and indolidine alkaloids can be prepared [61–64]. Using a similar method, *N*-(3-iodopropyl)indole (**50**) activated by a formyl group, produces the cyclic compound (**51**), as shown in eq. 3.16b.

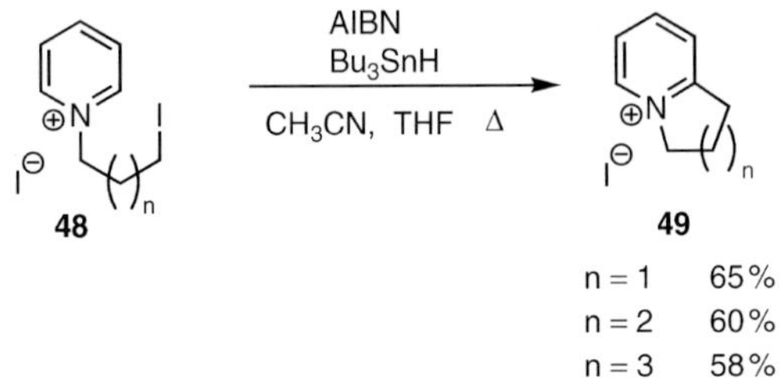

(3.16a)

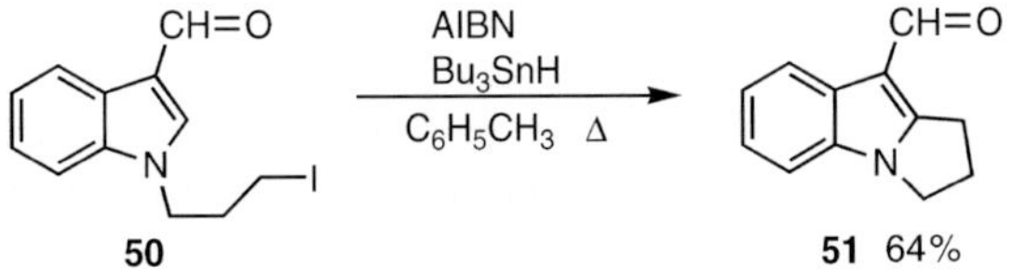

(3.16b)

Formation of 5- and 6-membered rings via skillful procedure

Generally, alkenyl halides are used for the radical cyclization. However, other functional groups can also be used for the Bu_3SnH-mediated radical cyclization. Eq. 3.17a shows the addition of $Bu_3Sn^{\cdot}$ to a formyl group in α-allyloxyacetaldehyde (**52**), the formation of a carbon-centered radical, and 5-*exo-trig* ring closure to form a tetrahydrofuran derivative (**53**), while eq. 3.17b shows the addition of $Bu_3Sn^{\cdot}$ to an acetyl group in bis-(α,β-unsaturated methyl ketone) (**54**), the formation of ketyl radical, 5-*exo-trig* ring closure, and finally, the elimination of Bu_3Sn group to form *trans*-1,2-diacetonylcyclopentane (**55a**) [66–77].

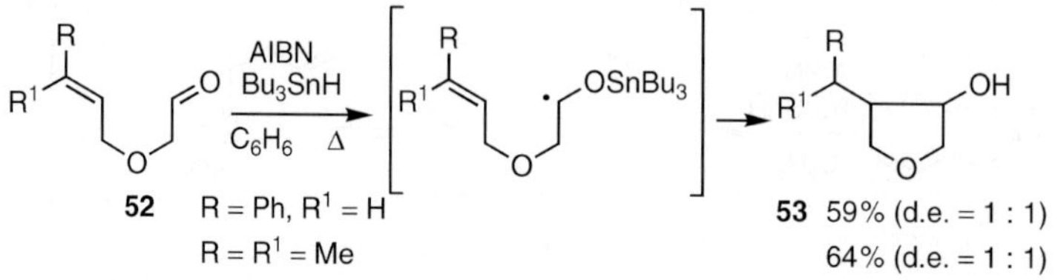

(3.17a)

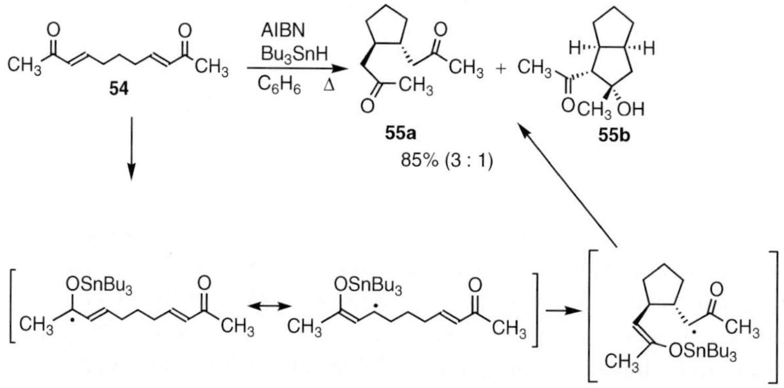

(3.17b)

In the reaction of *O*-benzoyl hydroxamic acid (**56**) with Bu_3SnH, $Bu_3Sn^{\cdot}$ adds to the carbonyl oxygen and subsequent β-elimination occurs to form an amidyl radical through the weak N–O bond cleavage, where finally, γ-lactam (**57**) via 5-*exo-trig* manner is formed (eq. 3.18).

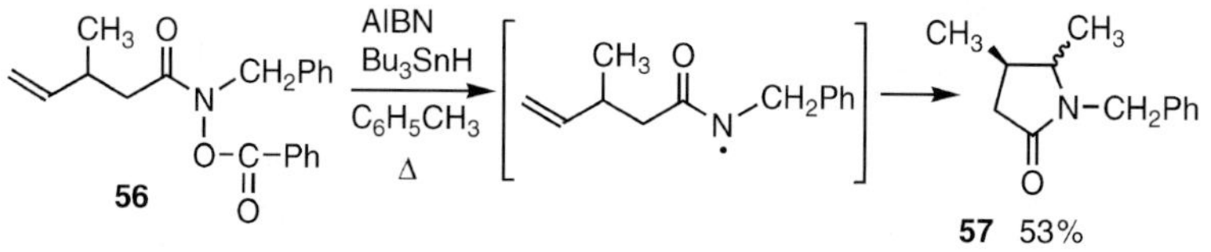

(3.18)

Xanthates, which are easily prepared from the reaction of alcohols, carbon disulfide, and methyl iodide in the presence of a base, or from the reaction of alcohols with *N,N′*-thiocarbonyldiimidazole, can generally be used for the deoxygenation of alcohols to the corresponding hydrocarbons. So, eq. 3.19 shows the deoxygenative cyclization of

5-alken-1-ol and 6-alken-1-ol derivatives (**58**) to form cyclopentane and cyclohexane derivatives (**60**), respectively.

(3.19)

Eq. 3.20 shows the addition of $Bu_3Sn^{\cdot}$ to thiocarbonyl sulfur of compound (**61**) to form carbon-centered radical, which cyclizes to olefinic group via 5-*exo-trig* manner to give thiollactone (**62**), after hydrolysis.

(3.20)

Eq. 3.21 shows the introduction of a hydroxymethyl group at the β-position of allylic alcohol with the silicon-tethered radical approach developed by Stork [78–83]. Allyl silyl ether (**64**) was first prepared by the treatment of allylic alcohol (**63**) with dimethylbromomethylsiliyl chloride, and subsequently treated with Bu_3SnH/AIBN, and finally treated with hydrogen peroxide in the presence of KF, to give a 1,3-diol derivative (**66**), as shown in eq. 3.21.

(3.21)

Acyl radical has a nucleophilic character and cyclizes via SOMO–LUMO interaction. Eq. 3.22 shows the cyclization of acyl radicals formed from the reaction of selenol esters (**67**) with Bu_3SnH/AIBN or Bu_3SnH/Et_3B, to give the cyclic ketones (**68**) via 5-*exo-trig* or 6-*exo-trig* manner through the transition state [II] [84–90].

(3.22)

Phenylseleno carbonate (**69**) bearing acetylenyl silyl ether is converted into γ-lactone (**70**) with Ph_3SnH/AIBN, through 5-*exo-dig* ring closure, intramolecular hydrogen atom transfer from the silicon atom to a vinyl radical, and 5-*endo-trig* ring closure. This lactone is the precursor of juruenolide C, an *anti*-fungal agent (eq. 3.23).

(3.23)

The same treatment of *Se*-phenyl selenocarbamate (**71**) with $(Me_3Si)_3SiH$/AIBN gives the γ-lactam (**72**) via a carbamoyl radical, the rate constant of which for cyclization is about $10^8\ s^{-1}$, as shown in eq. 3.24.

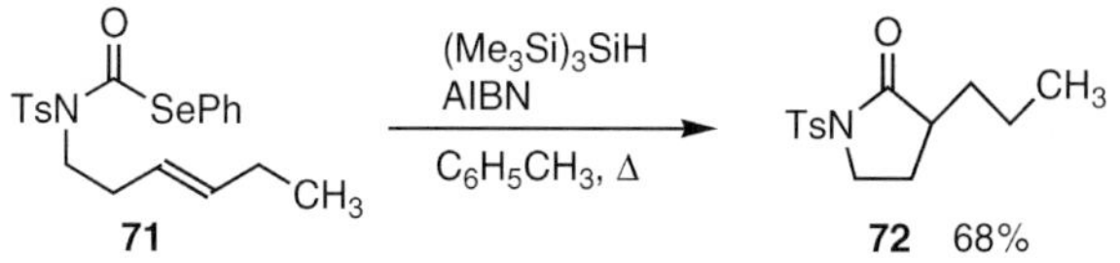
O
TsN
SePh
CH3
71
(Me3Si)3SiH
AIBN
C6H5CH3, Δ
O
TsN
CH3
72 68%

(3.24)

Carbon-centered radicals can cyclize at N–N multi bonds such as an azo group or azide group via 5-*exo-trig* or 6-*exo-trig* manner to give pyrrolidine or piperidine derivatives as shown in eq. 3.25 [91, 92]. Aminosugars such as nojirimicins can be prepared by this method.

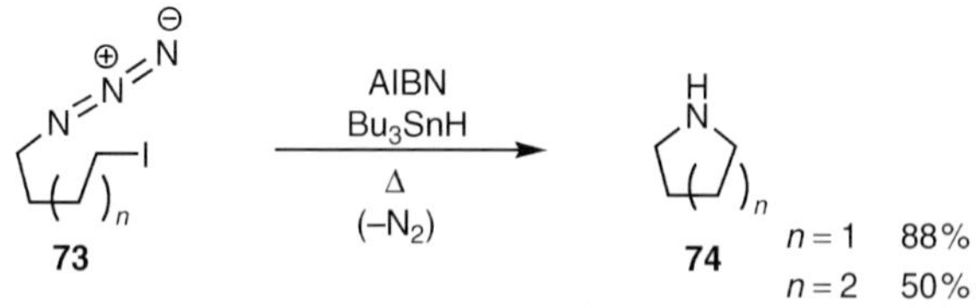
73
AIBN
Bu3SnH
Δ
(–N2)
H
N
74
n = 1 88%
n = 2 50%

(3.25)

α-Haloacetamides and α,α,α-trihaloacetamides have electrophilic character and can be easily reduced by single-electron transfer (SET) reagents. For example, as shown in eq. 3.26, treatment of α,α,α-trichloroacetamide (**75**) with Cu^{1+} forms the corresponding electrophilic α-amide radical, Cl^-, and Cu^{2+}, and then the formed α-amide radical cyclizes onto a vinyl group via 5-*endo-trig* manner and the formed electron-rich carbon-centered radical is oxidized by Cu^{2+} to carbocation, together with Cu^{1+} [93–97]. Thus, Cu^{1+} acts as an electron transfer catalyst.

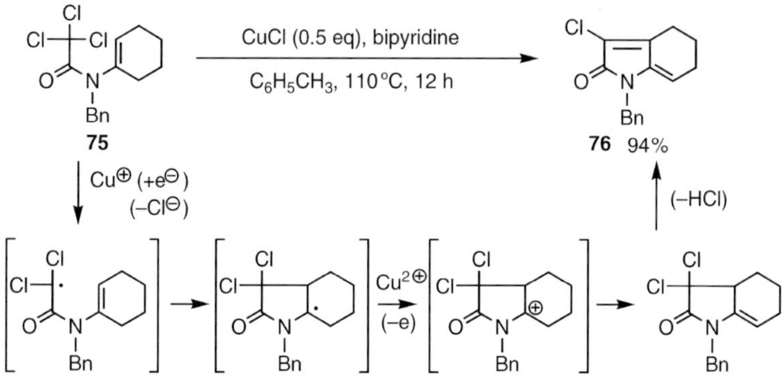
CuCl (0.5 eq), bipyridine
C6H5CH3, 110°C, 12 h
75
76 94%
Cu⊕ (+e⊖)
(–Cl⊖)
(–HCl)
Cu2⊕
(–e)
Bn

(3.26)

A variety of 3-substituted furans, including the natural perillene and dendrolasin, are obtained in good yields via reductive annulation of 2,2,2-trichloroethyl propargyl ethers using Cr^{2+} generated from $CrCl_3$ with Mn /TMSCl (eq. 3.27).

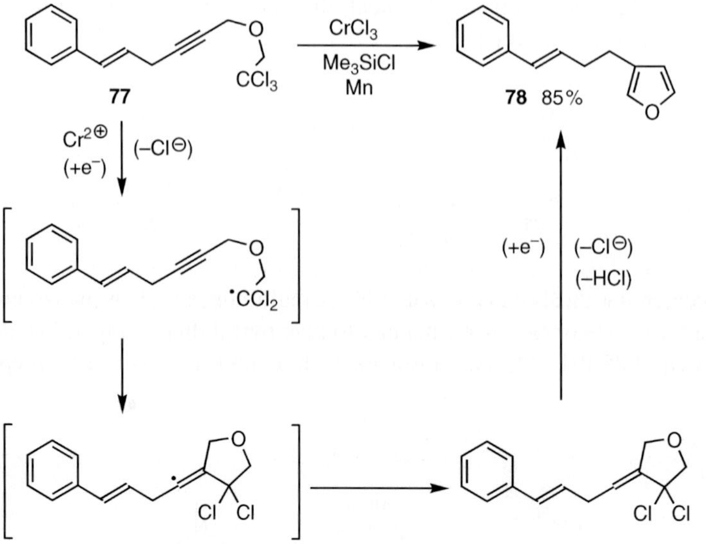

(3.27)

Experimental procedure 4 (eq. 3.27).

A solution of 2,2,2-trichloroethyl propargyl ether (1 mmol) in THF (2 ml) was added to a stirred, room-temperature suspension of dry $CrCl_3$ (15 mol%), Mn powder (4 mmol), and freshly distilled TMSCl (4 mmol) in THF (8 ml) under an argon atmosphere. After the addition, the reaction mixture was heated at 60 °C. After 12 h, the reaction mixture was cooled and quenched with water, and extracted three times with ether. After removal of the solvent, the residue was chromatographed on silica gel to give a product in 85% yield [97].

SmI_2 is an effective SET reagent and the end of the reaction can be noted by the disappearance of the greenish-blue color of the SmI_2 solution. Treatment of ω-iodo-α,β-unsaturated ester (**79**) with SmI_2 gives a carbocyclic sugar (**80**) via 5-*exo-trig* manner (eq. 3.28) and treatment of *O*-benzyl ketoxime (**81**) with SmI_2 forms the corresponding aminocarbocyclic sugar (**82**) via a ketyl radical generated by SET to the ketone group (eq. 3.29) [98–100].

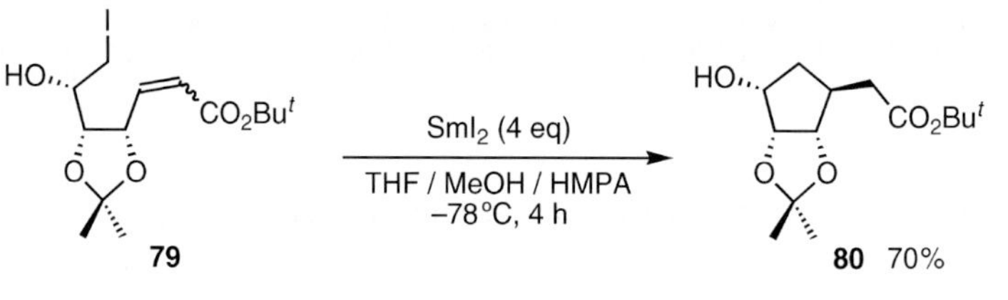

(3.28)

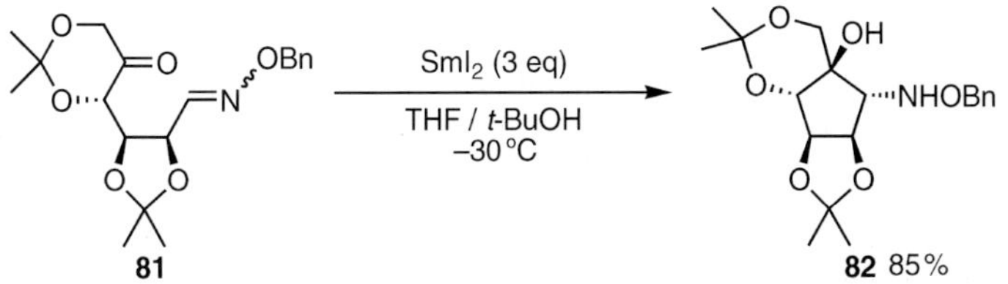

(3.29)

This would be a good method for the preparation of carbocyclic sugar analogs, since some sugar analogs have potent biological activities.

Reaction of cyclic γ-cyanoketones (**83**) with SmI_2 in the presence of *t*-BuOH as a proton source in HMPA-THF produces the α-hydroxycycloalkanones (**84**) after hydrolysis (eq. 3.30) [101–103].

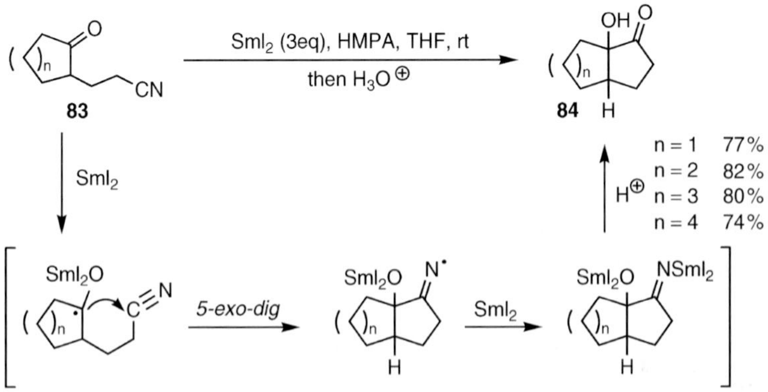

(3.30)

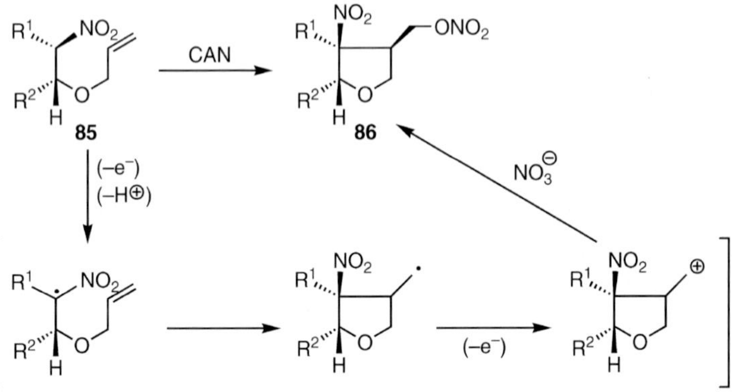

(3.31)

One-electron oxidation of the β-allyloxy nitroalkane (**85**) by cerium ammonium nitrate (CAN) produces a tetrahydrofuran derivative (**86**) (eq. 3.31).

Formation of 4-membered ring

Generally, the construction of a 4-membered ring is difficult with polar reaction methods. Therefore, 4-membered ring compounds are prepared mainly by [2 + 2] photocycloaddition reactions of olefins. On the other hand, formation of a 4-membered ring via 4-*exo-trig* manner with radical reactions presents difficulties because of the ring strain and consequently, the preparation of 4-membered ring compounds with radical 4-*exo-trig* manner is not useful. However, when the SOMO–LUMO or SOMO–HOMO interaction in radical cyclization is promoted and the cyclized radical formed via 4-*exo-trig* manner is stabilized, it can be a useful method for the construction of a 4-membered ring [104–114]. For example, a radical bearing a cyclic *gem*-dialkoxy group at the β-position cyclizes via 4-*exo-trig* manner to generate 4-membered ring (**88**) quantitatively as shown in eq. 3.32.

AIBN, Bu_3SnH, C_6H_6, Δ; **87** → **88** (CO_2CH_3); $n = 1$ 0%; $n = 2$ 100% (3.32)

The reaction takes place through SOMO–LUMO interaction to form a stable α-ester radical, and the cyclic *gem*-dialkoxy group (1,3-dioxane group) plays an important role in making the conformation rigid in the transition state. However, the dioxolane group does not induce 4-*exo-trig* ring closure because of the ring strain in the transition state. β-Lactam (**90**) shown in eq. 3.33 is formed through SOMO–HOMO interaction to generate the stable benzylic radical via 4-*exo-trig* manner. β-Lactone (**92**) can be prepared via 4-*exo-trig* manner from α-acyloxymethyl radicals formed with Barton decarboxylation, as shown in eq. 3.34a. The cyclized radical is highly stabilized by two benzylic phenyl groups. This stabilizing effect is a major driving force for cyclization to a 4-membered ring.

AIBN, Bu_3SnH, $C_6H_5CH_3$, Δ; **89** → **90** 78% (57 : 21 : 20 : 2) (3.33)

Bu_3SnH, $h\nu$, $C_6H_5CH_3$, 80°C; **91** → **92** 59% (3.34a)

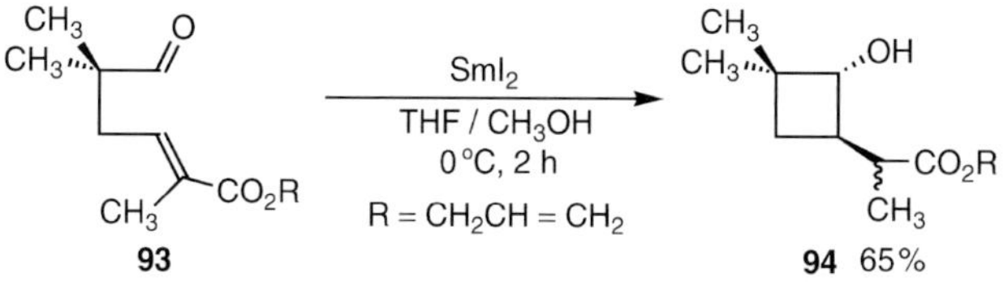

(3.34b)

SmI_2 is used for the preparation of a cyclobutanol derivative (**94**) from α-formyl α,β-unsaturated ester (**93**), via 4-*exo-trig* ring closure of the formed ketyl radical (eq. 3.34b).

Experimental procedure 5 (eq. 3.34b).

To the solution of SmI_2 (0.1 M, 1.0 mmol) in a mixture of THF (10 ml) and MeOH (2.5 ml) at 0 °C under a nitrogen atmosphere was added aldehyde (100 mg, 0.5 mmol) in THF (2 ml). After the mixture was stirred for 2 h at 0 °C, sat NaCl aq. solution (2 ml) and citric acid (128 mg, 0.61 mmol) were added to the mixture, and the organic layer was extracted with ethyl acetate three times. After the organic layer was dried over Na_2SO_4, the solvent was removed. The residue was chromatographed on silica gel (eluent: ethyl acetate /hexane = 3/7) to give 67 mg of a cyclobutanol derivative in 65% yield as an oil [111].

The compound (**95**) bearing an active methylene group can be converted to β-lactams (**96**) via 4-*exo-trig* manner by oxidative reagent, $Mn(OAc)_3$ as shown in eq. 3.35. The radical formed via 4-*exo-trig* manner is benzylic radical which is somewhat stabilized by the resonance effect and is one of the major driving forces for cyclization to a 4-membered ring.

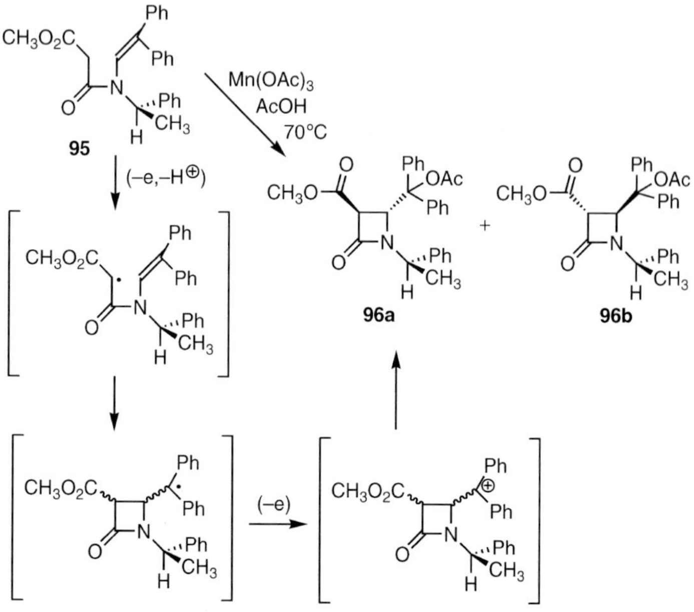

(3.35)

Cyclization of trihalo-enamide (**97**) with copper(I) proceeds effectively to form β-lactam (**98**) via 4-*exo-trig* manner. The copper(I)/TMEDA (tetramethylethylenediamine) system is the most effective for the formation of β-lactam (eq. 3.36).

CuCl, TMEDA
CH_3CN, Δ

97 **98** 85-86%

(3.36)

Formation of 3-membered ring

Study for the preparation of 3-membered ring compounds via 3-*exo-trig* is extremely limited because of the quite significant strain involved [115–117]. The product in eq. 3.37a seems to be formed via 3-*exo-trig* manner through the SET from SmI_2 to the iodide (**99**) and subsequent cyclization of the homoallylic radical thus formed. However, another mechanism, an anionic 3-*exo-trig* ring closure (intramolecular Michael addition), cannot be ruled out either. In the presence of SmI_2 and a proton source, δ-oxo-γ,γ-disubstituted-α,β-unsaturated ester (**101**) readily cyclize to give *trans*-cyclopropanol (**102**) via intermediate (III) as shown in eq. 3.37b.

SmI_2
THF / CH_3OH
r.t.

99 **100** 99%

(3.37a)

SmI_2
THF/ ^{t}BuOH
r.t.

101 **102** 90%

[III]

(3.37b)

The following reaction is not 3-*exo-trig* ring closure, but the formation of 3-membered ring compound (**104**) under photochemical conditions through Norrish II type reaction.

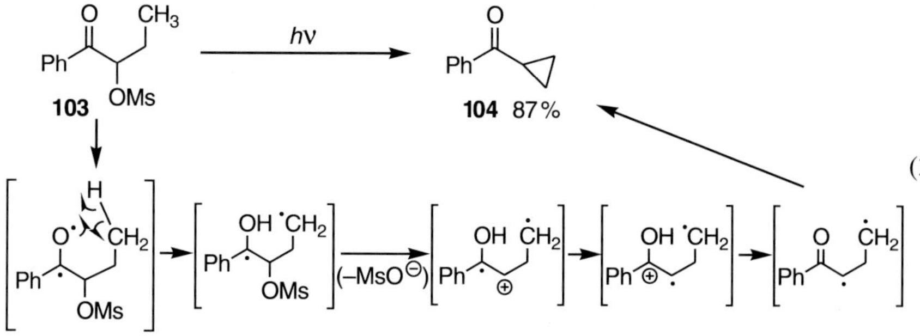

(3.38)

3.1.2 Cyclization onto olefinic groups and aromatic groups by sp^2 carbon-centered radicals

Formation of 5- and 6-membered rings

sp^2 Carbon-centered radicals formed by the treatment of aryl halides or vinyl halides with Bu_3SnH/AIBN, are σ radicals and much more reactive than sp^3 carbon-centered radicals which are π radicals. The high reactivity of sp^2 carbon-centered radical reflects the stronger bond dissociation energies of sp^2(C)–H and sp^2(C)–sp^3(C) bonds than those of sp^3(C)–H and sp^3(C)–sp^3(C) bonds. For example, the C–C bond dissociation energy of ethane is 88 kcal/mol, while that of the sp^2(C)–CH_3 bond in toluene is 102 kcal/mol. However, sp^2 carbon-centered radicals are also used for 5-*exo-trig* and 6-*exo-trig* ring closure. Thus, treatment of *o*-haloaryl allyl ether or *N*-*o*-haloaryl-*N*-allylamine (**105**) with Bu_3SnH/AIBN produces good yields of 2,3-dihydrobenzofuran or 2,3-dihydroindole (**107**) via 5-*exo-trig* manner (eq. 3.39a). This reaction can be also used for tandem intramolecular and intermolecular C–C bond formation as shown in eq. 3.39b [118–126].

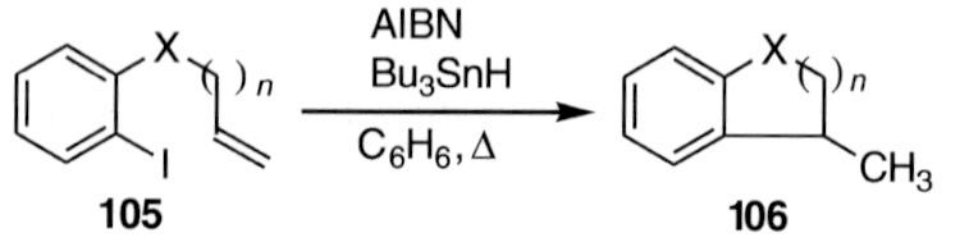

X		n	Yield
X	CH_2	n = 1	50%
X	O	n = 1	99%
X	NCH_3	n = 1	78%
X	O	n = 2	62%

(3.39a)

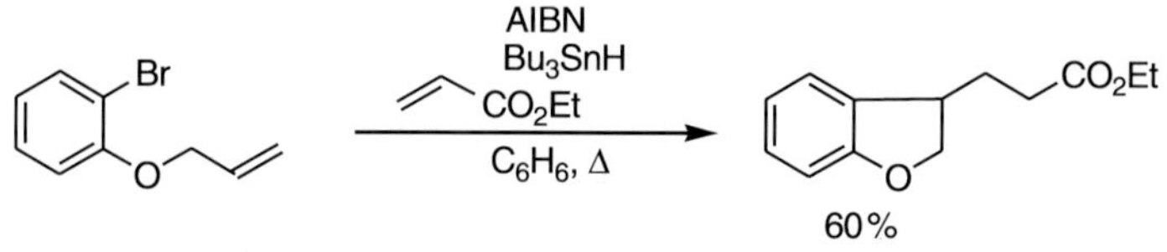

(3.39b)

Experimental procedure 6 (eq. 3.39b).

To a refluxing solution of bromide (2 mmol) and ethyl acrylate (4 mmol) in toluene (10 ml) was added dropwise a solution of Bu_3SnH (0.7 ml, 2.5 mmol) and AIBN (30 mg) in toluene (8 ml) over 40 min. After the addition, the mixture was further refluxed for

30 min. After removal of the solvent, the residue was chromatographed on silica gel to give ethyl 2-(2′,3′-dihydrobenzofuran-2′-yl)propanoate in 60% yield [123].

---

sp^2 Carbon-centered radical can be also used for 5-*exo-dig* cyclization with a propargyl group. sp^2 Carbon-centered radical is very reactive, so phenyl radical reacts onto an olefinic group bearing poor reactivity such as a uracil group, to form tricyclic isoindolinone (**108**) as shown in eq. 3.40 [127–131].

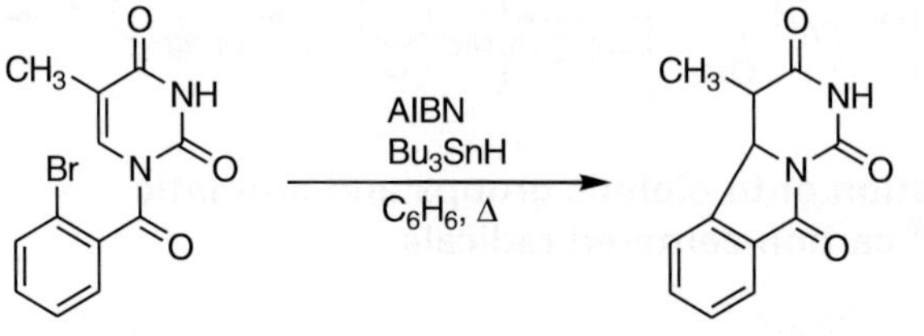

AIBN, Bu_3SnH, C_6H_6, Δ: **107** → **108** 80% (3.40)

Generally, sp^3 carbon-centered radicals do not cyclize onto an aromatic ring, due to the high stabilization of the aromatic ring. However, sp^2 carbon-centered radicals readily react at the aromatic ring via 6-*exo-trig* manner [132–139]. Eqs. 3.41a and 3.41b show the preparation of natural products, aporphine (**110**), and phenanthrene (**112**), respectively, with a radical reducing reagent, Bu_3SnH, through 6-*exo-trig* ring closure of an aryl radical onto electron-rich aromatic rings. In practice, the α-isobutyronitrile radical formed from AIBN or molecular oxygen in air acts as an hydrogen-abstracting agent of the cyclized intermediate radical formed (σ complex). [5]Helicene (**114**) can be prepared from 1,4-bis(2-*o*-iodophenylvinyl)benzene (**113**) with Bu_3SnH/AIBN, as shown in eq. 3.42a. Side-chained phenyl radicals cyclize onto the C_3 and C_2 positions of indoles (**115**) and (**117**), respectively (eqs. 3.42b and 3.42c). Eq. 3.43 show the preparation of penta-cyclic acridine (**120**) using the same method.

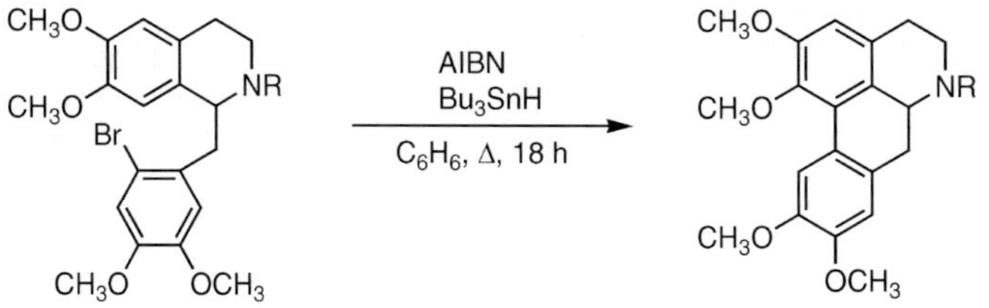

AIBN, Bu_3SnH, C_6H_6, Δ, 18 h: **109** → **110** 80% (3.41a)

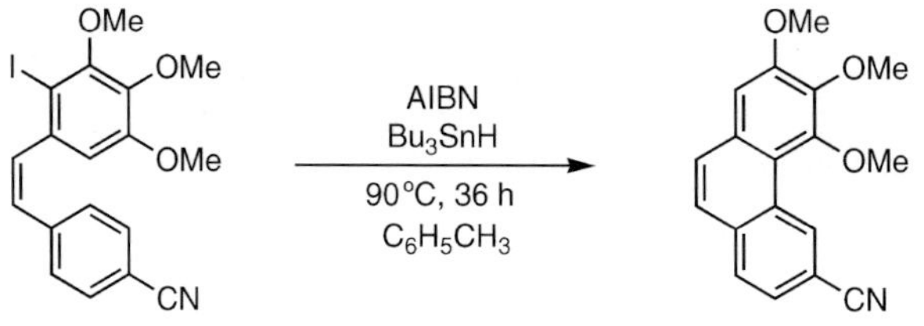

AIBN, Bu_3SnH, 90°C, 36 h, $C_6H_5CH_3$: **111** → **112** 85% (3.41b)

AIBN
Bu_3SnH
$C_6H_5CH_3$, 90°C

113 → **114** 58% (3.42a)

AIBN
Bu_3SnH
Δ

115 → **116** 58% (3.42b)

AIBN
Bu_3SnH
Δ

117 → **118** 90 % (3.42c)

AIBN
Bu_3SnH
$C_6H_5CH_3$, Δ
50%

119 → **120** ← **121** (3.43)

AIBN
Bu_3SnH
$C_6H_5CH_3$, Δ
56%

Experimental procedure 7 (eq. 3.43).

A mixture of Bu_3SnH (1.0 eq.) and AIBN (0.1 eq.) in toluene was added dropwise to a refluxing solution of 9-(*o*-bromoanilino)acridine (1.0 eq.) in toluene over 1 h. After 12 h, the solvent was removed and the residue was chromatographed on silica gel (eluent: ethyl acetate) to provide the penta-cyclic acridine in 50% yield [137].

AIBN
Bu_3SnH
100°C, 72 h, $C_6H_5CH_3$

122 → **123** 42%

AIBN
Bu_3SnH
80°C, 24 h, $C_6H_5CH_3$

124 → **125** 70%

(3.44)

Eq. 3.44 shows the intramolecular addition of an sp^2 carbon-centered radical onto the quinoline group under the same conditions.

Treatment of 2-bromo-*N*-alkenylpyridinium salts (**127**) with Bu_3SnH/AIBN in acetonitrile under refluxing conditions generates the cyclized pyridinium salts (**128**), through the generation of sp^2 carbon-centered radicals, followed by 5-*exo-trig* or 6-*exo-trig* ring closure. After the reduction of pyridinium salts (**128**), the corresponding alkaloid analogs are obtained as shown in eq. 3.45 [140–144].

126 → **127** (AIBN, Bu_3SnH, $C_6H_5CH_3$, Δ, 12 h) → **128**

(3.45)

$n = 1$ 74%
$n = 2$ 53%

Radical cyclization from alkenyl iodide (**129**) via 5-*exo-trig* manner is affected by Bu_3SnH/AIBN to give cyclic *exo*-diene (**130**) fused to 5-membered rings (eq. 3.46) [145, 146]. Treatment of *o*-bromobenzamide (**131**) bearing a uracil group with Bu_3SnH or $(Me_3Si)_3SiH$/AIBN generates the corresponding tricyclic isoindolinone (**132**) as shown in eq. 3.47a. Reactive vinyl radical cyclizes at the imino-nitrogen atom via 5-*exo-trig* manner to give 5-membered cyclic enamine (**134**) which is further trapped by PhCOCl to provide compound (**135**), as shown in eq. 3.47b.

129 → (AIBN, Bu_3SnH) → **130** 75%

(3.46)

131 → (AIBN, Bu_3SnH or $(Me_3Si)_3SiH$, C_6H_6, Δ) → **132** 80%

(3.47a)

(3.47b)

Experimental procedure 8 (eq. 3.47a).

To a refluxing solution of *o*-bromobenzamide (2.6 mmol) in benzene (50 ml) were added Bu_3SnH or $(Me_3Si)_3SiH$ (3.2 mmol) and AIBN (0.15 mmol). After 2 h, AIBN (0.15 mmol) was added again. The mixture was refluxed more 12 h. After the reaction, the solvent was removed and the residue was chromatographed on silica gel to give the tricyclic isoindolinone in 80% yield [129].

Skillful methods through the ring closure

Eq. 3.48 shows a specific radical reaction. Thus, the formed sp^2 carbon-centered radical cyclizes at the vinylic position via 5-*exo-trig* manner and subsequent β-cleavage to produce an *o*-hydroxystilbene skeleton (**137**), together with evolution of SO_2 [147, 148]. This reaction has the appearance of a 1,4-transfer of the phenylvinyl group from the sulfur atom to an sp^2 carbon atom.

(3.48)

By a similar method, Ar groups can be transferred via 1,4-Ar migration to form biaryls as shown in eqs. 3.49 to 3.52 [149–156]. Treatment of *N*-methyl-*N*-(2-bromophenyl) arenesulfonamide (**138**) with 1,1,2,2-tetraphenyldisilane ($Ph_4Si_2H_2$) in the presence of AIBN under heating conditions gives the corresponding biaryl (**139a**) by means of the intramolecular radical *ipso*-substitution. 1,1,2,2-Tetraphenyldisilane is the most effective reagent for 1,5-*ipso*-substitution on the sulfonamides among typical radical reagents, such as Ph_2SiH_2, Bu_3SnH, $(Me_3Si)_3SiH$, and $Ph_4Si_2H_2$ [154].

138 → (AIBN, Reagent, *m*-Xylene Δ) → **139a** + **139b** + **139c** (3.49)

Reagent	139a	139b	139c
Ph_2SiH_2	20%	23%	–
Bu_3SnH	41%	37%	18%
$(Me_3Si)_3SiH$	56%	22%	–
$Ph_4Si_2H_2$	60%	31%	–

140 → (1) Bu_3SnH, AIBN, C_6H_6, 0.05 M; 2) MeLi) → **141** 71% (3.50)

142 → (AIBN, Ph_3SnH, Xylene Δ) → **143** 58% (3.51)

144 → (excess $(t\text{-}BuO)_2$, C_6H_5Cl, Δ) → [acyl radical] → ($-SO_2$) → **145** 80% (3.52)

Preparation of biaryl (**141**) via intramolecular 1,5-Ar transfer from the silicone atom in silyl ether (**140**) to the sp^2 carbon with Bu_3SnH in the presence of AIBN is similarly possible, as shown in eq. 3.50. Phosphinate (**142**) also give a biaryl (**143**) with Bu_3SnH in the presence of AIBN through the migration of the Ar group from the phosphorus atom to the sp^2 carbon as shown in eq. 3.51. Generally in these Ar-transfer reactions with Bu_3SnH, the concentration of Bu_3SnH must be quite diluted (~0.03 M) to reduce the formation of the direct reduction products. 2-Hydroxydiarylketone (**145**) can be prepared by means of the hydrogen atom abstraction from the formyl group by $t-BuO^{\cdot}$ formed, and subsequent intramolecular 1,5-Ar transfer from the sulfur atom to the acyl carbon, by the reaction of aromatic aldehyde (**144**) with di-*t*-butyl peroxide as shown in eq. 3.52.

A more interesting reaction is the shift of an sp^2 carbon-centered radical to an sp^3 carbon-centered radical via 1,5-H shift as shown in eq. 3.53. The formed sp^2 carbon-centered radical abstracts a hydrogen atom via 1,5-H shift to generate an sp^3 carbon-centered radical which then cyclizes at the olefinic position via 5-*exo-trig* manner to produce a cyclopentane derivative (**147**) [157–160]. The rate constant of this 1,5-H shift is approximately $3 \times 10^7\ s^{-1}$, and is quite fast.

R R CH3 Br O OCH3 **146** — AIBN, Bu3SnH, C6H6, R = CO2Et → R R CH3 O CH3 OCH3 **147** 75%

1,5-H shift

(3.53)

Experimental procedure 9 (eq. 3.53).

To a refluxing benzene solution (0.01 M) of arylbromide (1.0 eq.) and AIBN (0.1 eq.) under an argon atmosphere was added dropwise using a syringe pump a benzene solution of Bu_3SnH (1.3 ~ 1.5 eq.) and AIBN (0.3 eq.) over 10 to 15 h. After the reaction, the solvent was removed. Ether was added to the residue, and DBU (2.0 eq.) and iodine solution of ether (1 M) were added, and the mixture was stirred for 20 min. After removal of the white solids by filtration, the solvent was removed and the residue was chromatographed on silica gel to give a cyclopentane derivative [157].

Since a vinyl radical is also an sp^2 carbon-centered radical, a similar 1,5-H shift from the sp^2 carbon-centered radical to the sp^3 carbon-centered radical occurs, as shown in eq. 3.54, which finally produces heliotridanes (**149**) via the subsequent 5-*exo-trig* ring closure.

Bu$_3$SnH, AIBN
C$_6$H$_6$, Δ, 3h
148
149a 67% (13 : 1) 149b
1,5-H shift

(3.54)

Spirocyclic pyrrolidin-2-ones (**151**) are prepared through the initial 1,5-radical translocation by an sp^2 carbon-centered radical, followed by cyclization as shown in eq. 3.55 [161].

Bu$_3$SnH, AIBN
C$_6$H$_5$CH$_3$, 110°C
150
151

n = 1	< 5%
n = 2	73%
n = 3	71%
n = 4	70%
n = 5	74%

1,5-H shift
5-*exo-trig*

(3.55)

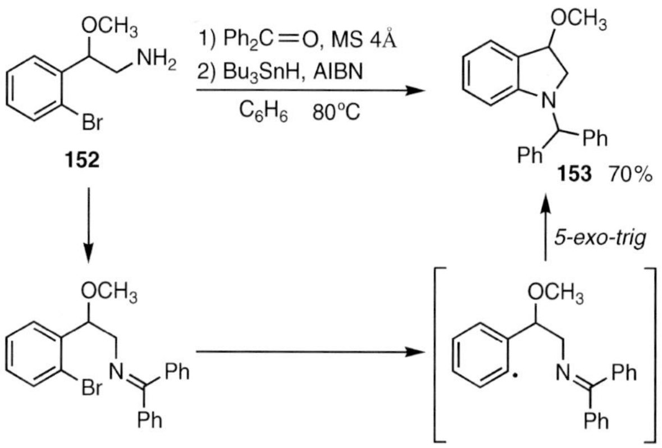

(3.56)

Treatment of ketimine formed from the reaction of *o*-bromophenethylamine (**152**) and benzophenone *in situ*, with Bu_3SnH/AIBN produces 2,3-dihydroindole (**153**) via 5-*exo-trig* manner which indicates the sp^2 carbon-centered radical reacts at the nitrogen atom of the imino group as shown in eq. 3.56 [162]. The treatment of aromatic diazonium salt with NaI, Cu^{1+}, SmI_2, or Co^{2+} also generates the corresponding sp^2 carbon-centered radicals together with N_2 gas via SET, which can be used for the cyclization [163–165].

3.1.3 Generation of sp carbon-centered radicals

Alkynyl radicals (sp carbon-centered radicals) are a highly energetic and highly reactive species. Therefore, the generation of an sp carbon-centered radical is rather difficult. The instability of the alkynyl radical is reflective of the strong bond dissociation energy of the C(sp)–H bond, which is estimated to be approximately 130 kcal/mol, nearly 20 kcal/mol higher than that of alkene, $C(sp^2)$–H. The study on alkynyl radicals is therefore extremely limited, and only one method for the generation of phenylethynyl radical from phenyliodoacetylene has been reported [166].

3.1.4 Cyclization to carbonyl groups by sp^3 or sp^2 carbon-centered radicals and other related reactions

The rate constants for the ring closure of a carbon-centered radical to a formyl group via 5-*exo-trig* and 6-*exo-trig* manners are close to those for the ring closure of 5-hexen-1-yl and 6-hepten-1-yl radicals ($1.4 \times 10^6\ s^{-1}$, $4.3 \times 10^4\ s^{-1}$ at 80 °C). This result is reasonable when we consider SOMO–LUMO orbital interactions between nucleophilic carbon-centered radicals and olefinic groups. There is, however, one big difference. The reverse reaction, i.e. β-cleavage of the formed cycloalkoxyl radical is extremely rapid, because the formed cycloalkoxyl radical (oxygen-centered radical) is unstable and much more reactive than the corresponding cycloalkylmethyl radical. Thus, radical ring closure to a formyl group does not generally work effectively. Radical ring closure to a ketone group becomes more difficult because of the steric hindrance. So how can we promote the radical ring closure to a carbonyl group and reduce the reverse ring-opening of the cycloalkoxyl radical? There are two possibilities; one is the design of compounds, i.e. retard of the ring-opening of cycloalkoxyl radicals formed by rigid conformation or structure, and the second is the rapid trap of the formed cycloalkoxyl radical by an oxygen-favored group such as a silicone group. Eq. 3.57 show two examples of the former [167].

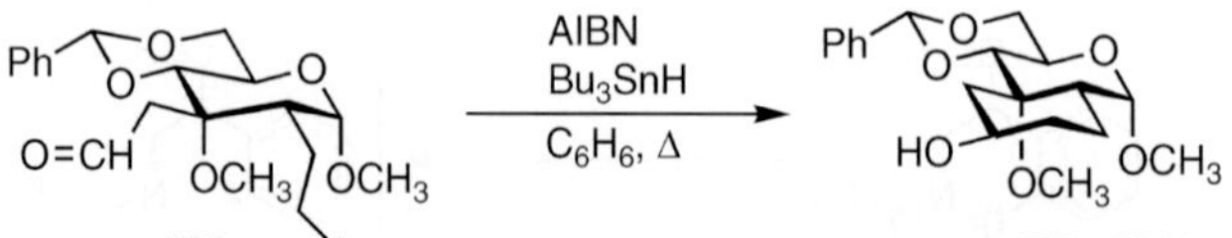

(3.57a)

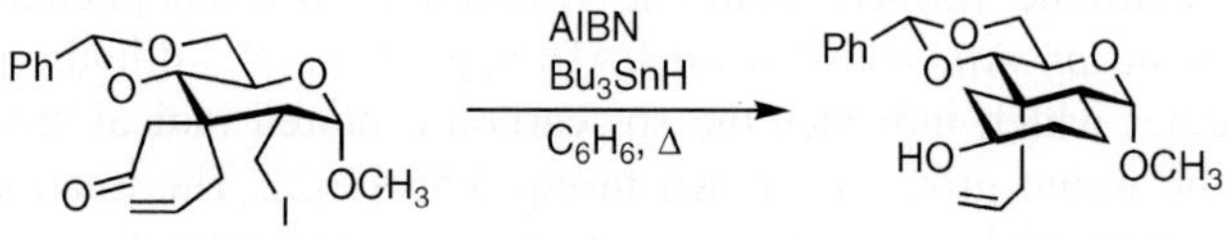

(3.57b)

The formation of a cyclohexyloxyl radical via 6-*exo-trig* manner is promoted by the rigid bicyclic structure derived from the sugar benzylidene group (eq. 3.57a). Moreover, eq. 3.57b indicates that radical ring closure to a formyl group is preferable to that onto an olefinic group.

An example of the latter possibility is the ring closure of formyl bromide (**158**) with $PhSiH_3$ instead of Bu_3SnH, where silyl radical has strong affinity to an oxygen-centered radical, to form cyclohexyl silyl ether (**159**) via 6-*exo-trig* manner (eq. 3.58).

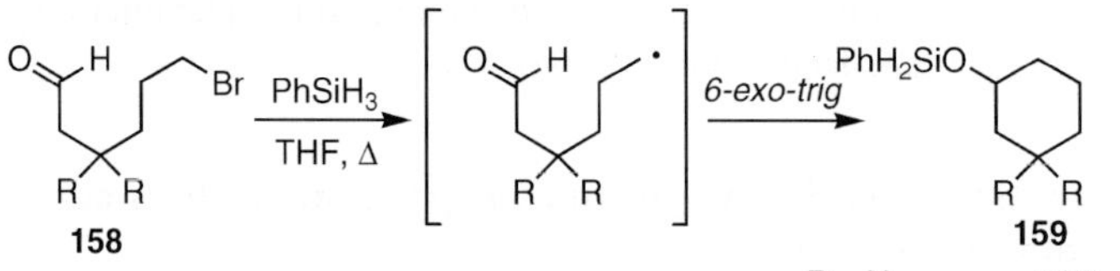

(3.58)

Treatment of bromoalkyl silyl ketone (**160**) with Bu_3SnH/AIBN generates cycloalkyl silyl ether (**161**), through ring closure of a carbon-centered radical onto ketone group to generate an oxygen-centered radical, Brook 1,2-rearrangement of a silyl group to form a carbon-centered radical, and finally abstraction of a hydrogen atom from Bu_3SnH, as shown in eq. 3.59. The tendency of 1,2-rearrangement of a silyl group is as follows: $Me_3Si < Me_2PhSi < MePh_2Si$. The experimental key point in this reaction is the slow, dropwise addition of Bu_3SnH, in order to keep the concentration of Bu_3SnH, a hydrogen donor, quite low [168–171].

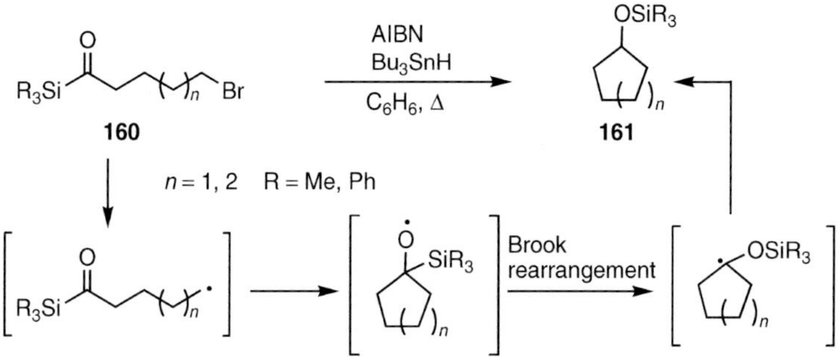

(3.59)

Treatment of an enal or enone compound with Bu_3SnH can be used for the formation of cyclopentanols and cyclohexanols through the formation of an *O*-stannyl ketyl radical and subsequent ring closure. The example in eq. 3.60 shows the 5-*exo-trig* ring closure of the formed ketyl radical onto the olefinic group to generate cyclopentanol (**163a**) and its cyclized lactone (**163b**). The example in eq. 3.61 shows 5-*exo-trig* and 6-*exo-trig* ring closure of the ketyl radicals formed onto the formyl group to generate cyclopentan-1,2-diol and cyclohexan-1,2-diol (**165**) respectively [172–178]. This type of reaction does not occur in diketone, due to steric hindrance.

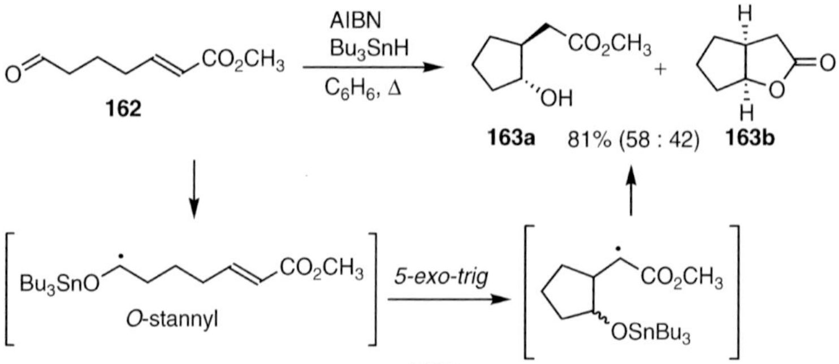

(3.60)

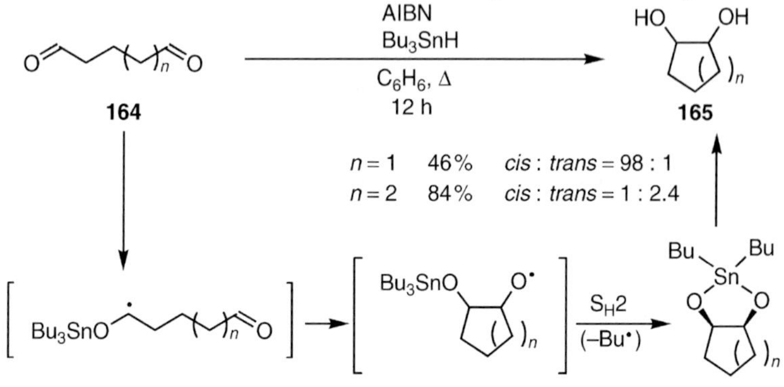

(3.61)

Experimental procedure 10 (eq. 3.61).

A benzene (36 ml) solution of dialdehyde (0.89 mmol), Bu_3SnH (311 mg, 10.7 mmol), and AIBN (15 mg, 0.089 mmol) was refluxed under a nitrogen atmosphere. After 3 h,

AIBN (0.089 mmol) was added again to the solution. After 12 h, the solvent was removed and the residue was chromatographed on silica gel to give the corresponding diol as an oil [175].

With this method, treatment of aldehydes (**166**) bearing a methoxy imino group with Bu_3SnH in the presence of AIBN generates the cyclic 1,2-aminoalcohols (**167**) with *trans* configuration (eq. 3.62).

AIBN, Bu_3SnH, C_6H_6, Δ, 17 h: **166** → **167**

n = 1 73% *trans* : *cis* = 18 : 1
n = 2 67% *trans* : *cis* = 3 : 1 (3.62)

Eq. 3.63 shows the preparation of *O*-benzyl cyclopentanone oxime (**169**) by means of the generation of an sp^3 carbon-centered radical, cyclization onto the imino carbon via 5-*exo-trig* manner, and elimination of the $Ph_3Ge^{\bullet}$ group (β-cleavage). Here, the same reaction can be observed with the Bu_3Sn group, instead of the Ph_3Ge group. Eq. 3.64 is the 5-*exo-trig* ring closure of sp^3 carbon-centered radical to imino carbon to generate *N*-benzyloxy cyclopentylamine (**171**) [179–187]. The rate constant for 5-*exo-trig* ring closure is 100 times faster than that for the ring closure of the 5-hexen-1-yl radical. This can be explained by the stabilization of the *N*-benzyloxy aminyl radical formed, similar to stable NO gas.

Hg-*h*ν, r.t. 90 min: **168** → **169** 95% (3.63)

Bu_3SnH, C_6H_6, Δ, 80°C: **170** → **171**

$k_{5\text{-}exo}\ (n = 1) = 4.2 \times 10^7 s^{-1}$
$k_{6\text{-}exo}\ (n = 2) = 2.4 \times 10^6 s^{-1}$ (3.64)

As previously mentioned, the rate for the 5-*exo-trig* ring closure of an sp^3 carbon-centered radical onto a carbonyl group is a little bit faster than that for the 5-*exo-trig* ring closure of an sp^3 carbon-centered radical onto an olefinic group. This result is supported by the following experiment shown in eq. 3.65. Treatment of the iodide (**172**) bearing both formyl and olefinic groups at the ε- and ε′-positions, with Bu_3SnH initiated by AIBN under benzene refluxing conditions, generates a cyclopentane derivative (**173b**) bearing a formyl group as a major product, which indicates that the reaction proceeds via a thermodynamically controlled pathway. However, the same treatment with Bu_3SnH initiated by Et_3B at room temperature generates a cyclopentanol derivative (**173a**) bearing an olefinic side chain as a major product, which indicates that the reaction proceeds via a kinetically controlled pathway. These results indicate that at a lower temperature, 5-*exo-trig* ring closure of an sp^3 carbon-centered radical onto a carbonyl group is kinetically more preferable than that of the sp^3 carbon-centered radical onto an olefinic group [188].

172 → **173a** + **173b** (3.65)

	173a	173b
Bu_3SnH, AIBN, C_6H_6, Δ	21%	63%
Bu_3SnH, Et_3B, r.t., $C_6H_5CH_3$ −78°C → 0°C	87%	0%

Radical ring closure of an sp^3 carbon-centered radical to a hydrazone group in compounds (**174**) can be carried out via 5-*exo-trig* manner. The silicon connection (Si-tethered) is removed by oxidative treatment with H_2O_2 to generate 2-amino-1,3-diols (**175**) (eq. 3.66) [189–191].

174 —(Bu_3SnH, AIBN, C_6H_6, Δ)→ [ ] —(KF, $KHCO_3$, H_2O_2; THF, MeOH (1 : 1))→ **175** (3.66)

R	Yield
R = Me	76%
R = *i*-Bu	68%
R = *i*-Pr	80%
R = Ph	57%

176 → Bu_3SnH, AIBN, C_6H_6, Δ → [Bu_3SnO radical] → 177

R = Ph, R^1 = H 59% (d.e. = 1 : 1)
R = R^1 = Me 64% (d.e. = 1 : 1)

(3.67)

178 → R-SOCl, Pr_2NEt, PhSSPh, CH_2Cl_2, −50°C → r.t. → 179 70%

R = 2,6-dimethylphenyl (CH_3, CH_3)

N-O-S(O)-R → (−$RSO_2^{\bullet}$) → [N•] → [cyclized radical]

(3.68)

Bu_3SnH-mediated cyclization of aldehyde (**176**) bearing an allyloxy group generates a cyclopentanol derivative (**177**) via the addition of $Bu_3Sn^{\cdot}$ onto the formyl group to form an *O*-stannyl ketyl radical, and subsequent 5-*exo-trig* ring closure (eq. 3.67). The imino radical generated from the thermal decomposition of *O*-sulfinyl oxime (**178**) formed *in situ*, cyclizes at the olefinic group via 5-*exo-trig* manner to give 5-membered cyclic imino compound (**179**), as shown in eq. 3.68.

Irradiation with a mercury lamp of bromoacetal (**180**) bearing an alkynyl group in the presence of Et_3N, produces the corresponding cyclic acetal (**181**) through the initial SET from Et_3N to bromoacetal, generation of an sp^3 carbon-centered radical via α-cleavage of the anion radical, ring closure via 5-*exo-dig* manner, and finally abstraction of a hydrogen atom from the solvent (eq. 3.69).

180 → Hg-*h*ν(254 nm), Et_3N (10 eq), CH_3CN, 30 min → 181 94%

180 → e^- (−$Br^{\ominus}$) → [radical] → [vinyl radical] → hydrogen atom abstraction → 181

(3.69)

(3.70)

(3.71)

Irradiation of ketone (**182**) with a mercury lamp generates biradicals which are formed via n-π^* electron transition. The reactive oxygen center of the biradical abstracts a hydrogen atom via 1,6-H shift, and the formed carbon-centered biradical couples intramolecularly to produce 5-membered compound (**183**) (eq. 3.70) Eq. 3.71 shows the same type of photochemical cyclization through the generation of a biradical, 1,9-H shift, and finally intramolecular coupling of the formed carbon-centered biradical [192–199].

Generally, 1,5-H and 1,6-H shifts through 6- and 7-membered transition states are favored, while 1,9-H shift through 10-membered transition state is disfavored. However in this case, the stabilized benzylic radical is formed via the 1,9-H shift, and so its transition state is not so disfavored thermodynamically.

Treatment of an isocycanobenzene derivative (**186**) bearing an alkenyl or alkynyl group at the *o*-position, with Bu_3SnH/AIBN generates an indole derivative (**187**) as shown in eq. 3.72 [200–203].

(3.72)

Experimental procedure 11 (eq. 3.72).

A mixture of isocyanobenzene derivative (0.85 mmol), Bu_3SnH (0.93 mmol), AIBN (0.04 mmol) in dry CH_3CN (5 ml) was refluxed for 1 h under an argon atmosphere. After the reaction, 3 M aq. HCl was added to the mixture and the obtained mixture was extracted with ether. The organic layer was washed with aq. sat. KF solution and was dried. After removal of the solvent, the residue was chromatographed on silica gel to provide indole derivative [200].

Recently, Barton reported that H_3PO_2 works as a radical reagent and a hydrogen atom donor. Based on this result, thioamide (**188**) was treated with H_3PO_2 instead of toxic Bu_3SnH, in the presence of AIBN to generate the corresponding indole skeleton (**189**) via 5-*exo-trig* manner, as shown in eq. 3.73 [200–203].

AcO S N H R **188** AIBN H_3PO_2 Et_3N *i*-PrOH, Δ AcO N H R **189** $H_2\dot{P}O_2$ H_3PO_2 $(-H_3PO_2S)$ AcO N H R $S-PO_2H_2$ OAc N H R $S-PO_2H_2$ OAc N H R $S-PO_2H_2$ (3.73)

AcO S N H Cbz N CO_2Me **190** AIBN H_3PO_2 NEt_3 AcO Cbz N N H CO_2Me **191** N N H CO_2Me (±)-Catharanthine (3.74)

This method can be used for the stereocontrolled synthesis of (+)-catharanthine, an anticancer alkaloid, shown in eq. 3.74

Divalent chalcogens are radicophilic. Thus, refluxing treatment of 2-(*o*-halophenyl)-propyl benzyl selenide or telluride (**192**) with $(Me_3Si)_3SiH$ in the presence of AIBN in benzene generates benzoselenophene or benzotellurophene (**193**), through the formation of an sp^2 carbon-centered radical, followed by S_Hi reaction on the selenium or tellurium atom to form a 5-membered skeleton together with a stable benzyl radical, and its dehydration as shown in eq. 3.75a [204–210]. The same refluxing treatment of thiolesters (**194**) bearing both iodophenyl and azide groups with Bu_3SnH/AIBN or $(Me_3Si)_3SiH$/AIBN system in benzene promotes 5- and 6-membered cyclization onto the azide group by the acyl radical (S_Hi reaction), to give benzolactones (**195**) together with dihydrobenzothiophene, as shown in eq. 3.75b. This type of substitution reaction does not occur at tri-coordinated or tetra-coordinated chalcogenides, because of steric hindrance.

CH_3, OH, Y–Bn, X
192
$(Me_3Si)_3SiH$, AIBN, C_6H_6, Δ
CH_3, Y
193
X = I, Br
Y = Se, Te
$(-H_2O)$
Y = Se $k = 3 \times 10^7 s^{-1}$ (80°C)
Y = Te $k = 5 \times 10^8 s^{-1}$ (80°C)
S_Hi
$(-Bn^\bullet)$
(3.75a)

I, S, O, $( )_n$–N_3
194
Bu_3SnH, AIBN, C_6H_6, Δ
O, NH, $( )_n$ + S
195
$n = 1$ 87%
$n = 2$ 87%
$( )_n$
$(-N_2)$
(3.75b)
$N{=}N{=}N$

1) $(COCl)_2$, C_6H_6
2) NaO–N(pyridine-2-thione), DMAP
3) BBr_3, CH_2Cl_2
CH_3O, HO, CH_3, R, $SeCH_2Ph$
196
HO, CH_3, Se, R
198
R = butyl 48%
R = 4,8,12-trimethyl tridecyl 62%
(3.76)
$(-PhCH_2SPy)$
$(-2CO_2)$
197

Treatment of 1-(2-benzylselenenyl-5-methoxyphenyl)-3-methyl-3-heptanol and 1-(2-benzylselenenyl-5-methoxyphenyl)-3,7,11,15-tetramethyl-3-hexadecanol (**196**) with oxalyl chloride and sodium salt of 2-mercaptopyridine *N*-oxide generates the oxalate diesters (**197**) (Barton oxalate ester). Refluxing treatment of oxalate esters (**197**), followed by deprotection with BBr_3, gives rise to moderate yields of the selenium-containing α-tocopherol analogues (**198**) (eq. 3.76).

p-Toluenesulfonyl iodide is a rather unstable yellowish solid, and easily decomposes. *p*-Toluenesulfonyl bromide (TsBr) is less unstable than *p*-toluenesulfonyl iodide, yet a *p*-toluenesulfonyl radical can be readily generated by the addition of AIBN, irradiation, or heating conditions, and can be used for the following cyclization with *N,N*-diallyl sulfonamide (**199**) via a radical chain process (eq. 3.77). The product contains all groups of the *p*-toluenesulfonyl bromide, which means that this reaction does not lose any functional groups, and is an atom(group)-transfer reaction [211–218].

TsX ⇌ (hν or Δ) Ts• + X• X = I, Br, SePh

199 → (AIBN, Ts-SePh, C_6H_6, Δ) **200a** + **200b** 97% (57 : 43) (3.77)

201 → (AIBN, TsBr, $C_6H_5CH_3$, 90°C, 4 h) **202** 73% (3.78)

Regio- and stereoselective toluenesulfonyl bromide-mediated radical cyclization of bis(allene) (**201**) in the presence of AIBN also goes ahead to give rise to the *trans*-fused 5-membered ring (**202**) containing vinyl sulfone and vinyl bromide as shown in eq. 3.78.

Experimental procedure 12 (eq. 3.78).

A solution of *p*-toluenesulfonyl bromide (94 mg, 0.4 mmol), bis(allene) (100 mg, 0.36 mmol) and AIBN (0.2 eq.) in 4 ml of toluene was degassed and heated in a pressure tube at 90 °C for 4 h. After the reaction, toluene was removed *in vacuo*, and the residue was chromatographed on silica gel (hexane/ethyl acetate = 2/1) to give a product in 73% yield [218].

Stereoselective synthesis of (+)-botryodiplodin was carried out by a radical cyclization of dibromoacetal (**203**) containing an allene group, with Bu_3SnH initiated by Et_3B, through the 5-*exo-trig* cyclization, and the subsequent debromination with bulky

$(Me_3Si)_3SiH$ for the stereoselective reduction (eq. 3.79) [219].

(3.79)

3.2 CYCLIZATION TO MEDIUM-SIZED AND LARGE-SIZED RINGS

Preparation of medium- and large-sized rings such as 12 ~ 20-membered rings can be constructed by radical cyclization with highly diluted Bu_3SnH using a syringe-pump, with *endo* cyclization [220–227]. The reason for this can be explained as follows: there is not much energy difference between the transition state of *exo* cyclization and that of *endo* cyclization, because the ring strain in the formation of these transition states is somewhat reduced due to their large ring-sized transition states. Consequently, tertiary or secondary carbon-centered radicals are predominantly formed via *endo* cyclization. In nature, there are many kinds of 12 ~ 20-membered lactones and cyclic ketones, and some of them show potent biological activity. Therefore, preparation of large ring-sized compounds with a radical ring closure method is very important. Eq. 3.80 shows the preparation of 10-, 14-, and 18-membered cyclic ketones through SOMO–LUMO interaction of the formed radicals, by the reaction of iodoalkyl vinyl ketones with Bu_3SnH in the presence of AIBN, under highly dilute conditions.

(3.80)

Preparation of highly functionalized medium-sized carbocycles (**210**) and (**212**), derived from carbohydrates, can also be carried out with the same procedure via 8-*endo-trig* and 7-*exo-trig* manners, as shown below (eqs. 3.81, 3.82). Moreover, preparation of β-*C*-disaccharide (**214**) proceeds selectively via 8-*endo-trig* radical cyclization of two temporarily tethered monosaccharide (**213**), by means of the addition of a carbohydrate-derived radical onto an anomeric exomethylene group, as shown in eq. 3.83.

(3.81)

(3.82)

(3.83)

The same 8-*endo-trig* cyclization can also be carried out with sp^2 carbon-centered radicals, as shown in eq. 3.84. Thus, the preparation of *trans*-pyrano[3,2-c][2]benzoxocines (**216**) was produced in good yields through 8-*endo-trig* manner of aryl radicals derived from D-mannose pyranosides (**215**), using the Bu_3SnH /AIBN system.

(3.84)

(3.85)

Eq. 3.85 shows the preparation of (*S*)-(+)-zearalenone, a natural product, via 14-*endo-trig* manner of the cinnamyl radical formed [228, 229].

Lactones can be prepared through the reverse mode, the SOMO–HOMO interaction, as shown in eq. 3.86 [230–233]. Some medium-sized lactones are obtained by this mode under highly dilute conditions with the Bu_3SnH/AIBN system; however, ring closure reactions via 7-*endo-trig*, 9-*endo-trig*, and 10-*endo-trig* are generally not preferable.

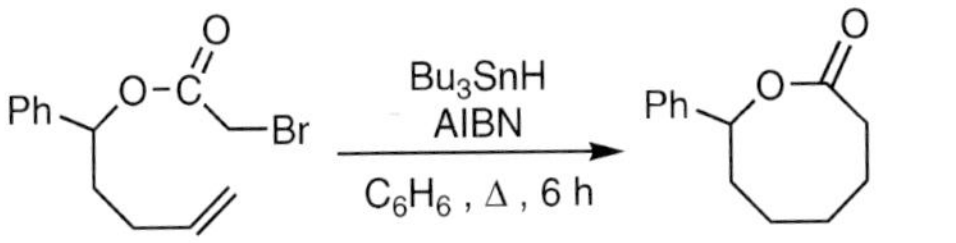

(3.86)

Experimental procedure 13 (eq. 3.86).

To a benzene solution of bromoacetate (0.015 M) was added dropwise a benzene solution of Bu_3SnH (1.4 eq.) and AIBN (0.1 eq.) over 5 h. After the reaction, the solvent was removed and the residue was chromatographed on silica gel to give lactone in 52% yield [230].

Treatment of 4-pentenyl iodoacetate (**221**) with bis(tributyltin) at room temperature under irradiation with a sun lamp generates γ-iodoheptanolactone (**222a**), which is an atom-transfer cyclization product (eq. 3.87) [234].

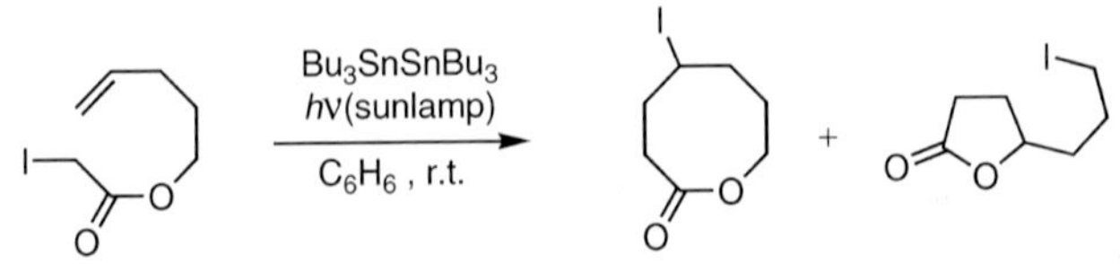

(3.87)

Experimental procedure 14 (eq. 3.87).

Boron trifluoride etherate (0.37 ml, 3 mmol) was added to a benzene (33 ml) solution of 4-pentenyl iodoacetate (1.0 mmol) under a nitrogen atmosphere at room temperature. *Bis*(tributyltin) (0.1 mmol) was added and the mixture was irradiated with a 300 W sun lamp at 20 °C. After 6 h, the solvent was removed and the residue was chromatographed on silica gel (hexane /ethyl acetate = 2/1) to produce the iodolactone in 80% yield [234].

Eq. 3.88 shows the preparation of a furanocembrane unit (**225**) through SOMO–LUMO interaction by an acyl radical, with Bu_3SnH [235]. The reaction shown in eq. 3.89 comprises of the initial addition of $Bu_3Sn^{\cdot}$ to a styrenyl group in compound (**226**) and subsequent 22-*endo-trig* ring closure of the formed benzylic radical to give compound (**227**) [236]. This rate constant for 22-*endo-trig* ring closure is not so small because it is $5.1 \times 10^3\ s^{-1}$ at 60 °C

PhSe O O 223 AIBN Bu_3SnH C_6H_6, Δ 224 40% TsOH $CHCl_3$, Δ 225 (3.88)

226 AIBN Bu_3SnH C_6H_6, Δ Bu_3Sn 227 74% (3.89)

More interestingly, as an oxidative condition with $Mn(OAc)_3$, various macrolide compounds (**230**) are obtained in high yields by the treatment of bis(acetylacetate) (**228**) and bis(olefin) (**229**) with $Mn(OAc)_3$ in acetic acid under highly diluted conditions (2 mM of substrate) [237,238]. This reaction proceeds as follows. The initial formation of a β-keto ester radical via single electron oxidation by $Mn(OAc)_3$ and its addition to olefin occur; the formed benzylic radical is oxidized by $Mn(OAc)_3$ to form a benzylic cation, then cyclization by oxygen of the enol form of the β-keto ester occurs, and again this reaction sequence is repeated in another β-keto ester group to ultimately form the macrolide compounds.

CH_3 O $(CH_2)_y$ O CH_3 228; Ph Ph $(CH_2)_x$ Ph Ph 229; $Mn(OAc)_3$ AcOH 100°C, 0.5 h; 230 (3.90)

x	y	ring-size	Yield (%)
1	2	11	0
2	2	12	71
2	3	13	64
2	4	14	66
4	3	15	75
4	4	16	74

Mn(OAc)$_3$, AcOH: 231 → 232 (3.91)

n	ring-size	Yield (%)
2	8	55
3	9	82
4	10	94
6	12	51
8	14	51

The same Mn(OAc)$_3$-mediated oxidation of 3,3-diphenyl-2-propenyloxyoligomethylene 3-oxobutanoate (**231**) in boiling acetic acid provides moderate to good yields of macrolides (**232**) via oxidative intramolecular radical cyclization, as shown in eq. 3.91.

As a reductive condition, treatment of α,α,α-trichloroacetate (**233**) with a Cu^{1+}-tris(pyridylmethyl)amine complex generates macrocyclic polyether (**234**) through initial SET from Cu^{1+} to trichloride, generation of α,α-dichloroacetate radical, 18-*endo-trig* ring closure, and abstraction of a chlorine atom from the α,α,α-trichloroacetate (**233**) by the formed carbon-centered radical as shown in eq. 3.92 [239].

233 → (Lig., CuCl, $ClCH_2CH_2Cl$, 80 °C) → 234 70% (3.92)

Lig: tris(2-pyridylmethyl)amine

3.3 RING EXPANSIONS

Cyclopropylmethyl (cyclopropylcarbinyl) radical can be observed by ESR at − 140 °C. However, on increasing the temperature to − 100 °C, β-cleavage occurs to form a 3-buten-1-yl radical. The rate constants for β-cleavage of cyclopropylmethyl radicals are shown in eq. 3.93, and are quite rapid [240–243]. The driving force for the rapid ring opening comes from the ring strain of the cyclopropyl ring. Similarly, ring opening of cyclopropyloxyl radicals is quite rapid for the same reasons. Here, the synthetic use of ring-opening reactions via β-cleavage will be mentioned, since radical ring-expansion reactions are based on β-cleavage reaction.

$k_1 = 1.3 \times 10^8 s^{-1}$ (25°C)
$E_a = 5.9$ kcal/mol (3.93a)

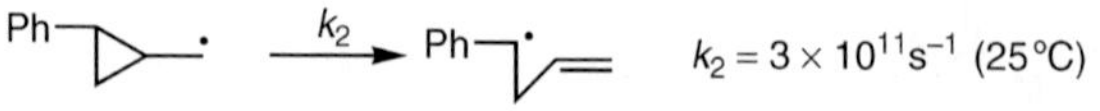

$k_2 = 3 \times 10^{11}\,s^{-1}$ (25 °C) (3.93b)

As a synthetic use of the ring-opening reaction of cyclopropyloxyl radicals, treatment of an α-bromomethyl compound (**236**) derived from α-methoxycarbonyl-cyclopentanone (**235**), with Bu_3SnH in the presence of AIBN generates one-carbon-expanded β-methoxycarbonylcyclohexanone (cyclic γ-keto ester) (**237**), through the formation of a primary carbon-centered radical, 3-*exo-trig* ring closure onto the carbonyl group, β-cleavage of the formed cyclopropyloxyl radical to give a stable α-ester radical, and finally abstraction of a hydrogen atom from Bu_3SnH (eq. 3.94) [244–253]. By this method, 6-membered and 7-membered cyclic γ-keto esters are obtained from 5-membered and 6-membered cyclic β-keto esters respectively.

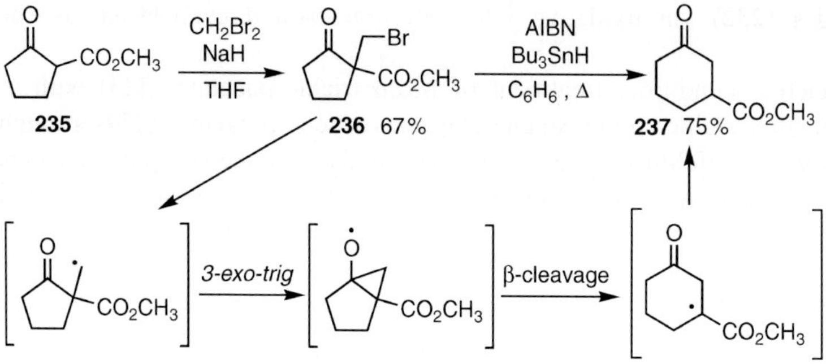

(3.94)

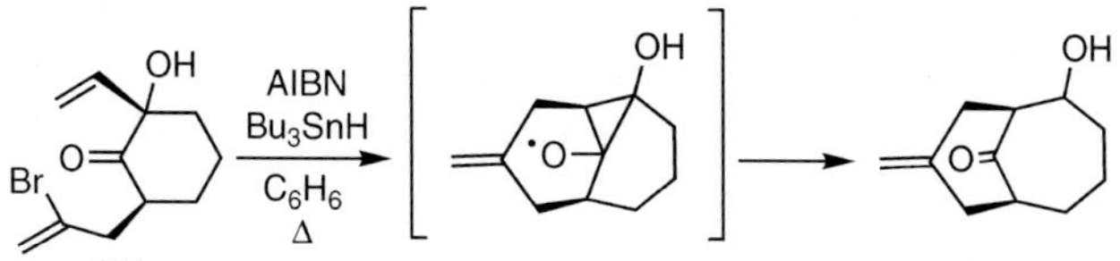

(3.95)

This reaction is further applied to the construction of a bridged bicyclic system (**239**), through 7-*endo-trig* ring closure of the vinyl radical, 3-*exo-trig* ring closure, and subsequent β-cleavage of the fused cyclopropyloxyl radical as shown in eq. 3.95.

One-carbon expansion of the ring-open-chained β-keto ester (**240**) also readily happens to form a good yield of γ-keto ester (**241**), (eq. 3.96), and the rate constant of the one-carbon expansion is approximately $10^4\ s^{-1}$ at 25 °C.

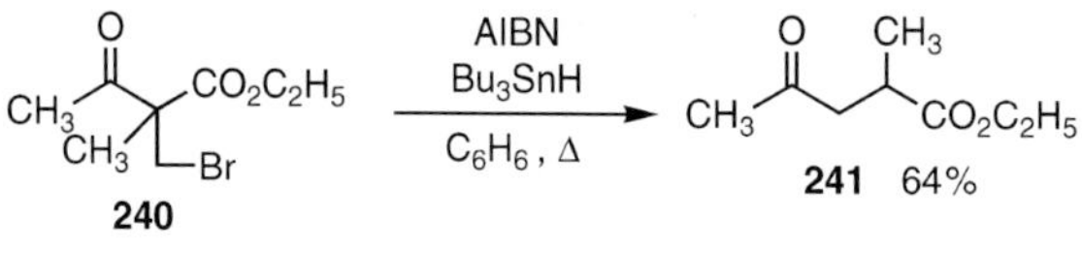

(3.96)

Scheme: 242 (2-bromomethyl-2-methoxycarbonylcycloheptanone) + MH, AIBN, $C_6H_5CH_3$, Δ → 243a (ring-expanded CO₂Me cyclooctanone) + 243b (2-methyl-2-methoxycarbonylcycloheptanone) (3.97)

MH	243a Yield (%)	243b Yield (%)
$Ph_4Si_2H_2$	75	trace
$(Me_3Si)_3SiH$	52	18
Bu_3SnH	20	63

Scheme: 244 + MH, AIBN, $C_6H_5CH_3$, Δ → 245a + 245b (3.98)

MH	245a Yield (%)	245b Yield (%)
$Ph_4Si_2H_2$	75	trace
$(Me_3Si)_3SiH$	76	trace
Bu_3SnH	42	36

The same ring-expansion reactions are also observed with less toxic $Ph_4Si_2H_2$ and $(Me_3Si)_3SiH$ as shown in eqs. 3.97 and 3.98. Under the same conditions, $Ph_4Si_2H_2$ and $(Me_3Si)_3SiH$ are more effective than Bu_3SnH, in terms of yield and selectivity of ring-expansion/direct reduction products.

Experimental procedure 15 (eq. 3.97).

AIBN (1.5 ~ 2.5 mmol) in toluene (10 ~ 15 ml) was added dropwise over 8 h by means of a dropping funnel to a refluxing solution of α-halomethyl cyclic β-keto ester (0.5 mmol) and $Ph_4Si_2H_2$ (0.6 mmol) in toluene (10 ml). After 12 h, the solvent was removed and the residue was purified by column chromatography (eluent: hexane/ethyl acetate = 5/1 ~ 10/1) to provide the ring-expanded cyclic γ-keto ester [253].

Moreover, in addition to the ring expansion of a cyclopropyloxyl radical, the present reaction can be applied to the ring expansion of a cyclopentyloxyl radical and a cyclohexyloxyl radical. Thus, 3-carbon and 4-carbon ring-expanded cyclic keto esters are obtained in good yields under highly diluted conditions as shown in eqs. 3.99b and 3.99c. However, 2-carbon ring-expanded cyclic keto ester cannot be formed, since the formation of 4-membered cyclobutyloxyl radical is highly disfavored (eq. 3.99a). Consequently, it is concluded that the driving force for these ring-expansion reactions is the formation of a stable α-ester radical, not reduction of the ring strain in cycloalkoxyl radicals. Eq. 3.100 shows the preparation of (*R*)-(−)-muscone (**253**) via the β-cleavage of a cyclopentyloxyl radical.

AIBN, Bu_3SnH, C_6H_6, Δ; **246** → **247** (3.99a)

AIBN, Bu_3SnH, C_6H_6, Δ; 5-*exo-trig*; **248** → **249** 69% (3.99b)

AIBN, Bu_3SnH, C_6H_6, Δ; 6-*exo-trig*; **250** → **251** 71% (3.99c)

NaH; 1) Bu_4NF 2) Ph_2PCl, I_2; **252**; R=TBS; $NaBH_3CN$, Ph_3SnCl; **253** (*R*)-(−)-Muscone (3.100)

$FeCl_3$ is a single-electron oxidant. Treatment of cyclopropyl silyl ether derivative (**254**) with $FeCl_3$ in DMF generates a cyclopropyloxyl radical initially, and subsequent β-cleavage and 5-*exo-trig* ring closure produce a bicyclic ketone (**255**) as shown in eq. 3.101. The product contains a chlorine atom, deriving from $FeCl_3$. A similar reaction using the $Fe(NO_3)_3$/1,4-cyclohexadiene system is also effective [254–256].

254 → $FeCl_3$, DMF → **255** 64% (3.101)

Experimental procedure 16 (eq. 3.101).

To a dry DMF (40 ml) solution of cyclopropyl silyl ether (2 mmol) was added dropwise a mixture of anhydrous $FeCl_3$ (4.4 mmol) and dry DMF (40 ml) over 40 min at 0 °C. The yellow solution obtained was further stirred for 1 h under an argon atmosphere at 0 °C. After the reaction, the mixture was poured into water (500 ml) and extracted with ethyl acetate three times (3 × 200 ml). The combined organic layer was washed with water (2 × 200 ml) and dried. After removal of the solvent, the residue was chromatographed on silica gel (ether/petroleum ether = 1/10) to give a bicyclic ketone in 64% yield [325].

Similarly to the cyclopropyloxyl radical, an epoxycarbinyl radical formed through the Barton–McCombie reaction rapidly gives rise to β-cleavage with $10^5\ s^{-1}$ rate constant. Eq. 3.102 shows the formation of a medium-ring sized cyclodecenone through C–O β-cleavage of an epoxycarbinyl radical to generate an oxygen-centered radical, then second β-cleavage to generate a stable α-ester radical [257–260].

AIBN
Bu_3SnH
C_6H_6, Δ
$CO_2C_2H_5$
256
$CO_2C_2H_5$
257 64%
C–O cleavage
$CO_2C_2H_5$
$CO_2C_2H_5$
$CO_2C_2H_5$

(3.102)

$Bu_3Sn^{\cdot}$ adds to an olefinic group, so treatment of α,β-unsaturated epoxide (**258**) with Bu_3SnH generates bicyclic *exo*-methylene alcohol (**259**), through an epoxycarbinyl radical, an alkoxyl radical via its C–O β-cleavage, 1,5-H shift, 5-*exo-trig* ring closure, and finally β-elimination of the Bu_3Sn group, as shown in eq. 3.103 [261–266].

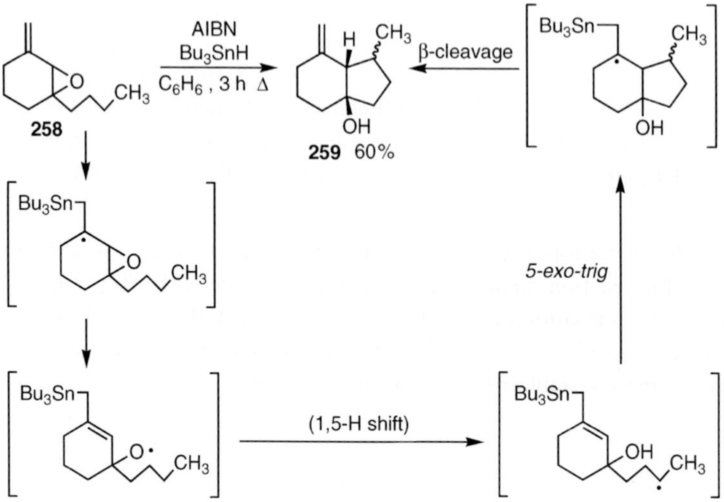

(3.103)

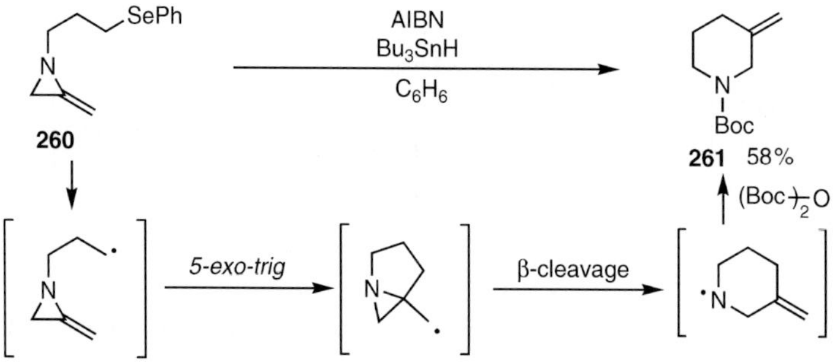

(3.104)

5-*Exo-trig* ring closure of a 3-(2-methyleneaziridin-1-yl)propyl radical leads to the generation of 3-methylenepiperidine (**261**) through the formation of a strained bicyclic aziridinylcarbinyl radical, and subsequent C–N β-cleavage to an aminyl radical, as shown in eq. 3.104.

Eq. 3.105 shows the reaction of 2-(3-bromopropyl)methylenecyclopropane (**262**) with $Bu_3Sn^{\cdot}$ to form *exo*-methylenecyclohexane (**263**), through 5-*exo-trig* ring closure to *exo*-methylene, and subsequent β-cleavage of the formed cyclopropylcarbinyl radical [267–270].

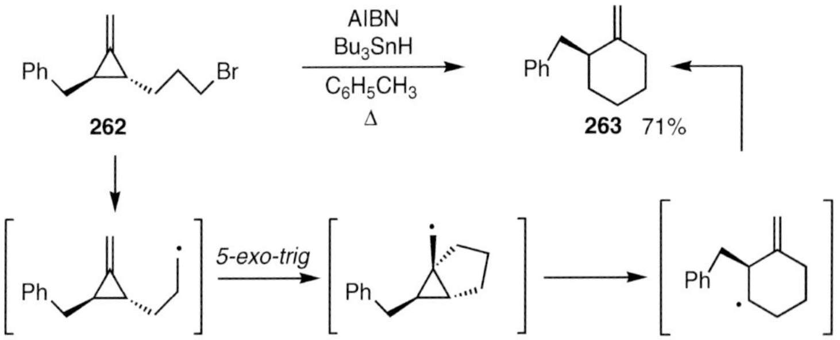

(3.105)

The Barton–McCombie reaction of bicyclo[3.1.0]-, [4.1.0]-, and [5.1.0]-imidazoylthiocarbonates (**264**) induces the formation of cyclopropylcarbinyl radicals, and subsequent β-cleavage, to generate one-carbon ring-expanded 6-, 7-, and 8-membered compounds (**265**) in good yields [271]. The driving force of these reactions is the reduction of ring strain (eq. 3.106).

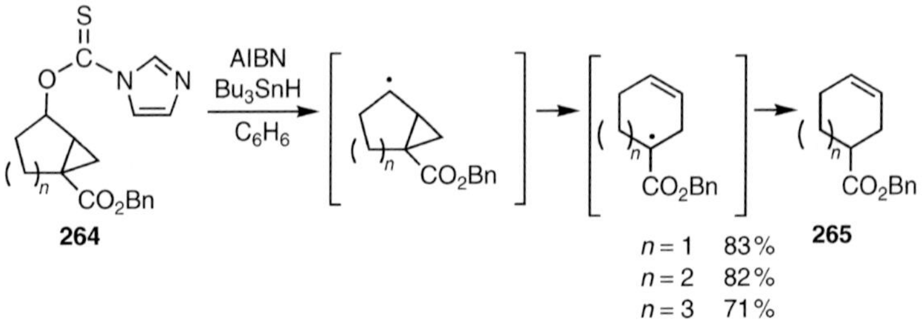

(3.106)

As a combined reaction of β-cleavage of a cyclopropylcarbinyl radical and intermolecular addition, treatment of vinylcyclopropane (**266**) and activated alkyne in the presence of PhSSPh and AIBN forms a cyclopentene skeleton (**267**), through the initial addition of a thiyl radical to the vinyl group, β-cleavage of the cyclopropylcarbinyl radical, addition of the carbon-centered radical to the alkyne, ring closure of a vinyl radical via 5-*exo-trig* manner, and finally subsequent β-elimination of the thiyl radical, as shown in eq. 3.107 [272–276]. Here, PhSSPh acts as a catalyst, since the thiyl radical is regenerated. Aliphatic disulfides such as

BuSSBu also work well.

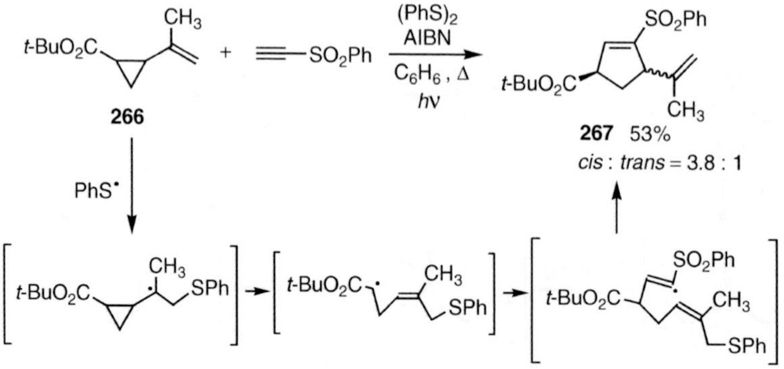

(3.107)

This reaction looks like [3 + 2] annulation with cyclopropane and alkyne derivatives.

What, then, about the cyclobutylcarbinyl radical ? Eq. 3.108 shows the treatment of tricyclic cyclobutylmethyl iodide (**268**) with Bu_3SnH in the presence of AIBN in refluxing benzene solution to generate bicyclic *exo*-methylene compound (**269**), via β-cleavage of the cyclobutylcarbinyl radical. This bicyclic *exo*-methylene (**269**) is the skeleton of a terpenoide natural product, guaiane alismol [277–280].

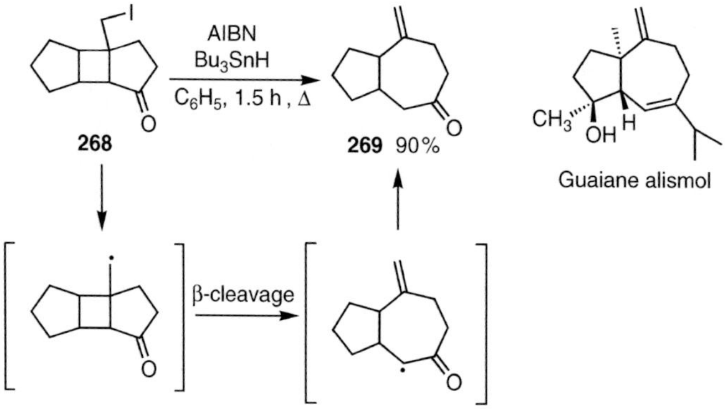

(3.108)

Treatment of tricyclic cyclobutylmethyl iodide with SmI_2 generates the same ring-expanded compounds via SET [281].

The cyclobutoxyl radical also induces β-cleavage. Thus, treatment of 2-(3-bromopropyl)cyclobutanone derivatives (**270**) with Bu_3SnH in the presence of AIBN forms the ring-expanded bicyclic ketones (**271**), through 5-*exo-trig* ring closure onto a carbonyl group, and subsequent β-cleavage of the cyclobutoxyl radical formed, as shown in eq. 3.109 [282–287].

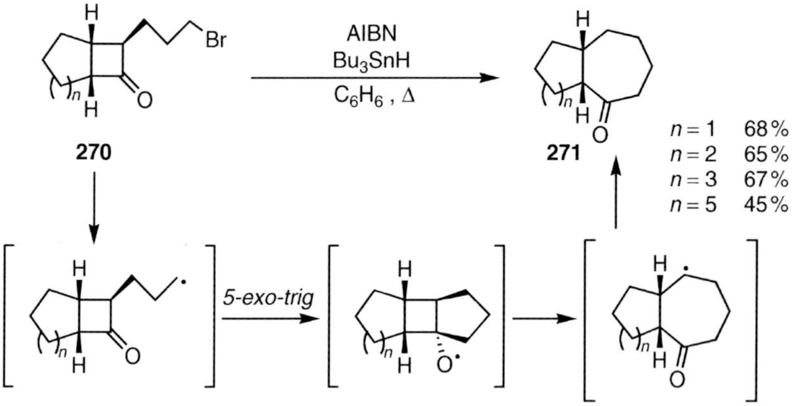

270 → 271 (AIBN, Bu_3SnH, C_6H_6, Δ; via 5-exo-trig)

$n = 1$	68%
$n = 2$	65%
$n = 3$	67%
$n = 5$	45%

(3.109)

272 → 273 (Bu_3SnH, AIBN, C_6H_6, Δ; via 5-exo-trig, (–X•))

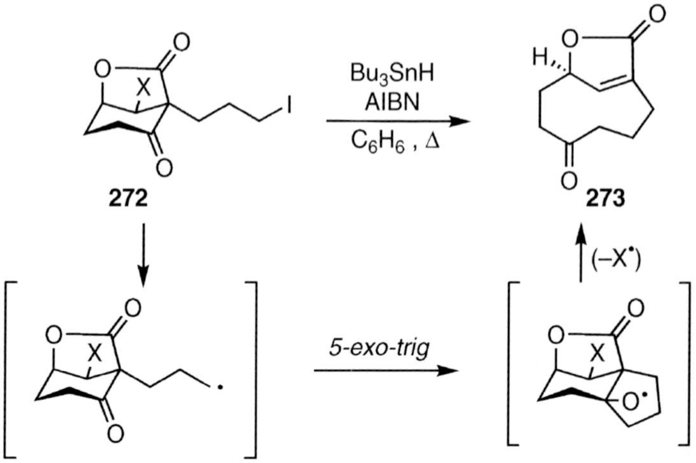

(3.110)

A highly strained cyclopentyloxyl radical formed from compound (**272**) *in situ* induces β-cleavage as shown in eq. 3.110.

SmI_2 can be also used for the same ring-expansion reaction. Thus, radical ring-expansion reaction of bicyclo[4.2.0]octanone (**274**) promoted by SmI_2 provides a mixture of cyclooctanol (**275a**) and lactone (**275b**) (eq. 3.111). In this case, CO_2Me, CN, and Ph groups play an important role for the ring expansion.

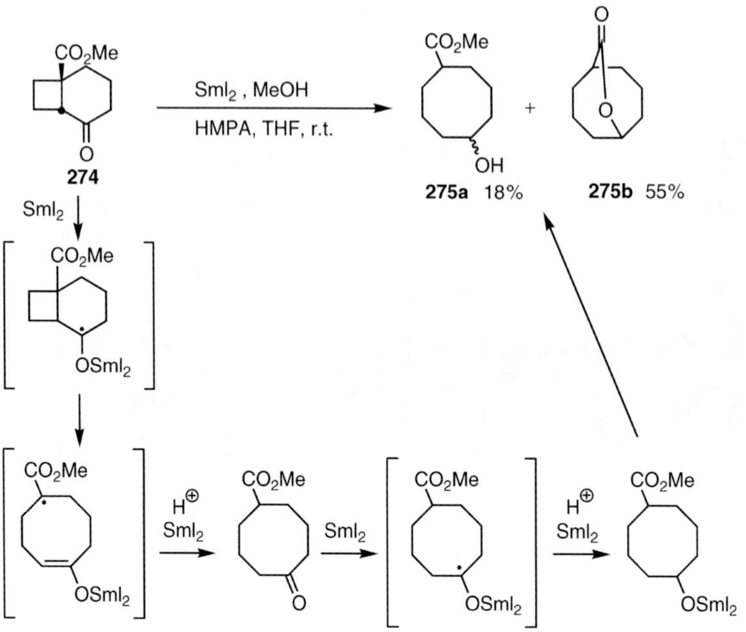

(3.111)

In a similar way to the cyclobutylcarbinyl radical, cyclobutylaminyl and cyclopropylaminyl radicals also induce rapid β-cleavage reactions as shown in eq. 3.112 [288–292].

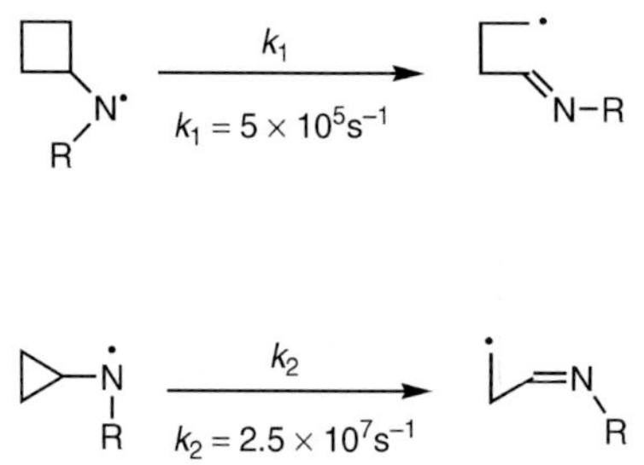

(3.112a)

(3.112b)

3.4 CYCLIZATION TO SPIRO SKELETON

Hydantocidin, which bears a spiro skeleton, is a natural product and a herbicide. Eq. 3.113 shows the typical preparation method of spiro-diketone (**277**) from the reaction of α-diazoketone (**276**) with $Rh_2(OAc)_4$ via a carbenoide species. However, this spiro skeleton can be also constructed by radical reactions. Treatment of β-keto ester (**278**) with $Mn(OAc)_3$ in the presence of electron-rich olefin generates spiro-cyclic thioketal (**279**) through the formation of a β-keto ester radical, intermolecular radical addition to olefin, and oxidative spiro-cyclization as shown in eq. 3.114 [293, 294].

Hydantocidin

276 $\xrightarrow{Rh_2(OAc)_4}$ 277 40%

(3.113)

278 $\xrightarrow{Mn(OAc)_3,\ AcOH}$ 279 48%

(–e, –H⊕)

(–e)

(3.114)

280 $\xrightarrow{Mn(OAc)_3}$ 281 51%

Mn(OAc)$_3$

(–e, –H⊕)

(–e, –H⊕)

(3.115)

Mn(OAc)$_3$-promoted oxidative reaction of 2-hydroxy-1,4-naphthoquinone (**280**) with a β-enamino carbonyl compound generates a spiro-lactam (**281**) as shown in eq. 3.115.

Spiro cyclization can be carried out by the treatment of organic halides or selenides with Bu_3SnH. Thus, treatment of vinyl iodide (**282**) with Bu_3SnH in the presence of AIBN produces a spiro-ketal (**283**), through the formation of a vinyl radical, abstraction of a hydrogen atom from the methine position via 1,5-H shift (from sp^2 carbon-centered radical to sp^3 carbon-centered radical), and 5-*exo-trig* ring closure (eq. 3.116) [295–298]. This reaction requires highly diluted conditions (0.01 M) of Bu_3SnH, to suppress the formation of direct reduction product. Treatment of cycloalkanone (**284**) bearing a β-iodo-α,β-unsaturated ester or a nitrile side chain, with Bu_3SnH in the presence of AIBN forms spiro-cyclic ketone (**285**) through 1,5-H shift (from sp^2 carbon-centered radical to sp^3 carbon-centered radical) and subsequent 5-*exo-trig* ring closure (eq. 3.117). By this method, the preparation of spiro-nucleosides may be possible. Preparation of spiro-dilactone (**287**) via the intramolecular free radical Michael addition onto the enol ester can be carried out as shown in eq. 3.118. The product is the skeleton of a natural product, altenuic acid II.

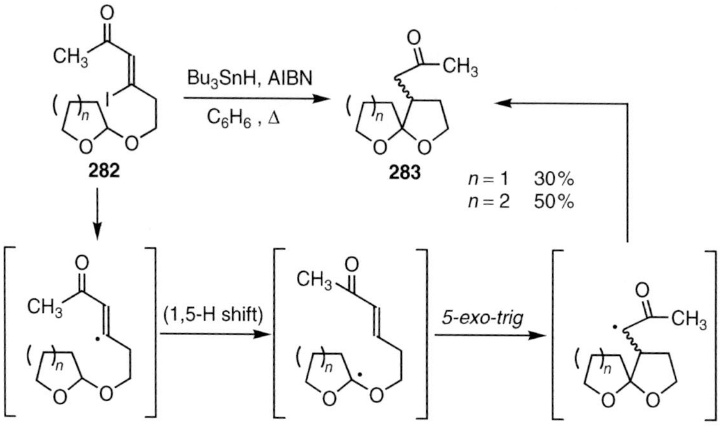

282 → 283 (Bu_3SnH, AIBN, C_6H_6, Δ); via (1,5-H shift) and 5-exo-trig; $n = 1$ 30%, $n = 2$ 50% (3.116)

284 → 285 (AIBN, Bu_3SnH, C_6H_6, Δ); via 1,5-H shift and 5-exo-trig; $E = CO_2CH_3$ 70% (2 : 1), $E = CN$ 76% (1.8 : 1) (3.117)

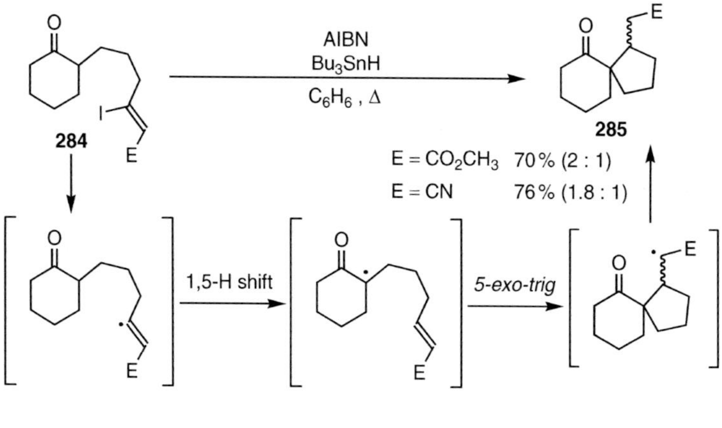

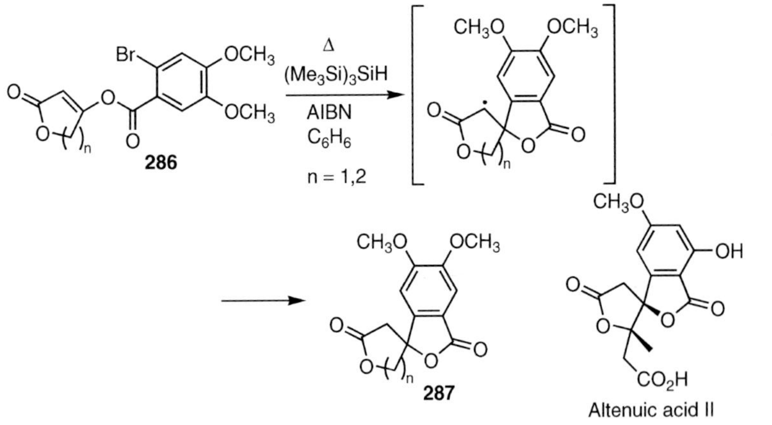

(3.118)

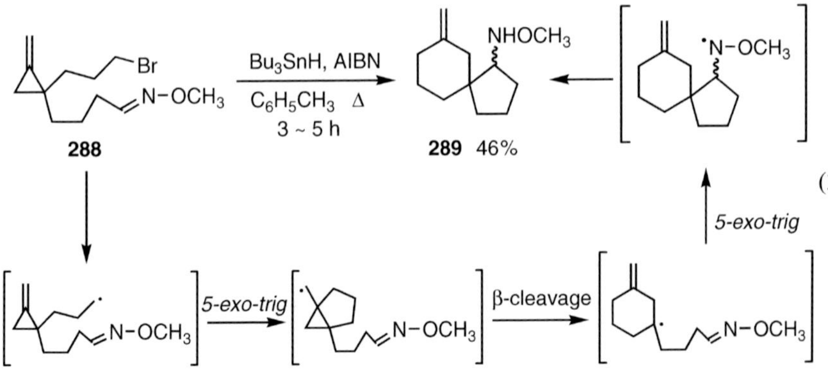

(3.119)

Eq. 3.119 shows the preparation of a spiro compound (**289**) through 5-*exo-trig* ring closure to form a cyclopropylcarbinyl radical, followed by β-cleavage, and subsequent 5-*exo-trig* ring closure to imino-carbon [299–301].

$PhI(O_2CCF_3)_2$-induced intramolecular cyclization of α-(2-arylethyl)cyclic-1,3-diketone (**290**) generates a high yield of a spiro-benzannulated compound (**291**), via SET [302].

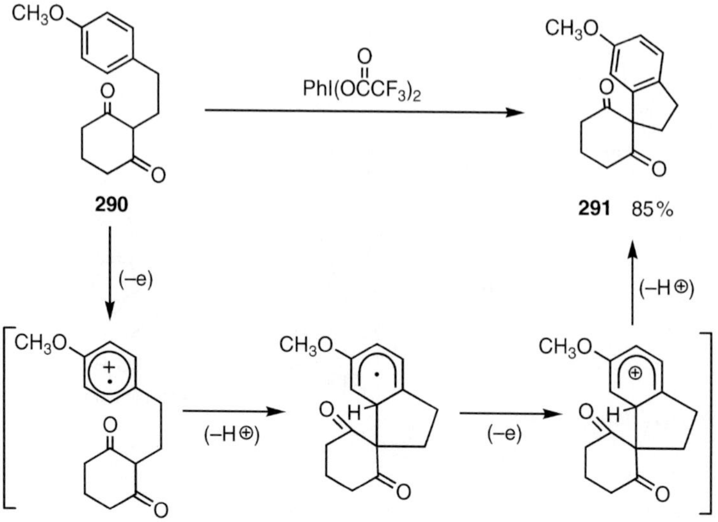

(3.120)

3.5 TANDEM CYCLIZATION TO POLYCYCLIC RINGS

The formation of a bicyclic skeleton from the reaction of alcohol with Fe^{3+} comprises of an alkoxyl radical, followed by β-cleavage to form a carbon-centered radical, and subsequent ring closure. For example, eq. 3.121 shows the reaction of TMS-protected cyclopropanol (**292**) with $Fe(NO_3)_3$ in 1,4-cyclohexadiene to give bicyclo[3,3,0]octan-3-one (**293**), through the generation of a cyclopropoxyl radical, its β-cleavage, and subsequent 5-*exo-trig* ring closure [303, 304].

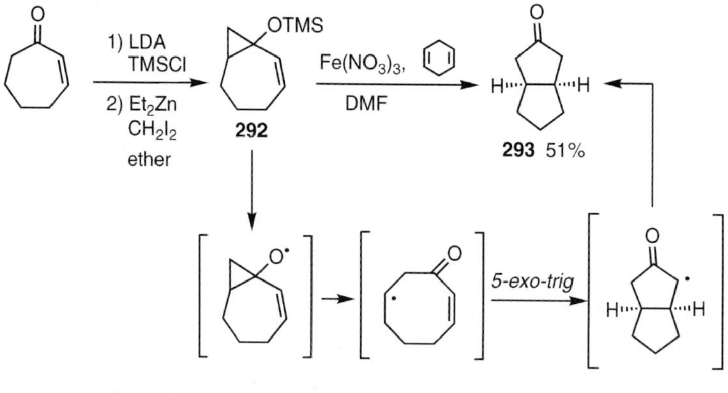

(3.121)

One-pot construction of a polycyclic skeleton with tandem radical cyclization is possible. Thus, treatment of iodoazide (**294**) with $(Me_3Si)_3SiH$ gives a tetracyclic alkaloid (**295**), through the 5-*exo-trig* ring closure by an sp^2 carbon-centered radical, followed by the second 5-*exo-trig* ring closure onto the azide group, together with the formation of molecular N_2. This tetracyclic compound is the skeleton of a natural product, aspidospermidine. This type of reaction is a radical specific reaction (eq. 3.122) [305–312].

(Me$_3$Si)$_3$SiH, AIBN, C$_6$H$_6$, 80°C, Δ; 294; 295 83%; Aspidospermidine; (−N$_2$)

(3.122)

Et$_3$B, Bu$_3$SnH, C$_6$H$_6$; E = CO$_2$CH$_3$; 296; 297a 83% (4 : 3) 297b

(3.123)

The same cascade reaction of the iodo-triene (**296**) with Bu$_3$SnH produces tricyclic esters (**297a**) and (**297b**) via triple 5-*exo-trig* ring closure as shown in eq. 3.123.

Eq. 3.124 shows the preparation of a polyhydroxyindolidine skeleton that is expected to be a glycosidase inhibitor and have *anti*-AIDS activity, from the reaction of compound (**298**) with Ph$_3$SnH. This reaction proceeds as follows. Initially comes the 5-*endo-trig* manner controlled thermodynamically, and the formed radical [**V**] is stabilized by captodative effect with amino and ester groups. Then, 6-*endo-trig* manner controlled thermodynamically occurs to form a bicyclic amide (**299**) [313]. This second 6-*endo-trig* cyclization is also controlled by a 5-membered ring formed through the initial cyclization. This tandem cyclization method may be useful for the construction of alkaloids, since naturally occurring alkaloids such as polyhydroxyindolizidines have recently been recognized as a glycosidase inhibitor.

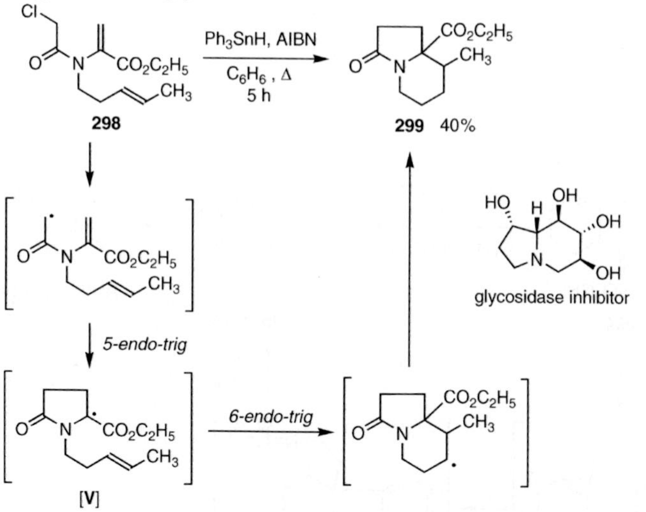

(3.124)

Another alkaloid, the paniculatine skeleton, can be constructed by the following procedure. Treatment of the α-iodo cyclic ketone (**300**) bearing TMS-protected acetylene and vinyl silyl groups, with Bu_3SnH provides a tricyclic ketone (**301**), through the initial 5-*exo-dig* ring closure onto the acetylene group, followed by the second cyclization via 5-*exo-trig* ring closure onto the vinyl silyl group as shown in eq. 3.125 [314]. Here, TMS plays an important role to increase the electron density of the acetylenic group against the electron-deficient α-ester radical.

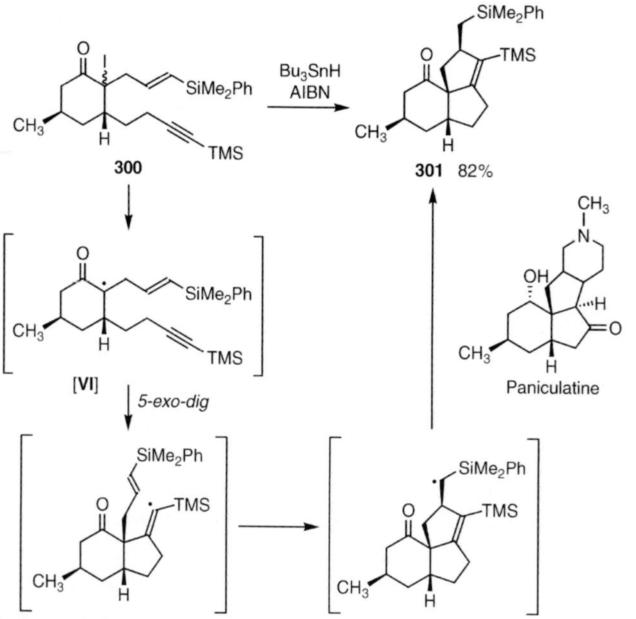

(3.125)

An acyl radical formed from *Se*-Ph selenoester, can be used for the construction of a polycyclic skeleton in one-pot reaction [315, 316]. In this way, triquinane (+)-

pentalenene was obtained by the reaction of α,β-unsaturated *Se*-Ph selenoester (**302**) with Bu_3SnH in the presence of AIBN, through sequential 5-*exo-trig* and 5-*exo-dig* radical cyclization involving a ketene intermediate as shown in eq. 3.126.

AIBN, Bu_3SnH, C_6H_6

302 → **303** 45%

(3.126)

AIBN, Bu_3SnH, C_6H_6

304 → (6-*endo-trig*) × 3 → **305** 90%

(3.127)

AIBN, Bu_3SnH, C_6H_6, Δ

306 → 7 × (6-*endo-trig*) → **307** 17%

(3.128)

Eq. 3.127 shows the cascade radical cyclization of an acyl radical leading to a steroid ring (**305**) with high regio- and stereo-selectivities and high yield, via triple 6-*endo-trig* ring closure, from *tri*-olefinic *Se*-Ph selenoester (**304**). The reaction in eq. 3.128 is the ultimate radical cyclization of *hepta*-olefinic *Se*-Ph selenoester (**306**) with Bu_3SnH to form *hepta*-cyclic ketone (**307**) with high stereoselectivity (all *trans* form) in one-pot reaction, through septet 6-*endo-trig* ring closure [317–324]. This is a very impressive reaction.

Experimental procedure 17 (eq. 3.128).

A mixture of *Se*-Ph selenoester (23.9 μmol) and AIBN (1 mg) in dry benzene (6 ml) was degassed and substituted by argon gas. Benzene (2 ml) solution of Bu_3SnH (29.8 μmol) and AIBN (1 mg) was added dropwise to the refluxing mixture over 8 h. After the addition, the mixture was refluxed further 3 h. After removal of the solvent, the residue was chromatographed on silica gel (eluent: ether and petroleum ether) to obtain a hepta-cyclic compound in 17% yield [322].

As an oxidative cascade reaction with $Mn(OAc)_3$, treatment of *tetra*-olefinic β-keto ester (**308**) with $Mn(OAc)_3$ in acetic acid generates a steroidal skeleton (**309**) via quartet *6-endo-trig* ring closure, in one-pot reaction (eq. 3.129).

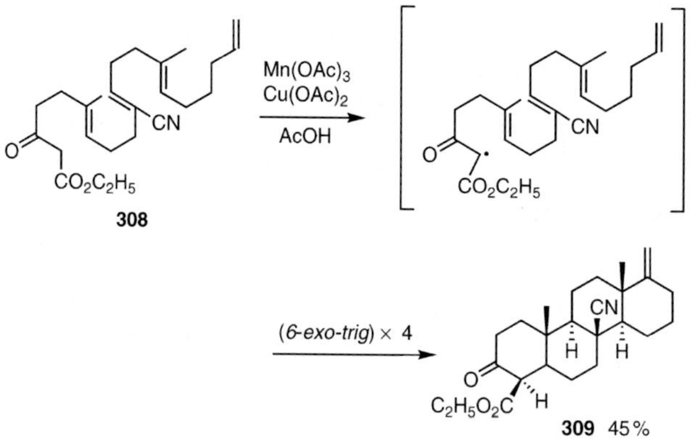

(3.129)

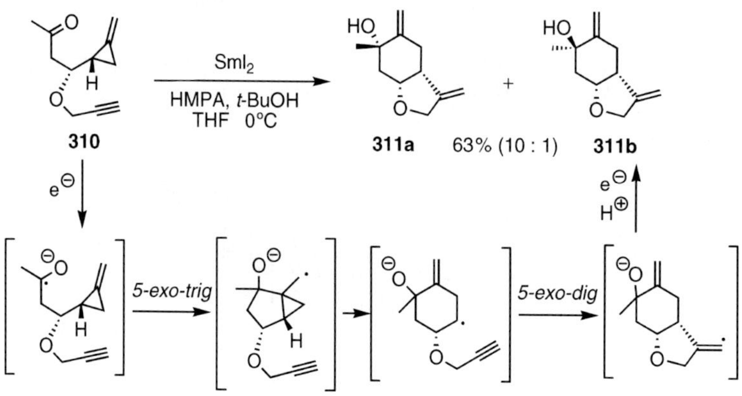

(3.130)

As a reductive cascade reaction with SmI_2, treatment of methylenecyclopropyl ketone (**310**) with SmI_2 in a mixture of HMPA and *t*-BuOH at 0 °C generates bicyclic alcohols (**311a**) and (**311b**) which are the skeleton of a natural product, paeonilactone, as shown in eq. 3.130.

These cascade reactions are very attractive in view of multi-C–C bond formation in one-pot and are a typically radical specific reaction.

REFERENCES

1. J Baldwin, *Chem. Commun.*, 1976, 734.
2. ALJ Beckwith and CH Schiesser, *Tetrahedron Lett.*, 1985, **26**, 373.
3. B Giese, B Kopping and C Chatgilialoglu, *Tetrahedron Lett.*, 1989, **30**, 681.
4. EW Della and SD Graney, *Org. Lett.*, 2002, **4**, 4065.
5. C Walling and A Cioffari, *J. Am. Chem. Soc.*, 1972, **94**, 6059.
6. ALJ Beckwith and G Moad, *Chem. Commun.*, 1974, 472.
7. ALJ Beckwith, CJ Easton and AK Serelis, *Chem. Commun.*, 1980, 482.
8. ALJ Beckwith, L Lawrence and AK Serelis, *Chem. Commun.*, 1980, 484.
9. ALJ Beckwith, G Phillipou and AK Serelis, *Tetrahedron Lett.*, 1981, **22**, 2811.
10. ALJ Beckwith, CJ Easton, T Lawrence and AK Serelis, *Aust. J. Chem.*, 1983, **36**, 545.
11. EW Della and AM Knill, *J. Org. Chem.*, 1995, **60**, 3518.
12. WR Dolbier, Jr. and XX Rong, *Tetrahedron Lett.*, 1996, **37**, 5321.
13. G Stork, R Mook, SA Biller and SD Rychnovsky, *J. Am. Chem. Soc.*, 1983, **105**, 3741.
14. DLJ Clive and PL Beaulieu, *Chem. Commun.*, 1983, 307.
15. M Ladlow and G Pattenden, *Tetrahedron Lett.*, 1984, **25**, 4317.
16. Y Ueno, M Moriya, K Chino, M Watanabe and M Okawara, *J. Chem. Soc. Perkin Trans. 1*, 1986, 1351.
17. DLJ Clive and DR Cheshire, *Chem. Commun.*, 1987, 1520.
18. A Spikrishna and G Sunderbabu, *Chem. Lett.*, 1988, 371.
19. F Villar, O Andrey and P Renaud, *Tetrahedron Lett.*, 1999, **40**, 3375.
20. SN Osipov and K Burger, *Tetrahedron Lett.*, 2000, **41**, 5659.
21. K Nozaki, K Oshima and K Utimoto, *Tetrahedron Lett.*, 1988, **29**, 6127.
22. MD Bachi and E Bosch, *J. Org. Chem.*, 1989, **54**, 1236.
23. GVM Sharma and SR Vepachedu, *Tetrahedron Lett.*, 1990, **31**, 4931.
24. ALJ Beckwith, I Blair and G Phillipon, *J. Am. Chem. Soc.*, 1974, **96**, 1613.
25. G Stork and NH Baine, *J. Am. Chem. Soc.*, 1982, **104**, 2321.
26. DJ Hart and YM Tsai, *J. Am. Chem. Soc.*, 1982, **104**, 1430.
27. JK Choi, DJ Hart and YM Tsai, *Tetrahedron Lett.*, 1982, **23**, 4765.
28. Y Ueno, K Chino, M Watanabe, O Moriya and M Okawara, *J. Am. Chem. Soc.*, 1982, **104**, 5564.
29. ALJ Beckwith, DM O'Shea and DH Roberts, *Chem. Commun.*, 1983, 1445.
30. Y Ueno, RK Khare and M Okawara, *J. Chem. Soc, Perkin Trans. 1*, 1983, 2637.
31. G Stork and R Mook, Jr., *J. Am. Chem. Soc.*, 1983, **105**, 3720.
32. G Stork and PM Sher, *J. Am. Chem. Soc.*, 1983, **105**, 6765.
33. A Padwa, H Nimmesgern and GSK Wong, *Tetrahedron Lett.*, 1985, **26**, 957.
34. A Padwa, H Nimmesgern and GSK Wong, *J. Org. Chem.*, 1985, **50**, 5620.
35. O Moriya, Y Urata, Y Ikeda, Y Ueno and T Endo, *J. Org. Chem.*, 1986, **51**, 4708.
36. M Pezechk, AP Brunetiere and JY Lallemand, *Tetrahedron Lett.*, 1986, **27**, 2715.
37. A Srikrishna and KC Rullaiah, *Tetrahedron Lett.*, 1987, **28**, 5203.
38. S Knapp, FS Gibson and YH Choe, *Tetrahedron Lett.*, 1990, **31**, 5397.
39. H Tanaka, H Suga, H Ogawa, AKM Abdul Hai and S Torii, *Tetrahedron Lett.*, 1992, **33**, 6495.
40. CK Sha, CY Shen, TS Jean, RT Chiu and WH Tseng, *Tetrahedron Lett.*, 1993, **34**, 7641.
41. EW Della and AM Knill, *Tetrahedron Lett.*, 1996, **37**, 5805.

42. E Lee, KS Li and J Lim, *Tetrahedron Lett.*, 1996, **37**, 1445.
43. Y Watanabe, S Ishikawa, G Takao and T Toru, *Tetrahedron Lett.*, 1999, **40**, 3411.
44. O Yamazaki, K Yamaguchi, M Yokoyama and H Togo, *J. Org. Chem.*, 2000, **65**, 5440.
45. S Mikami, K Fujita, T Nakamura, H Yorimitsu, H Shinokubo, S Matsubara and K Oshima, *Org. Lett.*, 2001, **3**, 1853.
46. Y Watanabe, Y Ueno, C Tanaka, M Okawara and T Endo, *Tetrahedron Lett.*, 1987, **28**, 3953.
47. Y Watanabe and T Endo, *Tetrahedron Lett.*, 1988, **29**, 321.
48. F Barth and CO- Yang, *Tetrahedron Lett.*, 1991, **32**, 5873.
49. SN Osipov and K Burger, *Tetrahedron Lett.*, 2000, **41**, 5659.
50. CP Chung and DJ Hart, *J. Org. Chem.*, 1983, **48**, 1784.
51. CS Wilcox and LM Thomasco, *J. Org. Chem.*, 1985, **50**, 546.
52. M Ladlow and G Pattenden, *J. Chem. Soc. Perkin Trans. 1*, 1988, 1107.
53. ALJ Beckwith and DH Roberts, *J. Am. Chem. Soc.*, 1986, **108**, 5893.
54. C Bonini, *Tetrahedron Lett.*, 1990, **31**, 5369.
55. MD Bachi, A Mesmaeker and N Mesmaeker, *Tetrahedron Lett.*, 1987, **28**, 2637.
56. H Ishibashi, T Sato, M Irie, S Harada and M Ikeda, *Chem. Lett.*, 1987, 795.
57. T Sato, Y Wada, M Nishimoto, H Ishibashi and M Ikeda, *J. Chem. Soc. Perkin Trans. 1*, 1989, 879.
58. T Sato, N Nakamura, K Ikeda, M Okada, H Ishibashi and M Ikeda, *J. Chem. Soc. Perkin Trans. 1*, 1992, 2399.
59. K Goodall and AF Parsons, *Tetrahedron Lett.*, 1997, **38**, 491.
60. T Itoh, H Ohara and S Emoto, *Tetrahedron Lett.*, 1995, **36**, 3531.
61. JA Murphy and MS Sherburn, *Tetrahedron Lett.*, 1990, **31**, 3495.
62. JA Murphy and MS Sherburn, *Tetrahedron Lett.*, 1990, **31**, 1625.
63. CJ Moody and CL Norton, *J. Chem. Soc. Perkin Trans. 1*, 1997, 2639.
64. SM Leit and LA Paquette, *J. Org. Chem.*, 1999, **64**, 9225.
65. EW Della and SD Graney, *Tetrahedron Lett.*, 2000, **41**, 7987.
66. FE Ziegler and ZI Zheng, *Tetrahedron Lett.*, 1987, **28**, 5973.
67. DS Middleton and NS Simpkins, *Tetrahedron*, 1988, **29**, 1315.
68. MD Bachi and E Bosch, *J. Chem. Soc. Perkin Trans. 1*, 1988, 1517.
69. A Srikrishna and G Sunderbabu, *Tetrahedron Lett.*, 1989, **30**, 3561.
70. T Morikawa, M Uejima and Y Kobayashi, *Chem. Lett.*, 1989, 623.
71. EJ Enholm and KS Kinter, *J. Am. Chem. Soc.*, 1991, **113**, 7784.
72. JL Esker and M Newcomb, *Tetrahedron Lett.*, 1993, **34**, 6877.
73. J Boivin, AM Schiano and SZ Zard, *Tetrahedron Lett.*, 1994, **35**, 249.
74. WR Bowman, MJ Broadhurst, DR Coghlan and KA Lewis, *Tetrahedron Lett.*, 1997, **38**, 6301.
75. AJ Clark and JL Peacock, *Tetrahedron Lett.*, 1998, **39**, 6029.
76. SY Chang, YF Shao, SF Chu, GT Fan and YM Tsai, *Org. Lett.*, 1999, **1**, 945.
77. D Bebbington, J Bentley, PA Nilsson and AF Parsons, *Tetrahedron Lett.*, 2000, **41**, 8941.
78. H Nishiyama, T Kitajima, M Matsumoto and K Itoh, *J. Org. Chem.*, 1984, **49**, 2298.
79. G Stork and M Kahn, *J. Am. Chem. Soc.*, 1985, **107**, 500.
80. M Koreeda and IA George, *J. Am. Chem. Soc.*, 1986, **108**, 8098.
81. G Stork and MJ Sofia, *J. Am. Chem. Soc.*, 1986, **108**, 6826.
82. K Tamao, K Maeda, T Yamaguchi and Y Ito, *J. Am. Chem. Soc.*, 1989, **111**, 4984.
83. GK Friestad, *Org. Lett.*, 1999, **1**, 1499.
84. D Crich and SM Fortt, *Tetrahedron Lett.*, 1988, **29**, 2585.
85. CJ Hayes and G Pattenden, *Tetrahedron Lett.*, 1996, **37**, 271.
86. A Stojanovic and P Renaud, *Synlett*, 1997, 181.
87. PA Evans and JD Rosenman, *J. Org. Chem.*, 1996, **61**, 2252.
88. JH Rigby, DM Danca and JH Horner, *Tetrahedron Lett.*, 1998, **39**, 8413.
89. PA Evans, T Manangan and AL Rheingold, *J. Am. Chem. Soc.*, 2000, **122**, 11009.
90. DLJ Clive and Elena-S Ardelean, *J. Org. Chem.*, 2001, **66**, 4841.
91. ALJ Beckwith, S Wang and J Warkentin, *J. Am. Chem. Soc.*, 1987, **109**, 5289.
92. S Kim, GH Joe and JY Do, *J. Am. Chem. Soc.*, 1994, **116**, 5521.

93. JH Udding, H Hiemstra, MNA Zanden and WN Speckamp, *Tetrahedron Lett.*, 1991, **32**, 3123.
94. AJ Clark, CP Dell, JM Ellard, NA Hunt and JP McDonagh, *Tetrahedron Lett.*, 1999, **40**, 8619.
95. DT Davies, N Kapur and AF Parsons, *Tetrahedron Lett.*, 1999, **40**, 8615.
96. AJ Clark, RP Filik, DM Haddleton, A Radigue and CJ Sanders, *J. Org. Chem.*, 1999, **64**, 8954.
97. DK Barma, A Kundu, R Baati, C Mioskowski and JR Falck, *Org. Lett.*, 2002, **4**, 1387.
98. Z Zhou and SM Bennett, *Tetrahedron Lett.*, 1997, **38**, 1153.
99. SM Bennett, RK Biboutou and BSF Salari, *Tetrahedron Lett.*, 1998, **39**, 7075.
100. S Bobo, IS de Gracia and JL Chiara, *Synlett*, 1999, 1551.
101. D Riber, R Hazell and T Skrydstrup, *J. Org. Chem.*, 2000, **65**, 5382.
102. Anne-C Durand, J Rodriguez and Jean-P Dulcere, *Synlett*, 2000, 731.
103. K Kakiuchi, Y Fujioka, H Yamamura, K Tsutsumi, T Morimoto and H Kurosawa, *Tetrahedron Lett.*, 2001, **42**, 7595.
104. H Ishibashi, N Nakamura, T Sato, M Takeuchi and M Ikeda, *Tetrahedron Lett.*, 1991, **32**, 1725.
105. ME Jung, ID Trifunovich and N Lensen, *Tetrahedron Lett.*, 1992, **33**, 6719.
106. K Ogura, N Sumitani, A Kayano, H Iguchi and M Fujita, *Chem. Lett.*, 1992, 1487.
107. ME Jung and M Kiankarimi, *J. Org. Chem.*, 1995, **60**, 7013.
108. A D'Annibale, S Resta and C Trogolo, *Tetrahedron Lett.*, 1995, **36**, 9039.
109. ME Jung and R Mraquez, *Tetrahedron Lett.*, 1997, **38**, 6521.
110. ME Jung, R Marquez and KN Houk, *Tetrahedron Lett.*, 1999, **40**, 2661.
111. D Johnston, CM McCusker and DJ Procter, *Tetrahedron Lett.*, 1999, **40**, 4913.
112. A D'Annibale, D Nanni, C Trogolo and F Umani, *Org. Lett.*, 2000, **2**, 401.
113. JS Bryans, NEA Chessum, AF Parsons and F Ghelfi, *Tetrahedron Lett.*, 2001, **42**, 2901.
114. K Castle, C-S Hau, JB Sweeney and C Tindall, *Org. Lett.*, 2003, **5**, 759.
115. H David, C Afonso, M Bonin, G Doisneau, MG Guillerez and F Guibe, *Tetrahedron Lett.*, 1999, **40**, 8557.
116. P Wessig and O Muhling, *Angew. Chem. Int. Ed.*, 2001, **40**, 1064.
117. H Villar and F Guide, *Tetrahedron Lett.*, 2002, **43**, 9517.
118. ALJ Beckwith and WB Gara, *J. Chem. Soc. Perkin Trans. 1*, 1975, 795.
119. K Shankaran, CP Sloan and V Snieckus, *Tetrahedron Lett.*, 1985, **26**, 6001.
120. K Jones, M Thompson and C Wright, *Chem. Commun.*, 1986, 115.
121. AN Abeywickrema and ALJ Beckwith, *Chem. Commun.*, 1986, 464.
122. KH Parker, DM Spero and KC Inman, *Tetrahedron Lett.*, 1986, **27**, 2833.
123. H Togo and O Kikuchi, *Tetrahedron Lett.*, 1988, **29**, 4133.
124. JP Dittami and H Ramanathan, *Tetrahedron Lett.*, 1988, **29**, 45.
125. DL Boger and RS Coleman, *J. Am. Chem. Soc.*, 1988, **110**, 1321.
126. DJ Hart and SC Wu, *Tetrahedron Lett.*, 1991, **32**, 4099.
127. S Takano, M Suzuki, A Kijima and K Ogasawara, *Tetrahedron Lett.*, 1990, **31**, 2315.
128. AJ Clark, K Jones, C McCarthyl and JM Storey, *Tetrahedron Lett.*, 1991, **32**, 2829.
129. W Zhang and G Pugh, *Tetrahedron Lett.*, 1999, **40**, 7591.
130. CG Martin, JA Murphy and CR Smith, *Tetrahedron Lett.*, 2000, **41**, 1833.
131. K Orito, Y Satoh, H Nishizawa, R Harada and M Tokuda, *Org. Lett.*, 2000, **2**, 2535.
132. AM Rosa, S Prabhakar and AM Lobo, *Tetrahedron Lett.*, 1990, **31**, 1881.
133. JC Estevez, MC Villaverde, RJ Estevez and L Castedo, *Tetrahedron Lett.*, 1991, **32**, 529.
134. K Jones, TCT Ho and J Wilkinson, *Tetrahedron Lett.*, 1995, **36**, 6743.
135. K Orito, S Uchito, Y Satoh, T Tatsuzawa and M Tokuda, *Org. Lett.*, 2000, **2**, 307.
136. DC Harrowven, BJ Sutton and S Coulton, *Tetrahedron Lett.*, 2001, **42**, 2907.
137. MJ Ellis and MFG Stevens, *J. Chem. Soc. Perkin Trans. 1*, 2001, 3180.
138. DC Harrowven, MIT Nunn and DR Fenwick, *Tetrahedron Lett.*, 2002, **43**, 3185.
139. DC Harrowven, MIT Nunn and DR Fenwick, *Tetrahedron Lett.*, 2002, **43**, 7345.
140. K Jones and A Fiumana, *Tetrahedron Lett.*, 1996, **37**, 8049.
141. AP Dobbs, K Jones and KT Veal, *Tetrahedron Lett.*, 1997, **38**, 5383.
142. DC Harrowven, MIT Nunn, NJ Blumire and DR Fenwick, *Tetrahedron Lett.*, 2000, **41**, 6681.

143. DC Harrowven, BJ Sutton and S Couton, *Tetrahedron Lett.*, 2001, **42**, 2907.
144. DC Harrowven, BJ Sutton and S Coulton, *Tetrahedron Lett.*, 2001, **42**, 9061.
145. CK Sha, ZP Zhan and FS Wang, *Org. Lett.*, 2000, **2**, 2011.
146. EN Prabhakaran, BM Nugent, AL Williams, KE Nailor and JN Johnston, *Org. Lett.*, 2002, **4**, 4197.
147. CRA Godfrey, P Hegarty, WB Motherwell and MK Uddin, *Tetrahedron Lett.*, 1998, **39**, 723.
148. A Ryokawa and H Togo, *Tetrahedron*, 2001, **57**, 5915.
149. DLJ Clive and S Kang, *Tetrahedron Lett.*, 2000, **41**, 1315.
150. WB Motherwell and S Vazquez, *Tetrahedron Lett.*, 2000, **41**, 9667.
151. A Studer, M Bossart and T Vasella, *Org. Lett.*, 2000, **2**, 985.
152. K Wakabayashi, H Yorimitsu, H Shinokubo and K Oshima, *Org. Lett.*, 2000, **2**, 1899.
153. H Senboku, H Hasegawa, K Orito and M Tokuda, *Tetrahedron Lett.*, 2000, **41**, 5699.
154. R Ryokawa and H Togo, *Tetrahedron*, 2001, **57**, 5915.
155. DLJ Clive and S Kang, *J. Org. Chem.*, 2001, **66**, 6083.
156. W Zhang and G Pugh, *Tetrahedron Lett.*, 2001, **42**, 5613.
157. DP Curran, AC Abraham and H Liu, *J. Org. Chem.*, 1991, **56**, 4335.
158. DP Curran and J Xu, *J. Am. Chem. Soc.*, 1996, **118**, 3142.
159. J Robertson, MA Peplow and J Pillai, *Tetrahedron Lett.*, 1996, **37**, 5825.
160. J Rancourt, V Gorys and E Jolicoeur, *Tetrahedron Lett.*, 1998, **39**, 5339.
161. JMD Storey, *Tetrahedron Lett.*, 2000, **41**, 8173.
162. JN Johnston, MA Plotkin, R Viswanathan and EN Prabhakaran, *Org. Lett.*, 2001, **3**, 1009.
163. ALJ Beckwith and GF Meijs, *J. Org. Chem.*, 1987, **52**, 1922.
164. AJ Clark and K Jones, *Tetrahedron Lett.*, 1989, **30**, 5485.
165. J Inanaga, O Ujikawa and M Yamaguchi, *Tetrahedron Lett.*, 1991, **32**, 1737.
166. G Martelli, P Spagnolo and M Tiecco, *J. Chem. Soc (B)*, 1970, 1413.
167. R Tsang and BF Reid, *J. Am. Chem. Soc.*, 1986, **108**, 2116 see also 8102.
168. YM Tsai, KH Tang and WT Jiaang, *Tetrahedron Lett.*, 1993, **34**, 1303.
169. YM Tsai, KH Tang and WT Jiaang, *Tetrahedron Lett.*, 1996, **37**, 7767.
170. SY Chang, WT Jiaang, CD Cherng, KH Tang, CH Huang and YM Tsai, *J. Org. Chem.*, 1997, **62**, 9089.
171. SY Chang, YF Shao, SF Chu, GT Fan and YM Tsai, *Org. Lett.*, 1999, **1**, 945.
172. EJ Enholm and G Prasad, *Tetrahedron Lett.*, 1989, **30**, 4939.
173. E Lee, JS Tae, YH Chong, YC Park, M Yun and S Kim, *Tetrahedron Lett.*, 1994, **35**, 129.
174. AF Parsons and RM Pettifer, *Tetrahedron Lett.*, 1997, **38**, 5907.
175. DS Hays and GC Fu, *J. Org. Chem.*, 1998, **63**, 6375.
176. J Tormo, DS Hays and GC Fu, *J. Org. Chem.*, 1998, **63**, 201.
177. P Devin, L Fensterbank and M Malacria, *Tetrahedron Lett.*, 1998, **39**, 833.
178. T Naito, K Nakagawa, T Nakamura, A Kasei, I Ninomiya and T Kiguchi, *J. Org. Chem.*, 1999, **64**, 2003.
179. FL Harris and L Weiler, *Tetrahedron Lett.*, 1987, **28**, 2941.
180. C Audin, JM Lancelin and JM Beau, *Tetrahedron Lett.*, 1988, **29**, 3691.
181. AD Mesmaeker, P Hoffmann, T Winkler and A Waldner, *Synlett*, 1990, 201.
182. DP Curran and H Liu, *J. Org. Chem.*, 1991, **56**, 3463.
183. J M Contelles, L Martinez, A M- Grau, C Pozuelo and ML Jimeno, *Tetrahedron Lett.*, 1991, **32**, 6437.
184. S Kim, Y Kim and KS Yoon, *Tetrahedron Lett.*, 1997, **38**, 2487.
185. U Iserloh and DP Curran, *J. Org. Chem.*, 1998, **63**, 4711.
186. GE Keck, SF McHardy and JA Murry, *J. Org. Chem.*, 1999, **64**, 4465.
187. DLJ Clive and R Subedi, *Chem. Commun.*, 2000, 237.
188. P Devin, L Fensterbank and M Malacria, *Tetrahedron Lett.*, 1999, **40**, 5511.
189. GK Friestad, *Org. Lett.*, 1999, **1**, 1499.
190. X Lin, D Stien and SM Weinreb, *Org. Lett.*, 1999, **1**, 637.
191. J Bentley, PA Nilsson and AF Parsons, *J. Chem. Soc. Perkin Trans. 1*, 2002, 1461.
192. GA Kraus and L Chen, *J. Am. Chem. Soc.*, 1990, **112**, 3464.
193. GA Kraus, PJ Thomas and MD Schwinder, *Tetrahedron Lett.*, 1990, **31**, 1819.

194. T Hasegawa, Y Horikoshi, S Iwata and M Yoshioka, *Chem. Commun.*, 1991, 1617.
195. PJ Wagner, Q Cao and R Pabon, *J. Am. Chem.Soc.*, 1992, **114**, 346.
196. G Pandey and GD Reddy, *Tetrahedron Lett.*, 1992, **33**, 6533.
197. Y Ueda, H Watanabe, J Uemura and K Uneyama, *Tetrahedron Lett.*, 1993, **34**, 7933.
198. J Cossy, JL Ranaivosata and V Bellosta, *Tetrahedron Lett.*, 1994, **35**, 8161.
199. S Sauer, A Schumacher, F Barbosa and B Giese, *Tetrahedron Lett.*, 1998, **39**, 3685.
200. T Fukuyama, X Chen and G Peng, *J. Am. Chem. Soc.*, 1994, **116**, 3127.
201. JD Rainier, AR Kennedy and E Chase, *Tetrahedron Lett.*, 1999, **40**, 6325.
202. MT Reding and T Fukuyama, *Org. Lett.*, 1999, **1**, 973.
203. H Tokuyama, M Watanabe, Y Hayashi, T Kurokawa, G Peng and T Fukuyama, *Synlett*, 2001, 1403.
204. MJ Laws and CH Schiesser, *Tetrahedron Lett.*, 1997, **38**, 8429.
205. CH Schiesser and SL Zheng, *Tetrahedron Lett.*, 1999, **40**, 5095.
206. L Engman, MJ Laws, J Malmstrom, CH Schiesser and LM Zuguro, *J. Org. Chem.*, 1999, **64**, 6764.
207. L Engman, MJ Laws, J Malmstrom, CH Schiesser and LM Zugaro, *J. Org. Chem.*, 1999, **64**, 6764.
208. Y Nishiyama, Y Hada, M Anjiki, S Hanita and N Sonoda, *Tetrahedron Lett.*, 1999, **40**, 6293.
209. N Al-Maharik, L Engman, J Malmstrom and CH Schiesser, *J. Org. Chem.*, 2001, **66**, 6286.
210. L Benati, R Leardini, M Minozzi, D Nanni, P Spagnolo, S Strazzari and G Zanardi, *Org. Lett.*, 2002, **4**, 3079.
211. CP Chuang and TJ Ngoi, *Tetrahedron Lett.*, 1989, **30**, 6369.
212. GL Edwards and KA Walker, *Tetrahedron Lett.*, 1992, **33**, 1779.
213. CP Chuang, *Tetrahedron Lett.*, 1992, **33**, 6311.
214. FE Gueddari, JR Grimaldi and JM Hatem, *Tetraheron Lett.*, 1995, **36**, 6685.
215. CP Chuang and SF Wang, *Synlett*, 1995, 763.
216. MP Bertrand, S Gastaldi and R Nouguier, *Tetrahedron Lett.*, 1996, **37**, 1229.
217. C Wang and GA Russell, *J. Org. Chem.*, 1999, **64**, 2066 see also 2346.
218. SK Kang, YH Ha, DH Kim, Y Lim and J Jung, *Chem. Commun.*, 2001, 1306.
219. F Villar, O Andrey and P Renaud, *Tetrahedron Lett.*, 1999, **40**, 3375.
220. NA Porter, VHT Chang, DR Magnin and BT Wright, *J. Am. Chem. Soc.*, 1988, **110**, 3554.
221. NA Porter and VHT Chang, *J. Am. Chem. Soc.*, 1987, **109**, 4976.
222. NA Porter, DR Magnin and BT Wright, *J. Am. Chem. Soc.*, 1986, **108**, 2787.
223. NA Porter, B Lacker, VHT Chang and DR Magnin, *J. Am. Chem. Soc.*, 1989, **111**, 8309.
224. J Marco-Contelles and E de Opazo, *Tetrahedron Lett.*, 2000, **41**, 5341.
225. B Vauzeilles and P Sinay, *Tetrahedron Lett.*, 2001, **42**, 7269.
226. J Marco-Contelles and E de Opazo, *J. Org. Chem.*, 2002, **67**, 3705.
227. A Nandi and P Chattopadhyay, *Tetrahedron Lett.*, 2002, **43**, 5977.
228. SA Hitchcock and G Pattenden, *J. Chem. Soc. Perkin Trans. 1*, 1992, 1323.
229. NJG Cox, SD Mills and G Pattenden, *J. Chem. Soc. Perkin Trans. 1*, 1992, 1313.
230. E Lee, CH Yoon and TH Lee, *J. Am. Chem. Soc.*, 1992, **114**, 10981.
231. E Lee, CH Yoon, TH Lee, SY Kim, TJ Ha, Y Sung, S Park and S Lee, *J. Am. Chem. Soc.*, 1998, **120**, 7469.
232. GA Russell and C Li, *Tetrahedron Lett.*, 1996, **37**, 2557.
233. B Ding, R Keese and H Stoecli-Evans, *Angew. Chem. Int. Ed.*, 1999, **38**, 375.
234. J Wang and C Li, *J. Org. Chem.*, 2002, **67**, 1271.
235. MP Astley and G Pattenden, *Synlett*, 1991, 333.
236. KJ Shea, R O'Dell and DY Sasaki, *Tetrahedron Lett.*, 1992, **33**, 4699.
237. T Yoshinaga, H Nishino and K Kurosawa, *Tetrahedron Lett.*, 1998, **39**, 9197.
238. S Jogo, H Nishino, M Yasutake and T Shinmyozu, *Tetrahedron Lett.*, 2002, **43**, 9031.
239. F De Campo, D Lastecoueres and J-B Verlhac, *J. Chem. Soc. Perkin Trans. 1*, 2000, 575.
240. B Maillard, D Forrest and KU Ingold, *J. Am. Chem. Soc.*, 1976, **98**, 7024.
241. JM Tanko, RE Drumright, NK Suleman and LE Brammer, Jr., *J. Am. Chem. Soc.*, 1994, **116**, 1785.

242. AA Martin-Esker, CC Johnson, JH Horner and M Newcomb, *J. Am. Chem. Soc.*, 1994, **116**, 9174.
243. F Tian and WR Dolbier, Jr., *Org. Lett.*, 2000, **2**, 835.
244. P Dowd and SC Choi, *J. Am. Chem. Soc.*, 1987, **109**, 3493 see also 6548.
245. P Dowd and SC Choi, *Tetrahedron*, 1989, **45**, 77.
246. ALJ Beckwith, DM O'Shea and SW Westwood, *J. Am. Chem. Soc.*, 1988, **110**, 2565.
247. C Chatgilialoglu, VI Timokhin and M Ballestri, *J. Org. Chem.*, 1998, **63**, 1327.
248. P Dowd and SC Choi, *Tetrahedron Lett.*, 1991, **32**, 565.
249. P Dowd and SC Choi, *Tetrahedron Lett.*, 1989, **30**, 6129.
250. H Nemoto, M Shiraki, N Yamada, N Raku and K Fukumoto, *Tetrahedron Lett.*, 1996, **37**, 6355.
251. MT Crimmins, S Huang and LE Guise-Zawacki, *Tetrahedron Lett.*, 1996, **37**, 6519.
252. JR Rodriguez, L Castedo and JL Mascarenas, *Org. Lett.*, 2001, **3**, 1181.
253. M Sugi and H Togo, *Tetrahedron*, 2002, **58**, 3171.
254. JS U, J Lee and JK Cha, *Tetrahedron Lett.*, 1997, **38**, 5233.
255. KI Booker-Milburn, A Barker and W Brailsford, *Tetrahedron Lett.*, 1998, **39**, 4373.
256. KI Booker-Milburm, B Cox, M Grady, F Halley and S Marrison, *Tetrahedron Lett.*, 2000, **41**, 4651.
257. DA Corser, BA Marples and RK Dart, *Synlett*, 1992, 987.
258. A Johns, JA Murphy, CW Patterson and NF Wooster, *Chem. Commun.*, 1987, 1238.
259. VH Rawal and HM Zhong, *Tetrahedron Lett.*, 1993, **34**, 5197.
260. JA Murphy and CW Patterson, *Tetrahedron Lett.*, 1993, **34**, 867.
261. S Kim, S Lee and JS Koh, *J. Am. Chem. Soc.*, 1991, **113**, 5106.
262. S Kim and JS Koh, *Tetrahedron Lett.*, 1992, **33**, 7391.
263. VH Rawal and V Krishnamurthy, *Tetrahedron Lett.*, 1992, **33**, 3439.
264. HS Dang and BP Roberts, *Tetrahedron Lett.*, 1992, **33**, 6169.
265. VH Rawal, V Krishnamurthy and A Fabre, *Tetrahedron Lett.*, 1993, **34**, 2899.
266. N Prevost and M Shipman, *Org. Lett.*, 2001, **3**, 2383.
267. C Destabel, JD Kilburn and J Knight, *Tetrahedron Lett.*, 1993, **34**, 3151.
268. M Santagostino and JD Kilburn, *Tetrahedron Lett.*, 1994, **35**, 8863.
269. EJ Kantorowski, B Borhan, S Nazarian and MJ Kurth, *Tetrahedron Lett.*, 1998, **39**, 2483.
270. S Kim, KH Kim and JR Cho, *Tetrahedron Lett.*, 1997, **38**, 3915.
271. PH Lee, B Lee, J Lee and SK Park, *Tetrahedron Lett.*, 1999, **40**, 3427.
272. KS Feldman and AKK Vong, *Tetrahedron Lett.*, 1990, **31**, 823.
273. KS Feldman, RE Ruckle and AL Romanelli, *Tetrahedron Lett.*, 1991, **30**, 5845.
274. DA Singleton, KM Church and MJ Lucero, *Tetrahedron Lett.*, 1990, **31**, 5551.
275. DA Singleton, CC Huval, KM Church and ES Priestley, *Tetrahedron Lett.*, 1991, **32**, 5765.
276. EJ Enholm and ZJ Jia, *Tetrahedron Lett.*, 1995, **36**, 6819.
277. GL Lange and C Gottardo, *Tetrahedron Lett.*, 1990, **31**, 5985.
278. GL Lange, C Gottardo and A Merica, *J. Org. Chem.*, 1999, **64**, 6738.
279. W Zhang and P Dowd, *Tetrahedron Lett.*, 1992, **33**, 7307.
280. GL Lange, C Gottardo and A Merica, *J. Org. Chem.*, 1999, **64**, 6738.
281. GL Lange and A Merica, *Tetrahedron Lett.*, 1999, **40**, 7897.
282. W Zhang and P Dowd, *Tetrahedron Lett.*, 1992, **33**, 3285.
283. P Dowd and W Zhang, *J. Am. Chem. Soc.*, 1991, **113**, 9875.
284. W Zhang, Y Hua, G Hoge and P Dowd, *Tetrahedron Lett.*, 1994, **35**, 3865.
285. G Lange and C Gottardo, *Tetrahedron Lett.*, 1994, **35**, 6607 see also 8513.
286. AG Schultz and X Zhang, *Chem. Commun.*, 2000, 399.
287. K Kakiuchi, K Minato, K Tsutsumi, T Morimoto and H Kurosawa, *Tetrahedron Lett.*, 2003, **44**, 1963.
288. M Newcomb, SM Park, J Kaplan and DJ Marquardt, *Tetrahedron Lett.*, 1985, **26**, 5651.
289. WR Bowman, DN Clark and RJ Marmon, *Tetrahedron Lett.*, 1991, **32**, 6441.
290. S Kim, GH Joe and JY Do, *J. Am. Chem. Soc.*, 1993, **115**, 3328.
291. L Benati, D Nanni, C Sangiorgi and P Spagnolo, *J. Org. Chem.*, 1999, **64**, 7836.
292. G Pattenden and DJ Schulz, *Tetrahedron Lett.*, 1993, **34**, 6787.

293. JM Mellor and S Mohammed, *Tetrahedron Lett.*, 1991, **32**, 7111.
294. CP Chuang and YL Wu, *Tetrahedron Lett.*, 2001, **42**, 1717.
295. DS Middleton and NS Simpkins, *Tetrahedron Lett.*, 1989, **30**, 3865.
296. DS Middleton and NS Simpkins, *Tetrahedron*, 1990, **46**, 545.
297. CDS Brown, NS Simpkins and K Clinch, *Tetrahedron Lett.*, 1993, **34**, 131.
298. CK Sha, CW Hsu, YT Chen and SY Cheng, *Tetrahedron Lett.*, 2000, **41**, 9865.
299. M Santagostino and JD Kilburn, *Tetrahedron Lett.*, 1995, **36**, 1365.
300. W Zhang, *Tetrahedron Lett.*, 2000, **41**, 2523.
301. GVM Sharma, AS Chander, VG Reddy, K Krishnudu, MHVR Rao and AC Kunwar, *Tetrahedron Lett.*, 2000, **41**, 1997.
302. M Arisawa, NG Ramesh, M Nakajima, H Tohma and Y Kita, *J. Org. Chem.*, 2001, **66**, 59.
303. KI Booker-Milburn and RF Dainty, *Tetrahedron Lett.*, 1998, **39**, 5049.
304. Z Cekovic and S Saicic, *Tetrahedron Lett.*, 1986, **27**, 5981.
305. DP Curran and DM Rakiewicz, *J. Am. Chem. Soc.*, 1985, **107**, 1448.
306. DP Curran and M-H Chen, *Tetrahedron Lett.*, 1985, **26**, 4991.
307. PJ Parsons, PA Willis and SC Eyley, *Chem.Commun.*, 1988, 283.
308. M Kizil, B Patro, O Callaghan, JA Murphy, MB Hursthouse and D Hibbs, *J. Org. Chem.*, 1999, **64**, 7856.
309. L Fensterbank, E Mainetti, P Devin and M Malacria, *Synlett*, 2000, 1342.
310. B Patro and JA Murphy, *Org. Lett.*, 2000, **2**, 3599.
311. K Takasu, S Maiti, A Katsumata and M Ihara, *Tetrahedron Lett.*, 2001, **42**, 2157.
312. S Zhou, S Bommezijn and JA Murphy, *Org. Lett.*, 2002, **4**, 443.
313. SR Baker, AF Parsons, JF Pons and M Wilson, *Tetrahedron Lett.*, 1998, **39**, 7179.
314. CK Sha, FK Lee and CJ Chang, *J. Am. Chem. Soc.*, 1999, **121**, 9875.
315. NM Harrington-Frost and G Pattenden, *Synlett*, 1999, 1917.
316. NM Harrington-Frost and G Pattenden, *Tetrahedron Lett.*, 2000, **41**, 403.
317. G Pattenden and L Roberts, *Tetrahedron Lett.*, 1996, **37**, 4191.
318. PA Zoretic, Z Shen and M Wang, *Tetrahedron Lett.*, 1995, **36**, 2929.
319. PA Zoretic, Z Chen, Y Zhang and AA Ribeiro, *Tetrahedron Lett.*, 1996, **37**, 7909.
320. PA Zoretic, X Weng, CK Biggers, MS Biggers, ML Caspar and DG Davis, *Tetrahedron Lett.*, 1992, **33**, 2637.
321. Q Zhang, RM Mohan, L Cook, S Kazanis, D Peisach, BM Foxman and BB Snider, *J. Org. Chem.*, 1993, **58**, 7640.
322. S Handa and G Pattenden, *J. Chem. Soc. Perkin Trans. 1*, 1999, 843.
323. HM Boehm, S Handa, G Pattenden, L Roberts, AJ Blake and W Li, *J. Chem. Soc. Perkin Trans. 1*, 2000, 3522.
324. RJ Boffey, WG Whittingham and JD Kilburn, *J. Chem. Soc. Perkin Trans. 1*, 2001, 487.
325. KI Booker-Milburn and DF Thompson, *J. Chem. Soc. Perkin Trans. 1*, 1995, 2315.

– 4 –

Intermolecular Radical Addition Reactions

4.1 ADDITION TO OLEFINS

As mentioned previously (Chapter 1.1.4), there are two types of radicals: nucleophilic radicals ($R_N^{\bullet}$, SOMO energy level is high, such as c-$C_6H_{12}^{\bullet}$ and t-$Bu^{\bullet}$) and electrophilic radicals ($R_E^{\bullet}$, SOMO energy level is low, such as $EtO_2CCH_2^{\bullet}$). The reactivity of these radicals with olefins depends on the electronic character and density of olefins in the addition reactions. Thus, based on the electron density of radicals, either electron-rich olefins (ED: electron-donating group) or electron-deficient olefins (EW: electron-withdrawing group) must be selected, to increase SOMO–HOMO orbital interaction or SOMO–LUMO orbital interaction. The yields can be increased when these orbital energy gaps are reduced. Mode **A** indicates that nucleophilic alkyl radical, $R_N^{\bullet}$, prefers to react with electron-deficient olefins, and Mode **B** indicates that electrophilic alkyl radical, $R_E^{\bullet}$, prefers to react with electron-rich olefins. Therefore, the reaction with the opposite combination, such as $R_N^{\bullet}$ with electron-rich olefins, or $R_E^{\bullet}$ with electron-deficient olefins, does not proceed effectively.

SOMO–LUMO orbital interaction with electron-deficient olefins (Mode **A**)

$$R_N^{\bullet} + CH_2{=}CH{-}EW \longrightarrow R_N{-}CH_2{-}\dot{C}H{-}EW$$

SOMO–HOMO orbital interaction with electron-rich olefins (Mode **B**)

$$R_E^{\bullet} + CH_2{=}CH{-}ED \longrightarrow R_E{-}CH_2{-}\dot{C}H{-}ED$$

4.1.1 Reductive conditions

Addition reactions with Mode **A** are popular, and most addition reactions are classified into this group. The most typical experimental procedure is the treatment of alkyl halides (RBr, RI) and electron-deficient olefins with Bu_3SnH in the presence of AIBN. Bu_3GeH or Ph_3GeH can be also used instead of Bu_3SnH; however, there are rather expensive and less reactive. Typical addition reactions are shown below, in the reactions of alkylhalides or nitroalkane with acryronitrile, acrylic acid, methyl acrylate, or butyl acrylate [1–17].

Et_3B, Bu_3SnH, $PhCH_2CH_2I$, r.t.; **1** → **2** 72% (4.1)

$CH_2{=}CHCN$, Bu_3SnH, Et_2O, Hg-$h\nu$; **3** → **4** 72% (4.2)

$CH_2{=}C(CO_2CH_3)P(O)(OCH_3)_2$, $(Me_3Si)_3SiH$, $h\nu$; **3** → **5** 80% (4.3)

$CH_2{=}CHCO_2Bu^t$, AIBN, Bu_3SnH, C_6H_6, Δ; β-gluco selenide **6** → **7** 76% (4.4)

$CH_2{=}CHCN$, AIBN, Bu_3SnH, C_6H_6, Δ; **8** → **9** 65% (4.5)

$CH_2{=}C(CH_3)CO_2CH_3$, AIBN, Bu_3SnH, C_6H_6, Δ; **10** → **11** 62% (4.6)

Generally, the reactions are carried out in refluxing benzene solution, since the yield in benzene is better than that in other solvents. Probably, the radicals formed may be somewhat stabilized by the weak orbital–orbital interaction between the radicals and benzene. However, from the environmental point of view, toluene or dioxane is recently used. As substrates, alkyl bromides or alkyl iodides are used, and the reactivity increases in the order: *prim*-alkyl < *sec*-alkyl < *tert*-alkyl. Sugar anomeric bromide (**3**) is generally not so stable, so the reaction is carried out under irradiation conditions with a mercury lamp at room temperature (eqs. 4.2 and 4.3). There are two types of anomeric glycosyl radicals as shown in Figure 4.1. One is the axial radical [**I**], and the other is the equatorial radical [**I′**]. The axial radical is more nucleophilic than the equatorial radical due to the stereoelectronic effect, where this effect comes

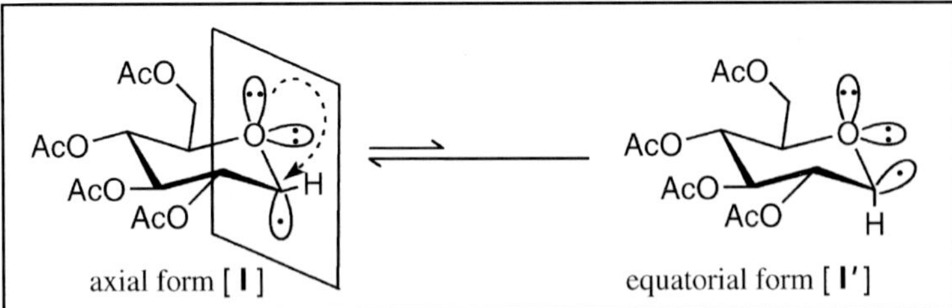

Figure 4.1

from the interaction between axial SOMO orbital and the axial lone-pair orbital on the neighboring oxygen atom (*anti*-periplanar relationship). Thus, electron density of the radical center in the axial form [**I**] is higher than that of the equatorial form [**I′**]. Therefore, the reactions of glucosyl bromide (**3**) with acrylonitrile and α-phosphonoacrylate in the presence of Bu_3SnH generates the addition products (**4**) and (**5**) with α-form alone, respectively. Sugar anomeric phenyl selenide (**6**) is stable, so the same reaction can be carried out under refluxing conditions as shown in eq. 4.4. Nitro compound (**10**) bearing a *tert*-alkyl group can be used for the addition reaction (eq. 4.6), however, nitro compounds bearing *sec*-alkyl and *prim*-alkyl groups are inert, and can not be used for the addition reaction. Xanthates derived from alcohols, and aromatic halides (ArI, ArBr) can be also used for the same addition reactions under heating conditions. Here, aryl radicals also behave as nucleophilic radicals, similarly to alkyl radicals.

Addition reactions with Mode B are not popular, but are occasionally useful. Eq. 4.7 indicates the reaction of ethyl bromoacetate and sugar vinyl ether with Bu_3SnH initiated by AIBN. The ethyl acetate radical is electrophilic and it reacts with electron-rich sugar vinyl ether through SOMO–HOMO orbital interaction to form a ribosyl anomeric radical, as shown below. Then, the formed ribosyl anomeric radical abstracts a hydrogen atom from Bu_3SnH through the less-hindered β-side, not through the α-side [18–20].

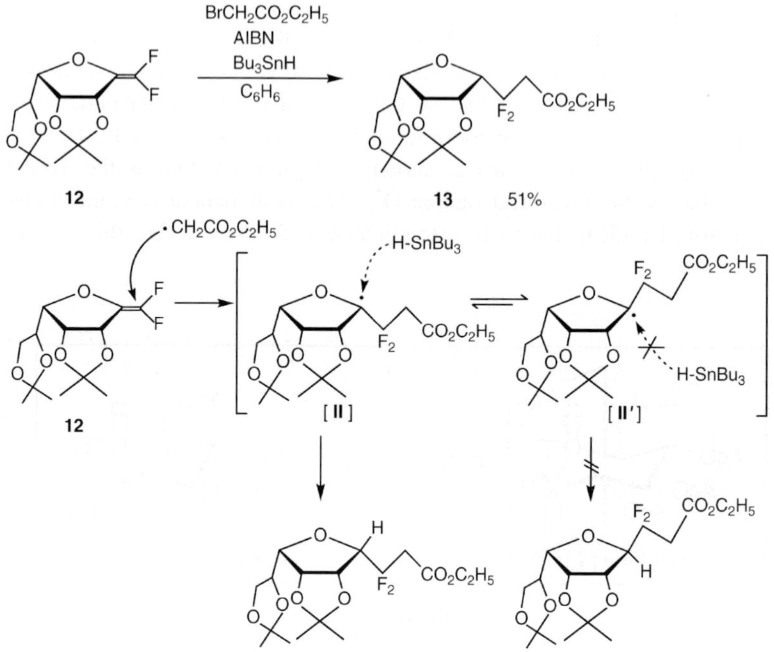

(4.7)

Treatment of α,α-dicyanoalkyl phenyl selenide (**14**) with vinyl ether initiated by AIBN under benzene refluxing conditions generates methyl ketone (**15**) (eq. 4.8). An electrophilic α,α-dicyanoalkyl radical is first formed, and then it adds to vinyl ether, followed by hydrolysis. Diethyl 3-iodoalkylphosphonate (**17**) can be formed through AIBN-initiated addition reaction of diethyl 1-iodomethylphosphonate (**16**) to alkene (eq. 4.9). This is an atom-transfer reaction. Both reactions (eqs. 4.8 and 4.9) do not require Bu_3SnH.

Eq. 4.10 shows addition–cyclization reaction of tributylstannyl α-iodoacetate (**18**) with 1-hexene. Here, $Bu_3Sn^{\cdot}$ is regenerated from the cyclization and, therefore, addition of Bu_3SnH is not required [21–24],

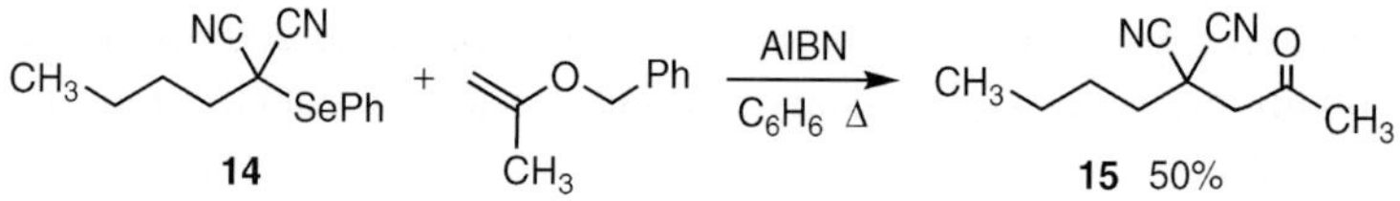

(4.8)

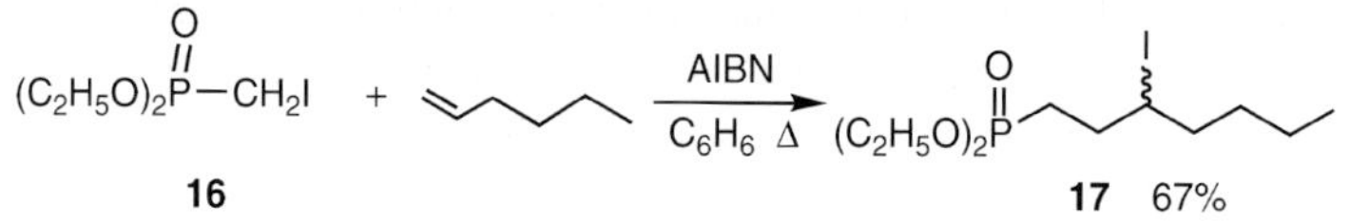

(4.9)

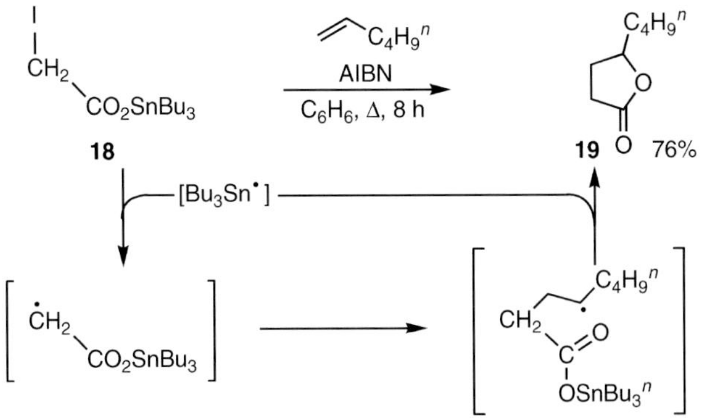

(4.10)

Se-Ph selenoester is a good precursor of an acyl radical where it behaves as a nucleophilic radical. Treatment of *Se*-Ph selenoester (**20**) and benzyl acrylate with Bu_3SnH generates a γ-keto ester (**21**) as shown in eq. 4.11a. 2-Indolylacyl radical can be generated from *Se*-Ph selenoester (**22**) by this method (eq. 4.11b) [25–30].

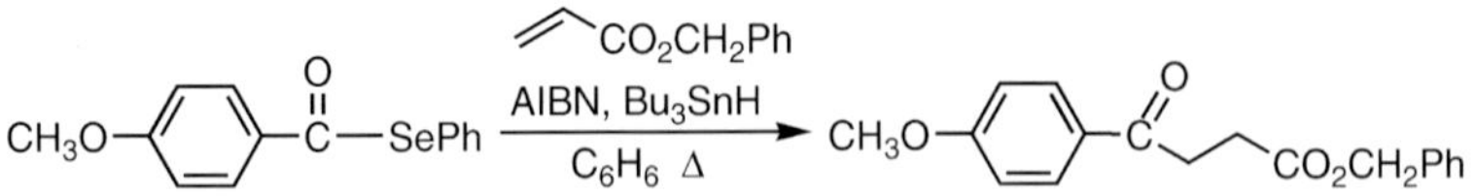

(4.11a)

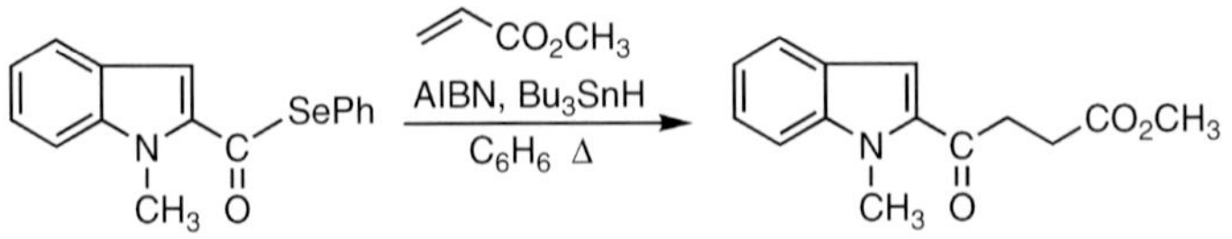

(4.11b)

Intermolecular alkyl radical addition to imino compound can be also carried out as shown in eq. 4.12 with $RI/Bu_3SnH/Et_3B$ system or RI/In (indium) system. Here indium is used as an electron transfer agent to RI [31–33].

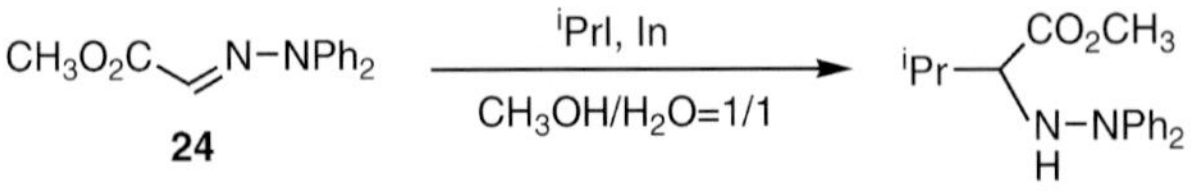

(4.12)

It is well known that Bu_3SnH is a highly toxic reagent. Thus, LD_{50}/kg in Me_3SnCl, Bu_3SnCl, and Bu_3SnH is 9–20 mg, 122 mg, and 127 mg, respectively. In 1989, $(Me_3Si)_3SiH$ was developed as a new, less-toxic radical reagent. The bond dissociation energy of Si–H in $(Me_3Si)_3SiH$ is 79 kcal/mol, while that of Sn–H in Bu_3SnH is 74 kcal/mol; however, the reactivity of $(Me_3Si)_3SiH$ is rather close to that of Bu_3SnH. This slightly stronger bond dissociation energy (5 kcal/mol) of $(Me_3Si)_3SiH$ suggests that the hydrogen-donating ability of $(Me_3Si)_3SiH$ is lower than that of Bu_3SnH, and this further suggests that $(Me_3Si)_3SiH$ is better than Bu_3SnH for carbon–carbon bond forming reactions. For example, when the reactions of 6-bromo-1-hexene with Bu_3SnH and of 6-bromo-1-hexene with $(Me_3Si)_3SiH$ were compared, % ratio of the direct reduction product, 5-*exo-trig* cyclization product, and 6-*endo-trig* cyclization product was 15, 83, and 1.2% in the former reaction, and 4, 93, and 2% in the latter reaction. Thus, reaction with Bu_3SnH generates the direct reduction product and cyclization products in 15 and 84% yields, respectively, while reaction with $(Me_3Si)_3SiH$ generates them in 4 and 95% yields, respectively.

SmI_2 can be used for SET reagent to carbonyl groups. Thus, eq. 4.13 shows the initial SET from SmI_2 to a carbonyl compound (**26**) to generate a ketyl radical, a nucleophilic radical, which then reacts with electron-deficient ethyl acrylate through SOMO–LUMO interaction to form γ-lactone (**28**) [34–36].

$R_1R_2C{=}O$ (**26**) + $R_3CH{=}C(R_4)CO_2C_2H_5$ (**27**) $\xrightarrow[\text{ROH, THF}]{SmI_2}$ **28**; **26** $\xrightarrow{e^-}$ [$R_1R_2\dot{C}{-}OSmI_2$] ⟶ [radical adduct, $R_1R_2C(OSmI_2)$–CHR_3–$\dot{C}R_4CO_2C_2H_5$] $\xrightarrow{e^-}$ [carbanion] $\xrightarrow{ROH}$ **28** (4.13)

Treatment of α,α-bromochlorocarboxylate ester (**29**) with CuBr/Fe generates the corresponding α-chloro ester radical which further reacts with electron-rich olefins such as 1-octene as shown in eq. 4.14 [37]. CuBr/Fe works as a SET reagent. This reaction is an atom-transfer reaction through the chain reaction. Atom-transfer reactions do not lose any atoms or functional groups during the formation of products, so they are completely combination reactions and, therefore, environmentally friendly reactions, if the reactions proceed quantitatively without using any toxic solvent or reagents. Here, α,α-dichloro esters do not react effectively under the same conditions.

$RCBrCl(CO_2CH_3)$ (**29**) + $CH_2{=}CH(CH_2)_5CH_3$ (**30**) $\xrightarrow[\text{DMF/CH}_2\text{Cl}_2]{\text{CuBr, Fe}}$ $RCCl(CO_2CH_3)CH_2CHBr(CH_2)_5CH_3$ (**31**, 60%); R = *n*-C_4H_9; **29** $\xrightarrow{e^-\ (-Br^\ominus)}$ [$R\dot{C}Cl(CO_2CH_3)$] ⟶ [$RCCl(CO_2CH_3)CH_2\dot{C}H(CH_2)_5CH_3$] ⟶ **31** (4.14)

$Na_2S_2O_4$ works also as a SET reagent. Treatment of R_fI (perfluoroalkyl iodide) with $Na_2S_2O_4$ generates electrophilic radical, $R_f^{\cdot}$, which reacts with electron-rich ketene dithioketal (**32**) through a chain pathway, to form R_f-substituted ketene dithioketal (**34**) via the elimination of hydrogen iodide. So, it looks like a substitution reaction of a hydrogen atom by R_f group [38–43].

$$\mathbf{32} + n\text{-}C_4F_9I\ (\mathbf{33}) \xrightarrow[\text{DMF/H}_2\text{O},\ R = CH_3]{Na_2S_2O_4,\ NaHCO_3} \mathbf{34}\ 44\%$$ (4.15)

Experimental procedure 1 (eq. 4.15).

To a mixture of DMF (10 ml) solution of $Na_2S_2O_4$ (1.74 g, 10 mmol), $NaHCO_3$ (0.84 g, 10 mmol) was added n-C_4F_9I (5 mmol). The mixture was warmed under stirring at 50 °C for 4 h. After the reaction, water (10 ml) was added to the mixture and the obtained mixture was extracted with ether. The organic layer was dried over $MgSO_4$ and evaporated. The residue was chromatographed on silica gel to give the product in 44% yield [67].

Triethylborane generates Et$^{\cdot}$ via the reaction with molecular oxygen under air, and its reacts with R_fI to generate $R_f^{\cdot}$ and EtI via S_H2 reaction pathway. So, Et_3B/R_fI/electron-rich olefin system can be used for the carbon–carbon bond forming reaction.

Secondary phosphine (**35**) reacts with *N*-vinylpyrrole (**36**) in the presence of AIBN to give the adduct (**37**) via a phosphorous-centered radical (eq. 4.16a). Treatment of secondary phosphines, R_2PH, with vinyl chalcogenides provides the corresponding tertiary phosphines in good yields with *anti*-Markovnikov regioselectivity. Treatment of a thiophosphite, but not a phosphite, with alkene (**38**) and triethylborane generates reductive addition product (**39**) under mild conditions, through the initial hydrogen atom abstraction from thiophosphite by the ethyl radical to form a phosphorous-centered radical, followed by the addition to alkene, and then β-cleavage, as shown in eq. 4.16b. Sodium salt of hypophosphorus acid acts as a double radical precursor, initiated by peroxide (eq. 4.17) [44–48].

$$(PhCH_2CH_2)_2PH\ (\mathbf{35}) + \mathbf{36} \xrightarrow[60 \sim 75\,°C]{AIBN} \mathbf{37}\ 91\%$$

$$[(PhCH_2CH_2)_2P^{\cdot}] \longrightarrow [\text{radical adduct}]$$ (4.16a)

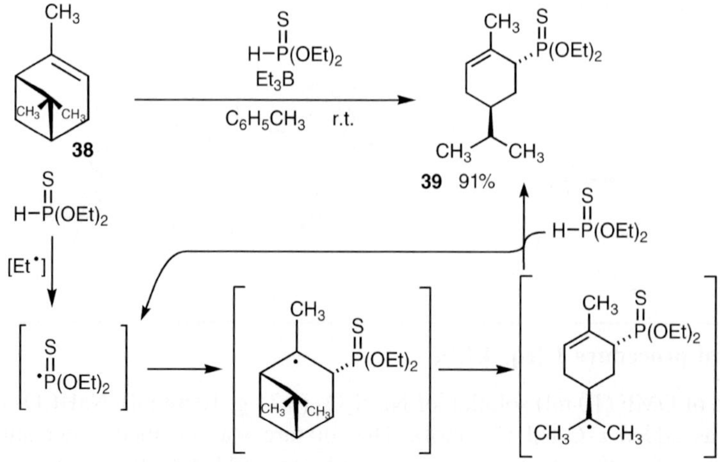

(4.16b)

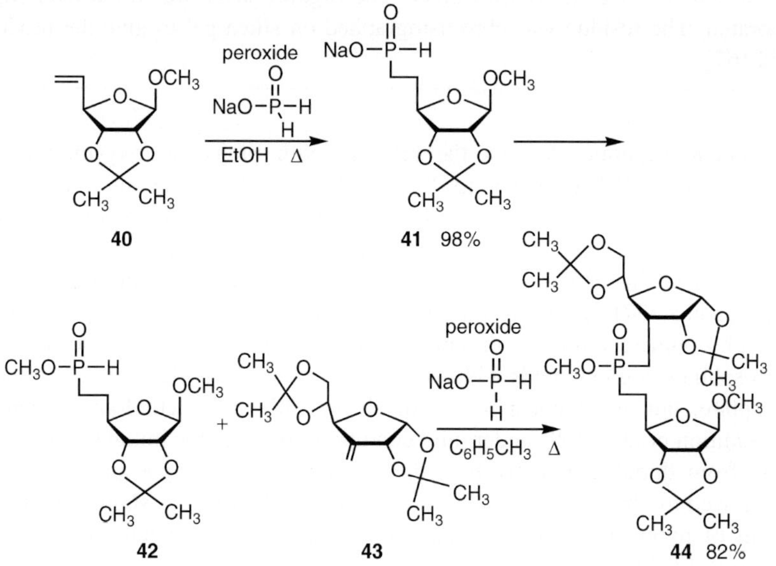

(4.17)

Experimental procedure 2 (eq. 4.16b).

To a solution of (1R)-(+)-α-pinene (136 mg, 1 mmol) and diethylthiophosphite (200 mg, 1.3 mmol) in toluene (5 ml) was added dropwise triethylborane (1 ml, 1 M solution in hexane) under aerobic conditions. After 15 min of additional stirring, the reaction mixture was poured into 2 M aq. solution of NaOH (15 ml), extracted with ether, dried over Na_2SO_4, and evaporated. Distillation of the residue gave the product in 91% yield (170 °C/0.1 mmHg) [45].

4.1.2 Oxidative conditions

Carbon–carbon bond formation reactions with peroxides are quite limited, in view of the synthetic features. However, the following atom-transfer reaction may be useful. Treatment of iodomethyl phenyl sulfone (**45**) with benzoyl peroxide generates an electrophilic benzenesulfonylmethyl radical, which adds to electron-rich 1-hexene to give the addition product (**46**) (eq. 4.18) [49–51]. The same treatment of bromoacetic acid and 1-hexene with benzoyl peroxide affords γ-butyl-γ-lactone through the cyclization of the formed γ-bromoctanoic acid.

PhSO$_2$CH$_2$I (**45**) + 1-hexene —(PhCO$_2$)$_2$, C$_6$H$_6$, Δ, 5 h→ **46** 70%
[Ph•]; PhI; [PhSO$_2$CH$_2$•]

(4.18)

Arenesulfonyl iodide and bromide are rather unstable compounds because of low bond dissociation energies of their SO$_2$–I and SO$_2$–Br. Therefore, treatment of *p*-tosyl bromide (**47**) with alkene or allene (**48**) produces 1,2-adduct (**49**) through the addition of the formed *p*-tosyl radical onto the allene as shown in eq. 4.19 [52]. Here, the *p*-tosyl radical attacks the central sp carbon of the allene group to generate the stable allylic radical, and then it reacts with *p*-tosyl bromide to give 1,2-adduct (**49**) and a *p*-tosyl radical again, i.e., chain pathway. So, this is also an atom(group)-transfer reaction.

TsBr (**47**) + **48** —AIBN (cat.), C$_6$H$_5$CH$_3$, Δ→ **49** 65%
[Ts•]; TsBr; Ts•

(4.19)

Treatment of arenesulfinate salts with Cu^{2+}, Mn^{3+}, or Ce^{4+} generates arenesulfonyl radicals, via single electron oxidation. Thus, reaction of alkene (**50**) with *p*-TsNa in the presence of Cu(OAc)$_2$ in AcOH gives *p*-tolyl allyl sulfone (**51**) through the addition of a toluenesulfonyl radical onto alkene, oxidation of the formed carbon-centered radical with Cu^{2+}, and then deprotonation (eq. 4.20a). This reaction requires acidic conditions for effective oxidation with Cu^{2+} or Mn^{3+} [53–58]. Eq. 4.20b is the same addition reaction with Ce^{4+}.

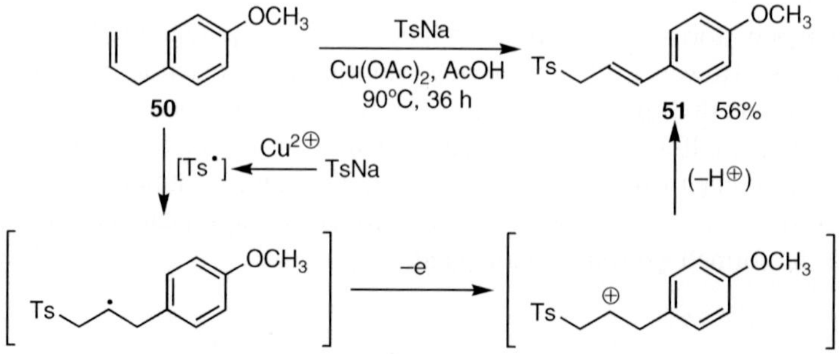

(4.20a)

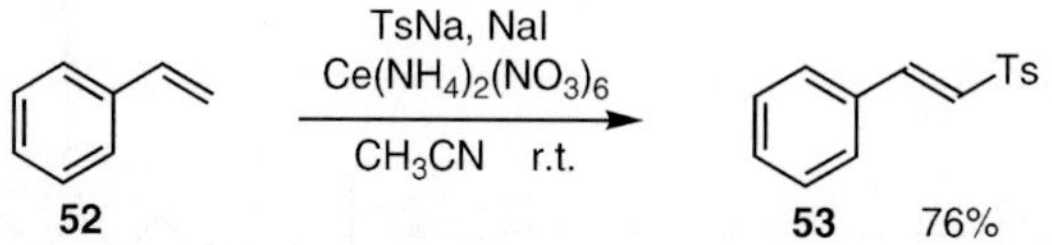

(4.20b)

Experimental procedure 3 (eq. 4.20b).

A mixture of styrene (1 mmol), sodium *p*-toluenesulfinate (1.2 mmol), and NaI (1.2 mmol) in dry CH_3CN (10 ml) was treated with CAN (2.5 mmol) in dry CH_3CN (15 ml) for 45 min under an argon atmosphere. After the reaction, the reaction mixture was washed with water, and extracted with CH_2Cl_2. The organic layer was washed with sat. aq. $Na_2S_2O_3$ solution and dried over Na_2SO_4. After removal of the solvent, the residue was chromatographed on silica gel to give the product [58].

1,3-Dicarbonyl compounds such as ethyl α-methylacetoacetate (**54**) can be efficiently alkylated in good yields with vinyl ethers in the presence of Cu^{2+} and Mn^{3+} as shown in eq. 4.21 [59].

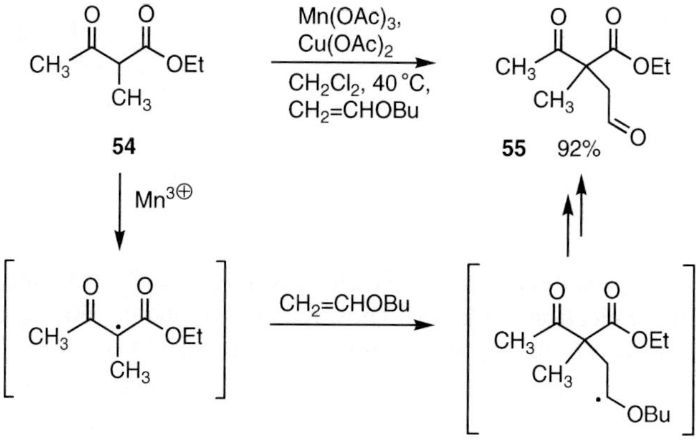

(4.21)

Experimental procedure 4 (eq. 4.21).

To ethyl α-methylacetoacetate (0.2 g, 1.38 mmol) in degassed dichloromethane (10 ml) was added butyl vinyl ether (0.77 g, 8.32 mmol) followed by $Mn(OAc)_3{\cdot}2H_2O$ (0.86 g, 3.19 mmol) and $Cu(OAc)_2{\cdot}H_2O$ (0.09 g, 0.41 mmol). The mixture was heated to reflux overnight (18 h). Then dichloromethane was added to the reaction mixture and the resulting mixture was filtered. The filtrate was washed with water (10 ml), and dried over $MgSO_4$. After removal of the solvent, the residue was purified by column chromatography on silica gel (eluent: dichloromethane/ethyl acetate = 7/3) to give ethyl α-methyl-α-(oxoethyl)acetoacetate in 92% yield [59].

Ce^{4+} can be also used for the same type of reaction, since it is a strong one-electron oxidant. Generation of sp^2 carbon-centered radicals such as aryl radicals, is not so easy, except for the reactions of aryl halides with Bu_3SnH or $Ph_4Si_2H_2$. However, treatment of arylhydrazines with Cu^{2+} generates aryl radicals through the initial oxidation to the arenediazonium ion with Cu^{2+}, and subsequent SET from Cu^+. Aryl radicals are much more reactive than alkyl radicals, and rapidly react with alkenes or imines as shown below (eq. 4.22) [60–63].

(4.22)

C_{60}, a typical fullerene, has high energy level of HOMO and low energy level of LUMO. This suggests that single electron oxidation and single electron reduction of C_{60} may occur smoothly. Practically, treatment of C_{60} with peroxide, $(RCO_2)_2$, induces SET from C_{60} to peroxide to generate 1,2-adduct, $R{-}C_{60}{-}O_2CR$, via the combination of $C_{60}^{+\cdot}$ $R^{\cdot}$, and RCO_2^-.

4.1.3 Photochemical conditions

Selenide (**58**) shown in eq. 4.23 is reactive, because its α-position is substituted by two electron-withdrawing groups. Irradiation of the selenide (**58**) with a mercury lamp in the presence of electron-rich 1-octene generates 1,2-adduct (**59**) with atom(group)-transfer addition reaction, through the initial homolytic bond cleavage of the C–SePh bond, addition of the formed carbon-centered radical onto 1-octene, and the reaction of the newly generated carbon-centered radical on the selenium atom of the starting selenide (chain reaction) [64–67].

$$\text{58} \xrightarrow[\text{C}_6\text{H}_6]{\text{Hg-}h\nu,\ \text{1-octene}} \text{59}\ \ 83\% \tag{4.23}$$

Experimental procedure 5 (eq. 4.23).

A benzene (2 ml) solution of selenide (0.33 mmol) and 1-octene (1 mmol) in a screw-cap pyrex test tube was bubbled with argon gas for 30 min., then the mixture was irradiated with a 450 W mercury lamp (Hanovia lamp) under an argon atmosphere (48 ~ 72 h). After the reaction, the solvent was removed and the residue was chromatographed on silica gel (eluent: acetonitrile/ethyl acetate = 1/5) to give 1,2-adduct in 83% yield [65].

Irradiation of C_{60} (**60**) with a mercury lamp in the presence of electron-rich keteneacetal generates 1,2-adduct (**61**), through the initial SET from keteneacetal to C_{60}, radical coupling reaction of the formed $C_{60}^{\bar{\cdot}}$ and keteneacetal cation radical, as shown in eq. 4.24 [68, 69].

$$\text{C}_{60}\ (\textbf{60}) + \text{CH}_2{=}\text{C(OTMS)OC}_2\text{H}_5 \xrightarrow[\text{C}_6\text{H}_6,\ \text{r.t.}]{\text{Hg-}h\nu} \textbf{61}\ \ 71\% \tag{4.24}$$

Benzophenone ($\lambda_{max} = 340$ nm, log $\epsilon = 2.5$, n-π^* electronic transition) can be used as a photochemical reagent and eq. 4.25 shows a radical Michael-addition reaction with benzophenone. The formed benzophenone biradical (triplet state, T_1) abstracts an electron-rich α-hydrogen atom from methyl 3-hydroxypropanoate (**62**) to generate an electron-rich α-hydroxy carbon-centered radical [**III**], then its radical adds to the electron-deficient β-carbon of α,β-unsaturated cyclic ketone (**63**) through the radical Michael addition. The electrophilic oxygen-centered radical in the benzophenone biradical abstracts an electron-rich hydrogen atom from methyl 3-hydroxypropanoate (**62**) [70]. So, an α-hydroxy carbon-centered radical [**III**] is formed, and an electron-deficient α-methoxycarbonyl carbon-centered radical [**III′**] is not formed.

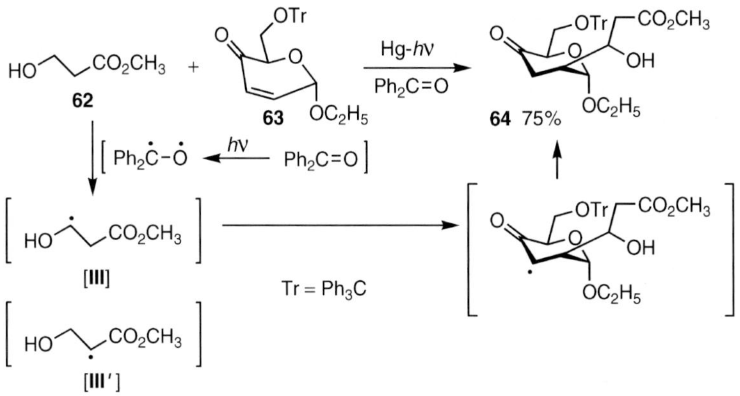

(4.25)

Tr = Ph_3C

4.1.4 Addition–elimination reactions

The Bu_3Sn group is a good radical leaving group. When it is substituted at the olefinic position, radical substitution reaction occurs. For example, treatment of alkyl bromide (**65**) and β-stannyl acrylate in the presence of AIBN gives β-alkyl substituted acrylate (**66**) through the addition–elimination reaction at the sp^2 carbon [71, 72]. Here, $Bu_3Sn^{\cdot}$ functions as a chain reaction mediator (eq. 4.26).

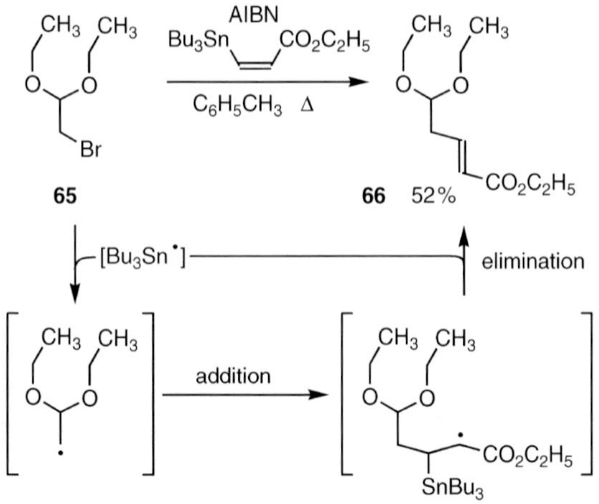

(4.26)

Since $ArSO_2$ group is also a good radical leaving group, irradiation of a mixture of anomeric sugar bromide (**67**) with benzenesulfonyl oxime ether in the presence of $Bu_3SnSnBu_3$ produces sugar oxime ether (**68**) as shown in eq. 4.27 [73–77].

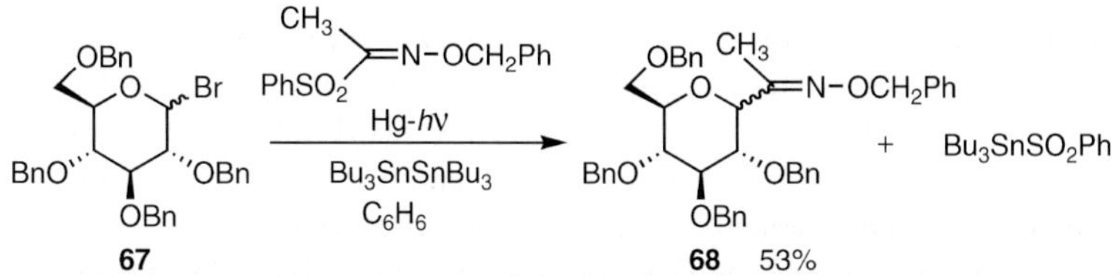

(4.27)

Here, rate constant for the addition reaction is about $10^5–10^6\ M^{-1}\ s^{-1}$.

Similarly, irradiation of a mixture of alkyl iodide (**69**) and trifluoromethanesulfonyl-acetylenic derivative in the presence of $Bu_3SnSnBu_3$ with a mercury lamp generates an alkyl-substituted acetylenic derivative (**70**) through the same addition–elimination reaction at the sp carbon (eq. 4.28) [78].

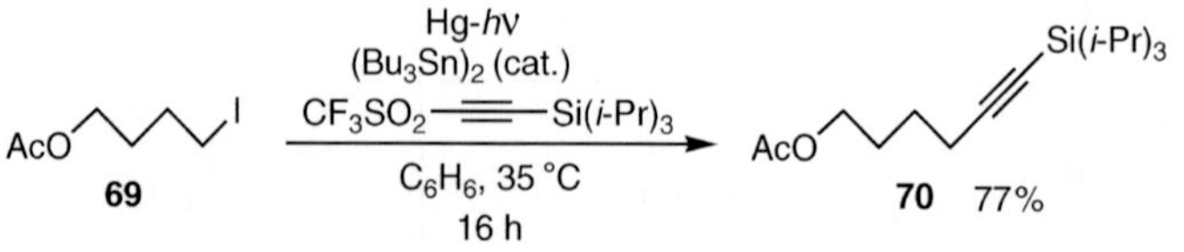

(4.28)

In the presence of peroxide, the same type of addition–elimination reaction with alkyl iodide (**71**) and ethyl vinyl sulfone can be carried out as shown in eq. 4.29 [79]. However, the reaction pathway is rather complicated.

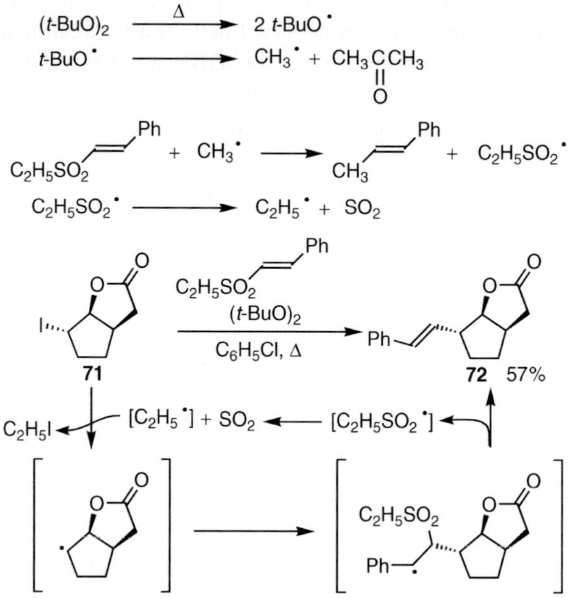

(4.29)

The methyl radical is formed from β-cleavage of the *t*-butoxyl radical derived from di-*t*-butyl peroxide (cat.), and it reacts with ethyl styrenyl sulfone to generate an ethanesulfonyl radical and *trans*-β-methylstyrene. The ethyl radical formed from β-cleavage of the ethanesulfonyl radical, at last, reacts with alkyl iodide (**71**) via S_H2 pathway to produce an alkyl radical and ethyl iodide. Here, the alkyl radical should be more stable than the ethyl radical. Therefore, alkyl radicals should be *sec*- or *tert*-alkyl radicals. The formed alkyl radical reacts with ethyl styrenyl sulfone to give addition–elimination product (**72**), together with ethanesulfonyl radical as a radical chain precursor.

Treatment of 2-aryl-1-nitroalkene (**73**) and 1-adamantyl iodide (**74**) with triethylborane under air produces *trans*-2-aryl-1-adamantylalkene (**75**) through the following reaction mechanism [80–82].

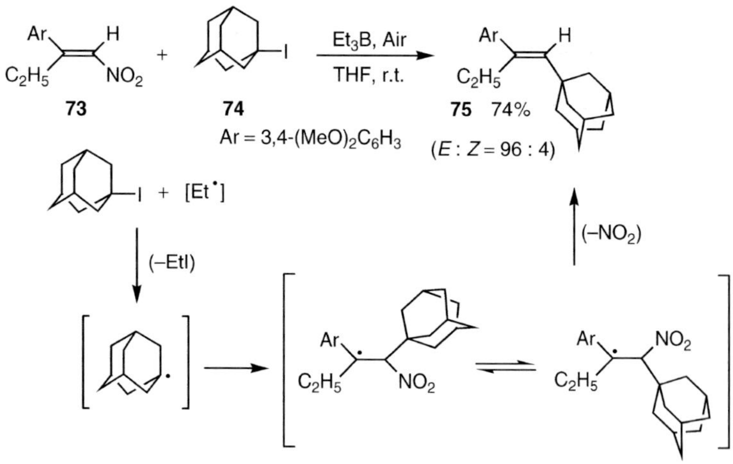

(4.30)

4.2 ALLYLATION REACTIONS

Radical allylations may proceed through S_H2 or S_H2' pathway, similarly to S_N2 or S_N2' pathway in polar reactions (eqs. 4.31a and 4.31b).

However, radical allylations mainly proceed through S_H2' pathway (eq. 4.31b). For example, treatment of adamantyl iodide and allyl-3,3-d_2-triphenyltin in the presence of AIBN generates mainly 3-adamantyl-1-propene-3,3-d_2, together with a minor amount of 3-adamantyl-1-propene-1,1-d_2.

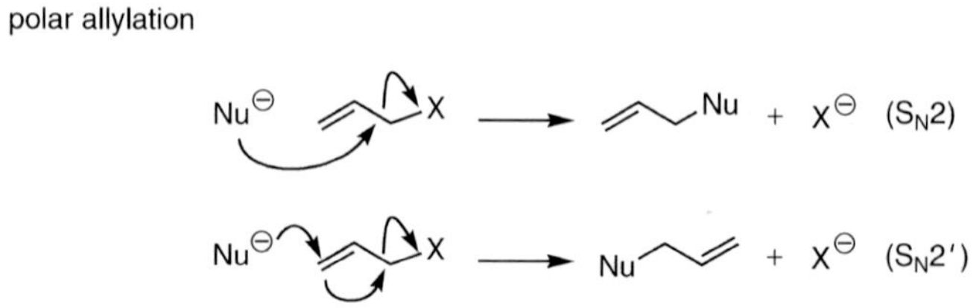

(4.31a)

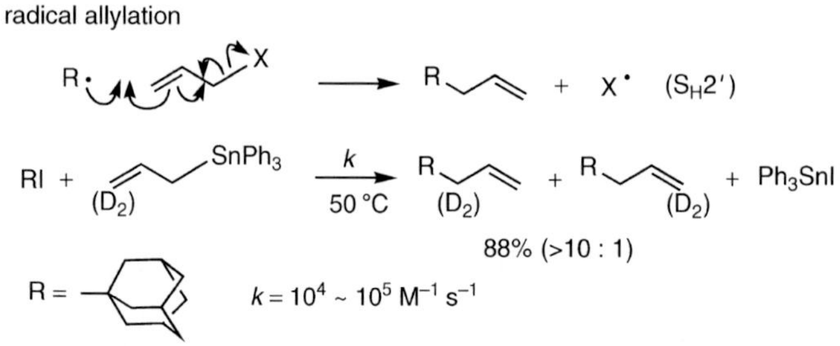

(4.31b)

The most well-known radical allylation reagent is allyltributyltin. Typical radical allylations are carried out by the irradiation of a benzene solution of alkyl bromide or selenide and allyltributyltin in the presence of AIBN with a mercury lamp. Eqs. 4.32a and 4.32b show the allylation of α-bromosulfone (**76**) and α-phenylselenyl-cyclopentanone derivatives (**78**) [83–95]. The olefinic group of allyltributyltin is electron-rich, and so an electrophilic radical such as α-sulfonyl radical or α-keto radical is preferable for efficient radical allylation. Moreover, perfluorinated allylic compound (**81**) can be obtained effectively from the reaction of perfluoroalkyl iodide (**80**) and allyltributyltin in the presence of AIBN under refluxing conditions of hexane (eq. 4.32c).

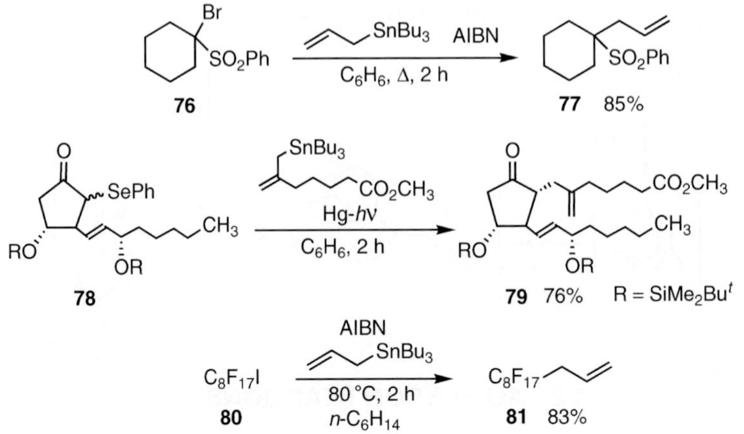

(4.32a)

(4.32b)

(4.32c)

Experimental procedure 6 (eq. 4.32a).

A mixture of α-bromosulfone (2 mmol), allyltributyltin (1.32 g, 4 mmol), and AIBN (65 mg, 0.4 mmol) in dry benzene (12 ml) was refluxed for 2 h under an argon atmosphere. After the reaction, the solvent was removed. The residue was dissolved in ether (30 ml) and the organic layer was washed with 10% KF aq. solution thrice (3 × 5 ml), and dried over Na_2SO_4. After removal of the solvent, the residue was chromatographed on silica gel to give α-allyl sulfone in 85% yield [91].

Hydroxy groups in sugars and nucleosides can be also converted to allyl groups via xanthate. Thus, eq. 4.33a shows the preparation of pyrimidine 3′-allyl-2′,3′-dideoxyribonucleosides through the Barton–McCombie type radical allylation, and here, the allyl group is introduced into the less-hindered side. Eq. 4.33b shows the preparation of 3,3-diallyl sugar (**86**) using β-cleavage of a cyclopropylmethyl radical, followed by allylation.

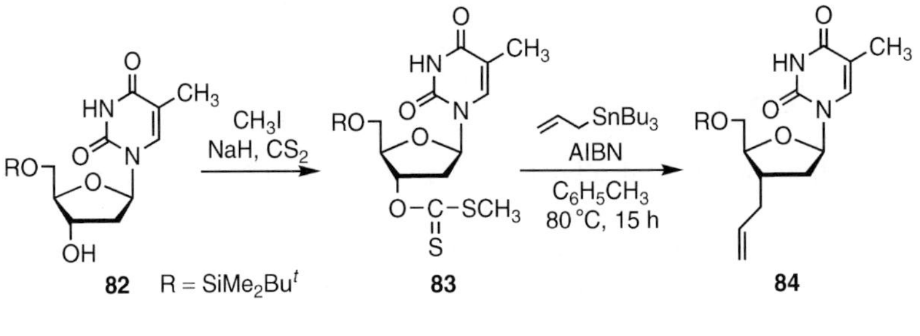

(4.33a)

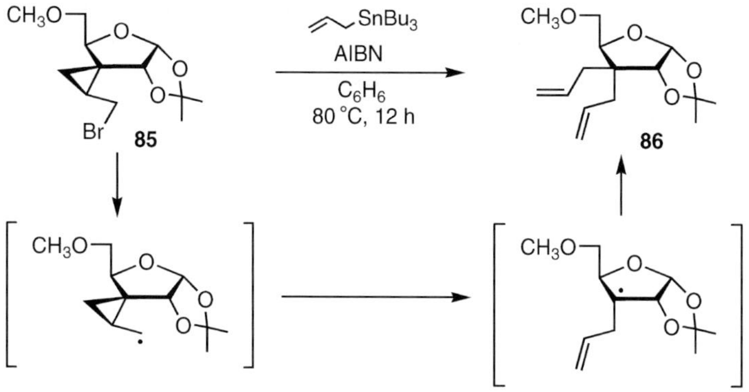

(4.33b)

Experimental procedure 7 (eq. 4.33b).

A mixture of cyclopropylmethyl bromide derivative (0.4 mmol), allyltributyltin (0.8 mmol), and AIBN (5 mol%) in dry benzene (3 ml) was degassed by bubbling argon gas for 30 min. The mixture was refluxed for 12 h. After removal of the solvent, aq. KF solution (5 ml) and ether (10 ml) were added to the residue and stirred for one hour. After filtration and separation, the organic layer was dried over Na_2SO_4. After removal of the solvent, the residue was chromatographed on silica gel using ethyl acetate/hexane to afford a diallyl product [95].

Allyltributyltin can be used in combinatorial radical allylation. Moreover, a polymer-supported reagent bearing an allyltin group was also developed and it can be used for the radical allylation of alkyl halides.

Allyl[*tris*(trimethylsilyl)]silane (**87**) and allyl alkyl sulfone (**93**), instead of toxic allyltributyltin, can be used for the allylation of alkyl halides under the same conditions as shown below (eqs. 4.34a–4.34c) [96–104].

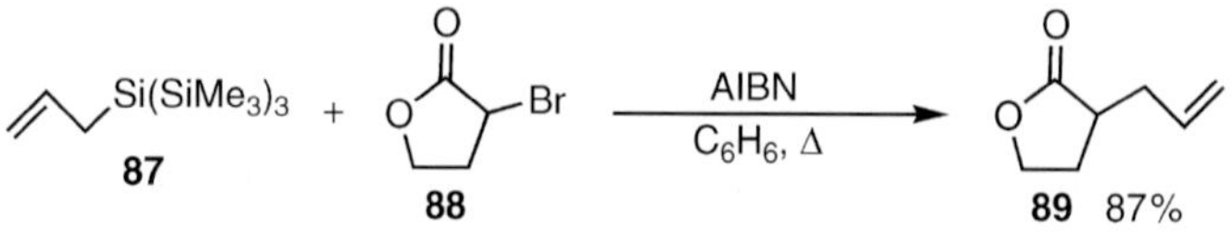

(4.34a)

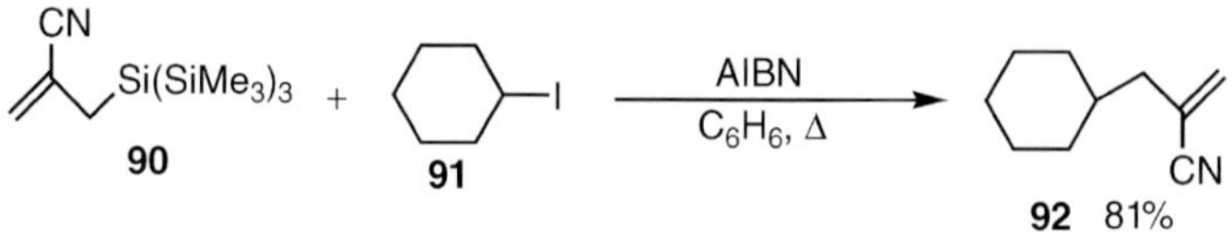

(4.34b)

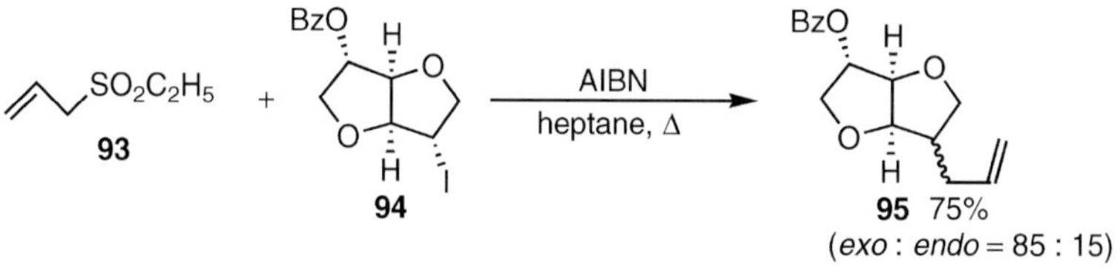

(4.34c)

Allyl halides and allyl sulfides can be also used for the same radical allylation; however, the reactivity is rather decreased. Radical addition of an allyl group to aromatic aldehydes was reported. This reaction proceeds through the addition of a *tris*(trimethylsilyl)silyl radical onto the formyl group in compound (**96**) to generate a benzylic radical, and subsequent S_H2' reaction onto allyl[*tris*(trimethylsilyl)]silane derivative (**97**) to form homoallyl silyl ether (**98**) (eq. 4.35) [105].

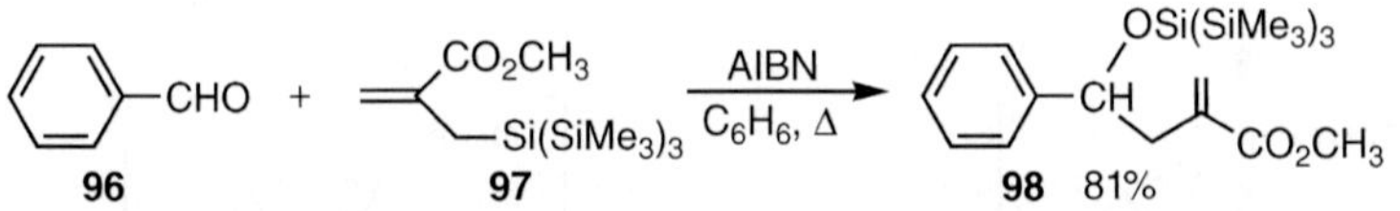

(4.35)

Under the oxidative conditions, allylation of active methine or methylene groups can be carried out with allylic sulfide/Mn^{3+}, Cu^{2+}. Eq. 4.36 shows the reaction of α-ethoxycarbonyl cyclopentanone (**99**) and 2-methyl-2-propenyl *t*-butyl sulfide (**100**) in the presence of $Mn(OAc)_3$ and $Cu(OAc)_2$ in acetic acid to form α-allyl-α-ethoxycarbonyl cyclopentanone (**101**) in good yield. Allyl sulfides are more effective than allyl sulfones [106, 107].

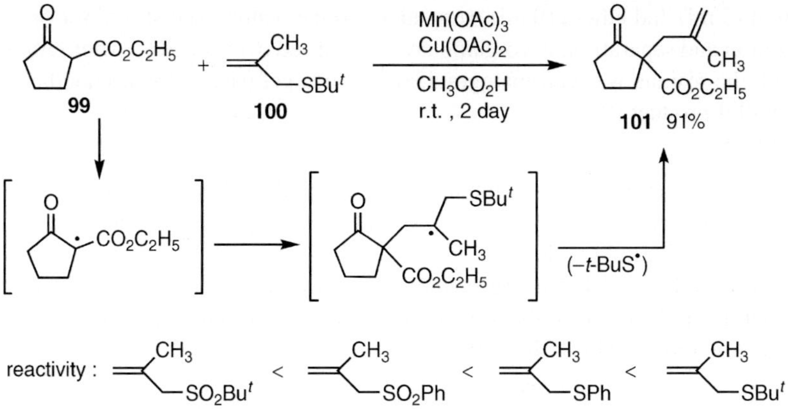

(4.36)

Treatment of *p*-TsBr with allene (**102**) in the presence of AIBN generates tosyl-substituted primary (*E*)-allylic bromide (**103**) via the addition of *p*-Ts˙ to the central sp carbon of the allene group (eq. 4.37) [108].

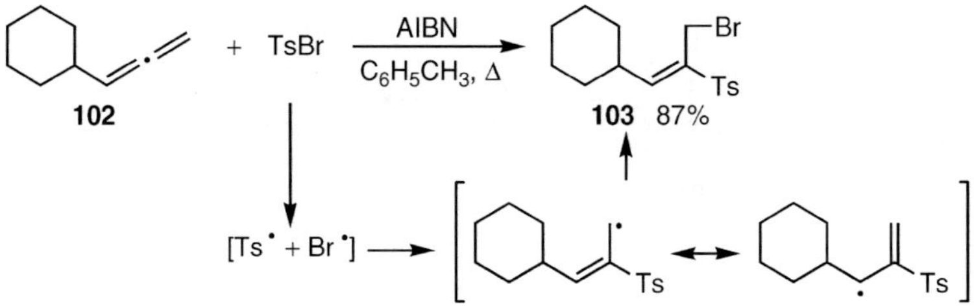

(4.37)

4.3 ADDITION–CYCLIZATION REACTIONS

Irradiation of a refluxing mixture of 4-iodo-1-butyne (**104**) and electron-deficient olefin (**105**) in the presence of $Bu_3SnSnBu_3$ with a tungsten lamp generates addition–cyclization products (**106a**) and (**106b**) as shown in eq. 4.38a. $Bu_3Sn^{\cdot}$ reacts with 4-iodo-1-butyne (**104**) to generate a 3-butynyl radical, then it reacts with methyl methacrylate via intermolecular addition and then cyclizes via 5-*exo-dig* manner to form an *exo*-methylene radical. This radical reacts again with the starting 4-iodo-1-butyne to form an *exo*-iodomethylene product (**106a**) and 3-butynyl radical through a radical chain reaction. So, this is an atom-transfer reaction. This methodology can be used for the construction of bicyclic and tricyclic compounds [109–117].

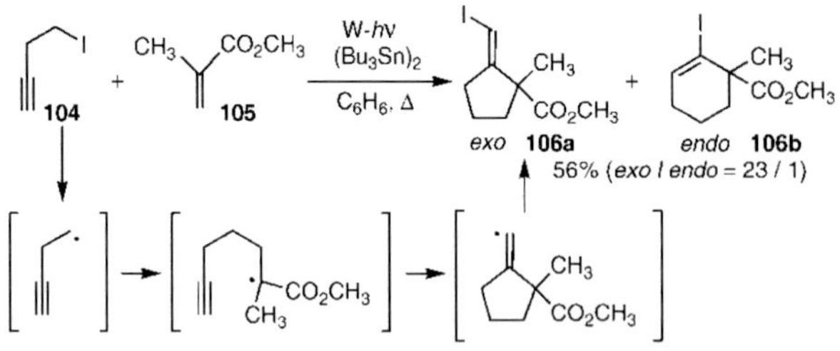

(4.38a)

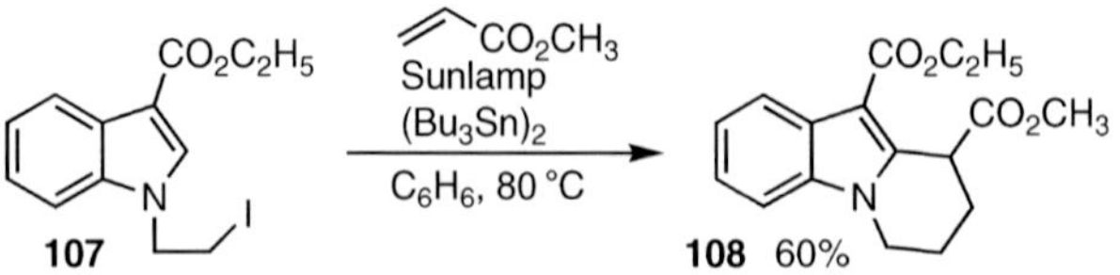

(4.38b)

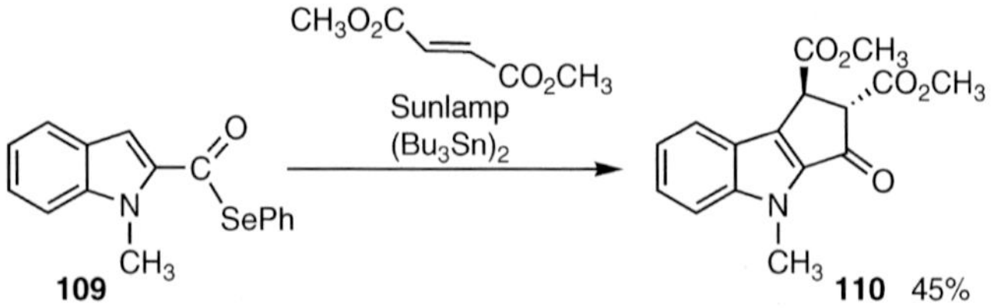

(4.38c)

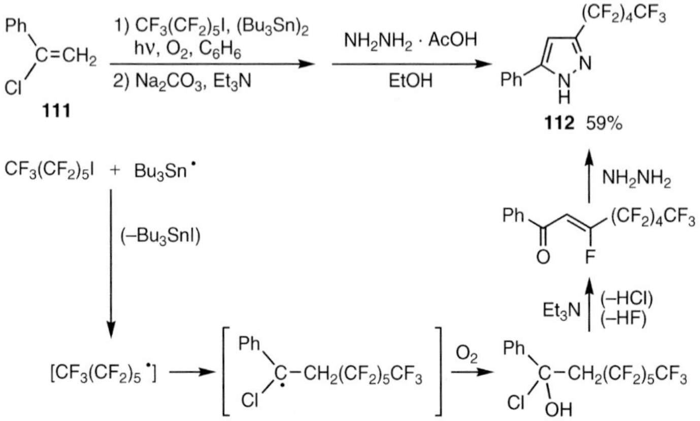

(4.38d)

The benzindolizidine skeleton (**108**) can be generated in moderate yield by a hexabutylditin-mediated consecutive radical addition and cyclization, from 1-(2-iodoethyl)indole

(**107**) and methyl acrylate (eq. 4.38b). The acyl radical formed from *Se*-phenyl selenoester (**109**) with hexabutylditin can be also used for consecutive radical addition with dimethyl fumarate (eq. 4.38c).

One-pot synthesis of pyrazole (**112**) bearing a perfluoroalkyl group is performed by the treatment of α-chlorostyrene (**111**), perfluoroalkyl iodide, and $Bu_3SnSnBu_3$, followed by reaction with hydrazine (eq. 4.38d).

Experimental procedure 8 (eq. 4.38d).

A solution of perfluoroalkyl iodide (0.4 mmol), α-chlorostyrene (1.2 mmol) and $Bu_3SnSnBu_3$ (0.44 mmol) in benzene (3 ml) was irradiated using a metal halide lamp (National Sky-beam MT-70) in Pyrex tube under O_2 atmosphere for 5 h. After removal of the solvent, ethanol and hydrazine acetate were added. The resultant solution was stirred under refluxing conditions for 2 h. After removal of the solvent, the residue was chromatographed on silica gel using a mixture of hexane and dichloromethane as an eluent, to give perfluoroalkylated pyrazole in 59% yield [117].

Co-cyclization of 1,6-diene (**113**) with a thiyl radical can be carried out involving the addition of thiyl radical to an olefinic group, cyclization via 5-*exo-trig* manner with a chair-like transition state, and termination by homolytic substitution at the sulfur (S_Hi). Preparation of *cis*-fused thiabicyclo[3.3.0]octane (**114**) is shown in eq. 4.39.

(*t*-BuS)$_2$, Et_3B, *n*-C_6H_{14}, hν, 10 °C, 5 h or (*t*-BuS)$_2$, *n*-C_6H_{14} hν, 10 °C, 5 h

113 → **114** 73%

t-BuS$^\bullet$; 5-*exo-trig* ; S_Hi (−*t*-Bu$^\bullet$) ; SBu-*t* ; S−Bu*t* ; H ; H H

(4.39)

Radical addition–cyclization of substrate (**115**) bearing two different radical acceptor such as acrylate and aldoxime ether moieties generates β-amino-γ-butyrolactones (**116a**) and (**116b**) (eq. 4.40) [118–120].

115 ($R = CH_2OSiPh_2Bu^t$) —C_2H_5I, Et_3B, $C_6H_5CH_3$, 20 °C→ **116a** + **116b** (NHOBn, C_2H_5, R) 53% (18 : 1)

(4.40)

Radical iodine atom transfer [3 + 2]-cycloaddition with alkene (**118**) using dimethyl 2-(iodomethyl)cyclopropane-1,1-dicarboxylate (**117**) forms cyclopentane derivative (**119**), through the formation of an electron-deficient homoallyl radical, followed by the addition to alkene, and cyclization via 5-*exo-trig* manner as shown in eq. 4.41.

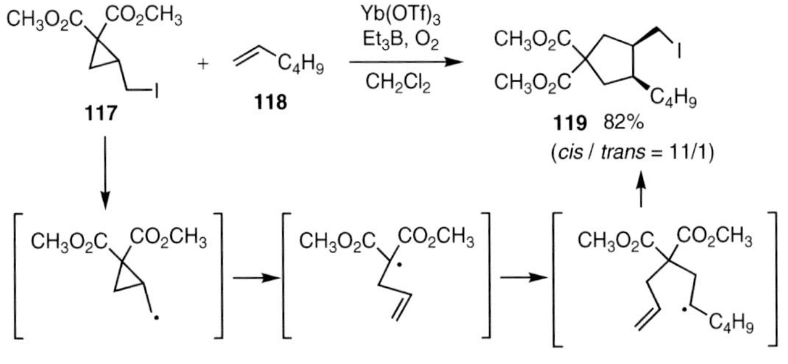

(4.41)

Experimental procedure 9 (eq. 4.41).

To a solution of dimethyl 2-(iodomethyl)cyclopropane-1,1-dicarboxylate (149 mg, 0.5 mmol), 1-hexene (0.13 ml, 1 mmol), and $Yb(OTf)_3$ (310 mg, 0.5 mmol) in dichloromethane (4 ml) was added Et_3B (0.5 ml, 1 *M* hexane solution). After being stirred for 5 h at 0 °C under an argon atmosphere, the mixture was poured into aq. NH_4Cl solution and extracted with ether. Ether extracts were washed with brine, dried over $MgSO_4$, and evaporated. Purification of the residue by silica-gel column chromatography gave a mixture of *cis*- and *trans*-cyclopentane derivative in 82% yield (*cis*/*trans* = 11/1) [159].

Free radical cyclization of 1,6-diene (**120**) using diethyl phosphite or diphenylphosphine oxide initiated by peroxide, produces an organophosphorus compound (**121**) via the addition of a phosphonyl radical to an olefinic group (eq. 4.42a). Radical addition of PH_3 to limonene (**122**) results in the formation of 4,8-dimethyl-2-phosphabicyclo[3.3.1]-nonane (**123**) (eq. 4.42b) [121, 122].

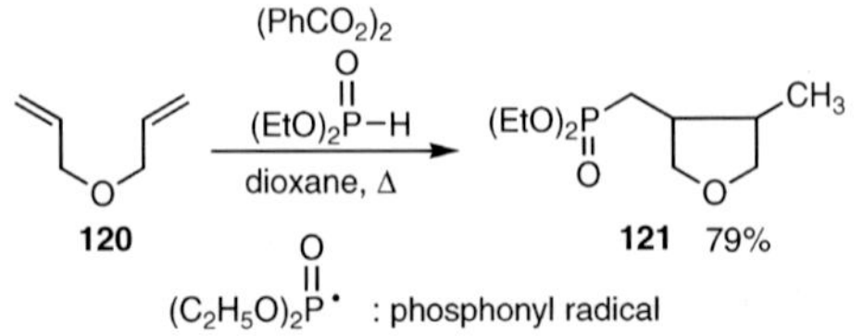

(4.42a)

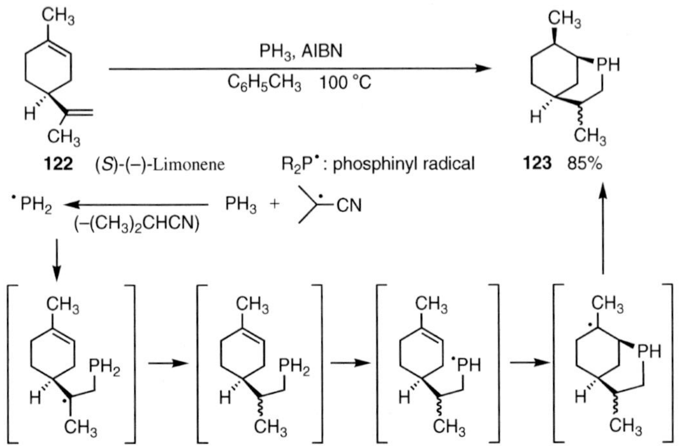

(4.42b)

[3 + 2] Reaction of the Barton ester (**124**) with vinyl sulfone can be carried out under irradiation with a tungsten lamp (eq. 4.43) [123].

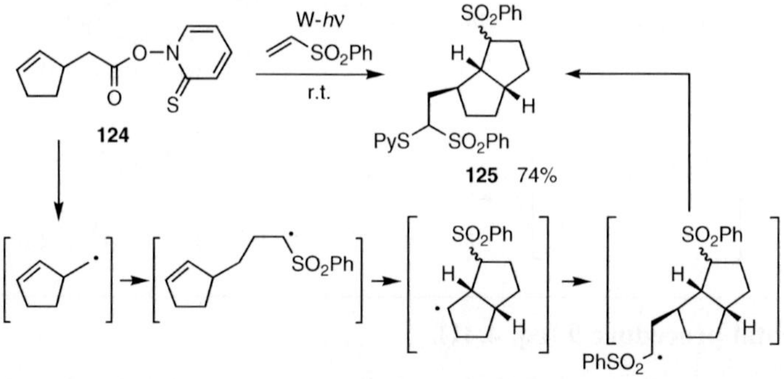

(4.43)

Bond dissociation energies of C–Br and C–I in $BrCCl_3$, CBr_4, CHI_3, etc. are lower, so treatment with AIBN promotes the radical reactions. Eq. 4.44 shows that the refluxing treatment of ene-yne-ketone (**126**) with bromotrichloromethane in the presence of AIBN generates cyclopentanone (**127**) bearing an *exo*-methylene group, through the addition of $CCl_3^{\bullet}$ to an electron-rich olefinic group, followed by 5-*exo-dig* cyclization and then bromine abstraction from bromotrichloromethane by the formed *exo*-vinyl radical [124]. This is an atom(group)-transfer reaction.

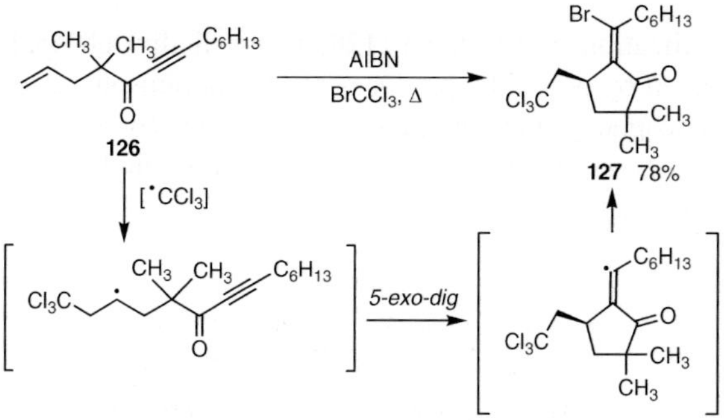

(4.44)

As an oxidative addition–cyclization reaction, treatment of 1,4-naphthoquinone (**128**) and diethyl phenylmalonate (**129**) with $Mn(OAc)_3$ in acetic acid generates benz[a]anthraquinone (**130**) as shown in eq. 4.45 [125–132].

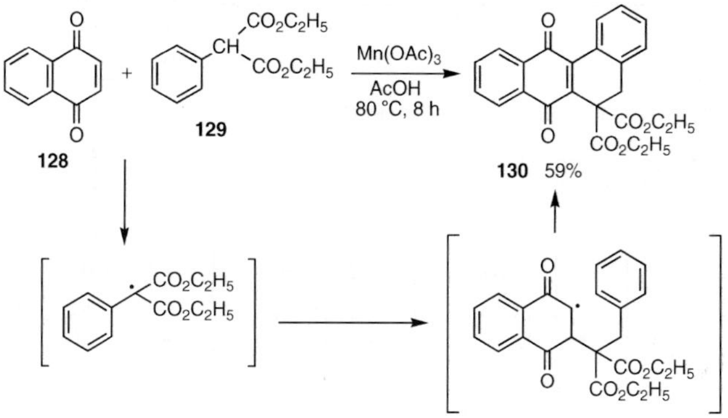

(4.45)

Experimental procedure 10 (eq. 4.45).

Acetic acid (10 ml) solution of 1,4-naphthoquinone (0.49 mmol), diethyl malonate (1.98 mmol), and $Mn(OAc)_3$ (798 mg, 2.99 mmol) was heated at 80 °C for 16 h. After the reaction, ethyl acetate (100 ml) was added to the solution and the mixture was washed with sat. aq. $NaHSO_3$ solution (25 ml), sat. aq. Na_2CO_3 solution (50 ml × 3), and water (25 ml × 3). The organic layer was dried over Na_2SO_4. The solvent was removed and the residue was chromatographed on silica gel (eluent: dichloromethane/hexane = 3/1) to give a product [126].

The following quinoline alkaloids (**132**) and (**134**) are prepared by the same addition–cyclization reaction with $Mn(OAc)_3$.

Mn(OAc)3, C6H13 (alkene), AcOH, 60 °C))))

131 → 132 50%

(−e) (−H⊕)

(−e)

(4.46a)

Mn(OAc)3, (CH3CO)2CH2, AcOH, r.t.

133 → 134 77%

$CH_3-\overset{O}{\overset{||}{C}}-CH_2-\overset{O}{\overset{||}{C}}-CH_3$ → (−e) (−H⊕) → $[CH_3-\overset{O}{\overset{||}{C}}-\dot{C}H-\overset{O}{\overset{||}{C}}-CH_3]$

(−e) (−H⊕)

(−e) (−H⊕)

(4.46b)

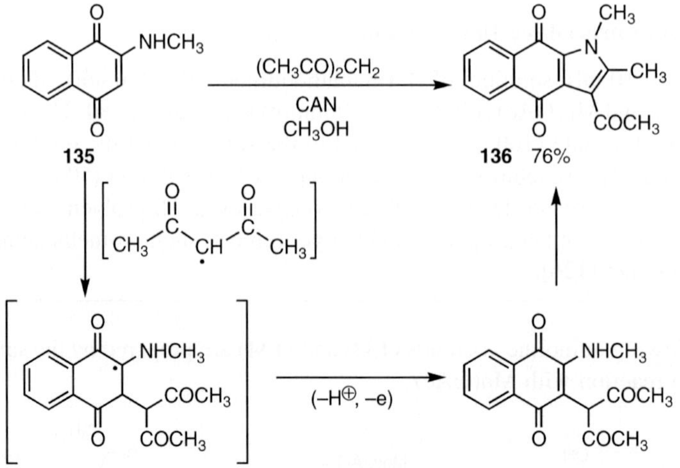

(4.47)

Treatment of 2-amino-1,4-naphthoquinone (**133**) and acetylacetone with $Mn(OAc)_3$ generates an oxidatively condensed alkaloid (**134**) (eq. 4.46b). The same oxidative reaction of 2-amino-1,4-naphthoquinone (**135**) and β-diketone with cerium(IV) ammonium nitrate (CAN) can be performed (eq. 4.47).

$Mn(OAc)_3$-mediated reaction of acetylacetone with 1,5-cyclooctadiene (**137**) leads to the formation of a *cis*-fused bicyclic compound (**138**), through the addition of a acetylacetonyl radical to 1,5-cyclooctadiene and subsequent intramolecular radical cyclization (eq. 4.48).

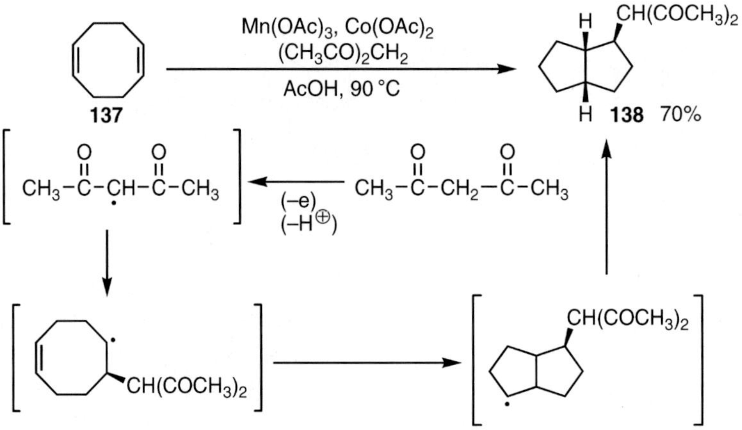

(4.48)

Ceric (IV) ammonium nitrate (CAN) works as the same oxidative radical reagent as $Mn(OAc)_3$. Diethyl 3-furylphosphonates was prepared in good yield under mild conditions by a simple two-step procedure using CAN-promoted oxidative addition of β-ketophosphonate (**139**) to vinyl acetate (**140**), followed by acid-catalyzed cyclization as shown in eq. 4.49.

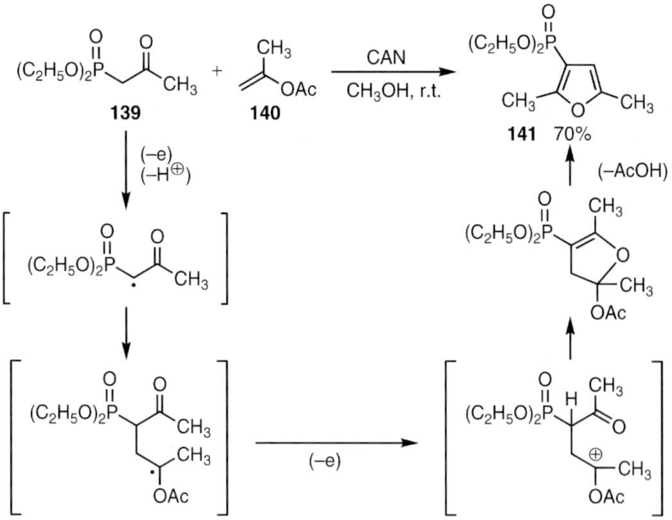

(4.49)

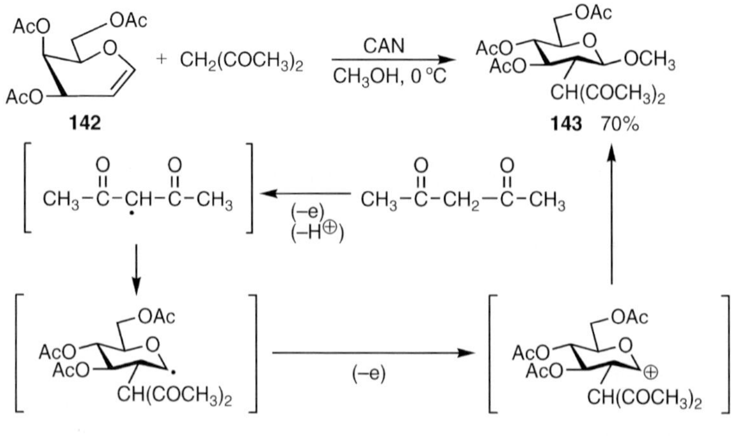

(4.50)

CAN-mediated radical addition of acetylacetone to glycal (**142**) affords *O*-methyl glycoside (**143**) (eq. 4.50).

Experimental procedure 11 (eq. 4.50).

A solution of CAN (2–4 eq.) in methanol (20–40 ml) was added to a solution of glycal (2 mmol) and dimethyl malonate (2.64 g, 20 mol) in methanol (5 ml) at 0 °C. After the disappearance of glycal (2–5 h) by TLC, ice-cold aq. diluted $Na_2S_2O_4$ solution (100 ml) was added to the mixture, and the obtained mixture was extracted with dichloromethane (40 ml × 4). The organic layer was dried over Na_2SO_4. After removal of the solvent, the residue was chromatographed on silica gel with a mixture of hexane and ethyl acetate [131].

As a reductive addition–cyclization reaction, treatment of allyl 4,6-di-*O*-acetyl-2,3-dideoxy-α-D-*erythro*-hex-2-enopyranoside (**144**) with $CF_3(CF_2)_7I$ and $Na_2S_2O_4$ as SET reagent at 5 °C generates 4,6-di-*O*-acetyl-1,2,3-trideoxy-4′-(2,2,3,3,4,4,5,5,6,6,7,7,

8,8,9,9,9-heptadecafluorononyl)-2′,3′,4′,5′-tetrahydro-3-iodo-α-D-glucopyranoso[1,2,-*b*]furan (**145**) (eq. 4.51), through the initial SET from $Na_2S_2O_4$ onto $CF_3(CF_2)_7I$ to form electron-deficient $CF_3(CF_2)_7^{\cdot}$, addition of $CF_3(CF_2)_7^{\cdot}$ to the terminal olefinic group of the substrate, 5-*exo-trig* cyclization to the another olefinic group, and finally abstraction of an iodine atom from $CF_3(CF_2)_7I$ [133]. So, this is an atom(group)-transfer reaction.

AcO, AcO, O, O, 144; $CF_3(CF_2)_7I$, $Na_2S_2O_4$, $NaHCO_3$, CH_3CN, H_2O, 0°C; AcO, AcO, O, H, O, I, H, 145, $CH_2(CF_2)_7CF_3$, 73% (59 : 14)

$2SO_2^{\cdot -}$ ← $S_2O_4^{2-}$; $CF_3(CF_2)_7I$ ($-SO_2$); [$CF_3(CF_2)_7^{\cdot}$] → [AcO, AcO, O, O, $(CF_2)_7CF_3$] → [AcO, AcO, O, O, $CH_2(CF_2)_7CF_3$]; $CF_3(CF_2)_7I$ [$-CF_3(CF_2)_7^{\cdot}$] (4.51)

4.4 REACTIONS WITH CARBON MONOXIDE

As shown in Hunsdiecker reactions and Barton decarboxylation reactions, radical decarboxylation of an alkylcarboxyl radical ($RCO_2^{\cdot}$) via β-cleavage is extremely rapid to form $R^{\cdot}$ and carbon dioxide, and the rate constant is $\sim 10^9\ s^{-1}$, while that of an arylcarboxyl radical ($ArCO_2^{\cdot}$) is $\sim 10^5\ s^{-1}$. Thus, decarboxylation is very rapid, and so it suggests that the synthetic use of the radical coupling reaction of alkyl radical and carbon dioxide to get carboxylic acid, i.e. the reverse reaction, gives disappointing results. However, decarbonylation of acyl radical ($RCO^{\cdot}$) via α-cleavage to form $R^{\cdot}$ and carbon monoxide is not so rapid, and the rate constant is about $\sim 10^3\ s^{-1}$. So, the synthetic use of the radical coupling reaction of alkyl radical and carbon monoxide may be possible. Practically, treatment of octyl bromide (**146**) with Bu_3SnH/AIBN under high-pressured carbon monoxide with an autoclave generates the corresponding aldehyde (**147**) via the formation of an acyl radical as shown in eq. 4.52 [134–136].

Br, 146; CO (50 atm), Bu_3SnH, AIBN, C_6H_6, 80 °C, 3 h; CHO, 147 63% + 36% (4.52)

The yields dramatically depend on the pressure of carbon monoxide and the concentration of Bu_3SnH, and the coupling reaction of the alkyl radical and carbon monoxide is the rate-determining step. The same treatment in the presence of electron-deficient olefin and allyltributyltin produces the corresponding ketones (**149**) and (**150**), respectively, through the addition reaction of acyl radical to olefins as shown in eq. 4.53 [137–141]. Generally, the acyl radical has a weak nucleophilic character.

CN; CO (80 atm); Bu_3SnH, AIBN; 80 °C, 2 h → **149** 46%

148

CO (30 atm); $SnBu_3$; AIBN; 80 °C, 12 h → **150** 60%

(4.53)

Eq. 4.54 shows the reaction of *n*-heptanol (**151**) with $Pb(OAc)_4$ under high-pressured carbon monoxide with an autoclave to generate the corresponding δ-lactone (**152**). This reaction proceeds through the formation of an oxygen-centered radical by the reaction of alcohol (**151**) with $Pb(OAc)_4$, 1,5-H shift, reaction with carbon monoxide to form an acyl radical, oxidation of the acyl radical with $Pb(OAc)_4$, and finally, polar cyclization to provide δ-lactone [142–146]. This reaction can be used for primary and secondary alcohols, while β-cleavage reaction of the formed alkoxyl radicals derived from tertiary alcohols occurs.

CH_3 OH **151** → CO (80 atm), $Pb(OAc)_4$, C_6H_6, 40 °C → CH_3 O O **152** 51%

[CH_3 O•] (1,5-H shift) [CH_3 OH] → [CH_3 O OH] → [CH_3 O ⊕OH]

(4.54)

Experimental procedure 12 (eq. 4.54).

To a benzene (40 ml) solution of alcohol (0.4 mmol) in a stainless reactor of autoclave was added $Pb(OAc)_4$ (295 mg, 0.6 mmol). The inside of the reactor was flashed twice with 10 atm of carbon monoxide. Then the reactor was pressured at 80 atm with carbon monoxide and heated at 40 °C with stirring for 3 days. After the reaction, the mixture was poured into 0.4 M aq. HCl solution, and extracted with ether thrice (20 ml × 3). The organic layer was dried over Na_2SO_4 and filtered. After removal of the solvent, the residue was chromatographed on silica gel (eluent: ethyl acetate/hexane = 1/9) to obtain δ-lactone in 51% yield [144].

Radical carbonylative ring-closure of 6-iodohexyl acrylate (**153**) with $(Me_3Si)_3$-SiH/AIBN in supercritical CO_2 as reaction media affords the 11-membered macrolide (**154**) (eq. 4.55).

(4.55)

(4.56)

Eq. 4.56 shows the formation of γ-lactam (**156**) from imino bromide (**155**), through the formation of an acyl radical with monoxide, 5-*exo-trig* ring closure at the nitrogen atom by the nucleophilic acyl radical, and finally abstraction of a hydrogen atom from Bu_3SnH [147].

Instead of Bu_3SnH, Mn^{3+} can be also used for the same carbonylation reaction with carbon monoxide in an autoclave [148].

4.5 ADDITION TO ACETYLENES

Radical additions of Bu_3SnH and $(Me_3Si)_3SiH$ onto acetylene derivatives (**157**), (**159**), and (**161**) initiated by AIBN or Et_3B smoothly proceed to give the corresponding addition products, vinyl silane (**158**) and vinyl stannanes (**160**) and (**162**), via β-silyl vinyl radical and β-stannyl vinyl radicals, respectively (eqs. 4.57–4.59) [149–158]. Reactions with terminal alkynes provide the (*E*)-isomer as a major adduct, due to the steric hindrance in the hydrogen-abstraction by vinyl radicals. However, (*E*) and (*Z*) selectivity generally depends on the substituent of alkynes, and radical reagents, Bu_3SnH and $(Me_3Si)_3SiH$.

(4.57)

initiator			
AIBN	90 °C	88%	$Z:E = 84:16$
Et_3B, O_2	25 °C	85%	$Z:E = 99:1$

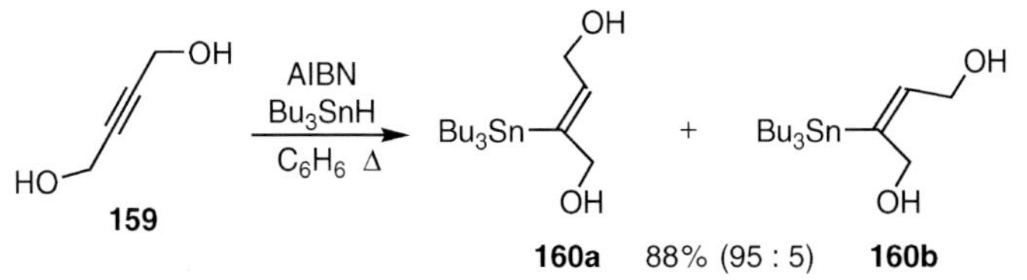

(4.58)

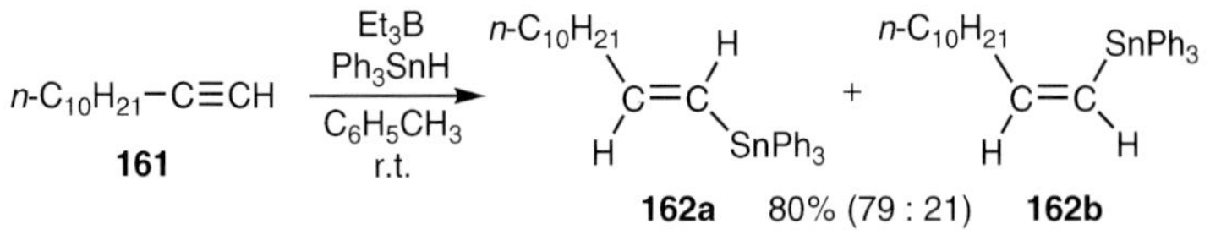

(4.59)

Experimental procedure 13 (eq. 4.57).

AIBN method: To a flask were added phenylacetylene (4 mmol), AIBN (0.96 mmol), and toluene (40 ml). After flash the flask with argon gas, $(Me_3Si)_3SiH$ (4.8 mmol) was added, and the obtained mixture was heated at 70 °C for 2–4 h. After the reaction, the solvent was removed, and the residue was distilled to give 2-[*tris*(trimethylsilyl)silyl]styrene (b.p. 170 °C/0.02 mmHg).

Et_3B method: To a mixture of phenylacetylene (4 mmol), $(Me_3Si)_3SiH$ (5.2 mmol) in dry toluene (40 ml) were added Et_3B (0.8 mmol, 1 M hexane solution) and air (10 ml) over 6–24 h with syringe pumps. After the reaction, the mixture was poured into water and extracted with ether thrice. The organic layer was dried over Na_2SO_4. After removal of the solvent, the residue was chromatographed on silica gel (eluent: pentane) to give the adduct [155].

Experimental procedure 14 (eq. 4.59).

Et_3B (0.1 mmol, 1 M hexane solution) was added to a mixture of 1-dodecyne (1 mmol) and triphenyltin hydride (1.2 mmol) in toluene (8 ml) at room temperature under an argon atmosphere. After 20 min the reaction mixture was poured into water and extracted with ethyl acetate thrice. The organic layer was washed with brine and dried over Na_2SO_4. After removal of the solvent, the residue was treated with preparative TLC to give a mixture of (*E*) and (*Z*)-1-(triphenylstannyl)-1-dodecene in 80% yield ($E/Z = 79/21$) [160].

The intermediate vinyl radicals, such as β-stannyl vinyl radicals and β-silyl vinyl radicals are a very reactive species because of the sp^2 carbon-centered radicals, so, these reactive intermediates can be used for the cyclization. Thus, treatment of ene-yne compounds (**163**) ans (**166**) with the Bu_3SnH/AIBN or $(Me_3Si)_3SiH$/AIBN system generates (*exo*-tributylstannylmethylene)cyclopentane (**164**) (eq. 4.60), α-(*exo*-tributylstannylmethylene)cyclopentanone (**167**) (eq. 4.61), respectively. Treatment of yne-oxime ether (**169**) with Bu_3SnH/AIBN provides methyl β-amino-γ-(*exo*-methylene)cyclopentanecarboxylate (**170**) (eq. 4.62). Treatment of propargyl *O*-glycoside (**171**) with Bu_3SnH/AIBN gives a bicyclic sugar (**172**) bearing an *exo*-(tributylstannyl)methylene group (eq. 4.63). Eq. 4.64 shows uncommon 7-*exo-trig* cyclization of a sugar ene-yne compound (**173**). In these reactions, the tributylstannyl group of the products is easily removed on silica-gel treatment or acid treatment.

AIBN, Bu_3SnH; silica gel, CH_2Cl_2

163 $E = CO_2CH_3$ → **164** 85% → **165** (4.60)

AIBN, Bu_3SnH, C_6H_6 Δ; $H^{\oplus}$, Et_2O

166 → **167** → **168** (4.61)

	167	168
$R_1 = R_2 = CH_3$	91%	88%
$R_1 = Ph, R_2 = H$	92%	92%
$R_1 = CO_2C_2H_5, R_2 = H$	88%	82%

AIBN, Bu_3SnH, C_6H_6 Δ; $H^{\oplus}$, Et_2O

169 → **170** 82% (4.62)

AIBN, Bu_3SnH

171 → **172** 75% ($E:Z = 3:1$) (4.63)

AIBN, Bu_3SnH, $C_6H_5CH_3$ 80 °C

173 → **174** 60% (4.64)

Experimental procedure 15 (eq. 4.60).

To a flask were added ene-yne diester (0.1 mol), Bu_3SnH (0.104 mol), AIBN (0.25 mmol) under an argon atmosphere. The mixture was heated at 75–85 °C for 0.5–1 h. After the reaction, the mixture was added into a mixture of dichloromethane (1000 ml) and silica gel (350 g), and stirred for 24–48 h. After the reaction, the mixture was filtered and washed with ethyl acetate. After the Celite filtration of the filtrate, the solvent was removed and the residue was distilled to give dimethyl 3-methylene-4-isopropyl-1,1-cyclopentanedicarboxylate in 85% yield (b.p. 80–85 °C/0.2 mmHg) [150].

REFERENCES

1. SD Burke, WF Fobare and DM Armistead, *J. Org. Chem.*, 1982, **47**, 3348.
2. N Ono, H Miyake, A Kamimura and A Kaji, *Tetrahedron Lett.*, 1982, **23**, 2957.
3. B Giese and J Dupuis, *Angew. Chem. Int. Ed. Engl.*, 1983, **22**, 622.
4. B Giese, JA G-Gomez and T Witzel, *Angew. Chem. Int. Ed. Engl*, 1984, **23**, 69.
5. N Ono, H Miyake, A Kamimura, I Hamamoto, R Tamura, and A Kaji, *Tetrahedron*, 1985, **41**, 4013.
6. P Pike, S Hershberger and J Hershberger, *Tetrahedron Lett.*, 1985, **26**, 6289.
7. N Ono, H Miyake and A Kaji, *Chem. Lett.*, 1985, 635.
8. G Stork, PM Sher and HL Chen, *J. Am. Chem. Soc.*, 1986, **108**, 6384.
9. B Giese and T Witzel, *Tetrahedron Lett.*, 1987, **28**, 2571.
10. B Giese and B Kopping, *Tetrahedron Lett.*, 1989, **30**, 681.
11. C Chatgilialoglu, B Giese and B Kopping, *Tetraheron Lett.*, 1990, **31**, 6013.
12. E Lee and DS Lee, *Tetrahedron Lett.*, 1990, **31**, 4341.
13. LA Paquette, DR Lagerwall and HG Korth, *J. Org. Chem.*, 1992, **57**, 5413.
14. H Junker, N Phung and W Fessner, *Tetrahedron Lett.*, 1999, **40**, 7063.
15. B Wu, BA Avery and MA Avery, *Tetrahedron Lett.*, 2000, **41**, 3797.
16. R SanMartin, B Tavasoli, K Walsh, DS Walter and T Gallagher, *Org. Lett.*, 2001, **2**, 4051.
17. L Grant, Y Liu, KE Walsh, DS Walter and T Gallagher, *Org. Lett.*, 2002, **4**, 4623.
18. WB Motherwell, BC Ross and MJ Tozer, *Synlett*, 1989, 68.
19. P Renaud, P Bjorup, PA Carrupt, K Schenk and S Schuber, *Synlett*, 1992, 211.
20. M Yoshida, M Ohkoshi, N Aoki, Y Ohnuma and M Iyoda, *Tetrahedron Lett.*, 1999, **40**, 5731.
21. GA Kraus and K Landgrebe, *Tetrahedron Lett.*, 1984, **25**, 3939.
22. MD Castaing, BD Jeso, GA Kraus, K Landgrebe and B Maillard, *Tetrahedron Lett.*, 1986, **27**, 5927.
23. P Balczewski, T Bialas and M Mikolajczyk, *Tetrahedron Lett.*, 2000, **41**, 3687.
24. MG Roepel, *Tetrahedron Lett.*, 2002, **43**, 1973.
25. M Yoshida, T Kimura and M Kobayashi, *Sulfur Lett.*, 1984, **2**, 29.
26. DL Boger and RJ Mathvink, *J. Org. Chem.*, 1989, **54**, 1777.
27. K Ogura, T Arai, A Kayano and M Akazome, *Tetrahedron Lett.*, 1998, **39**, 9051.
28. K Ogura, T Arai, A Kayano and M Akazome, *Tetrahedron Lett.*, 1999, **40**, 2537.
29. ML Bennasar, T Roca, R Griera and J Bosch, *Org. Lett.*, 2001, **3**, 1697.
30. M-L Bennasar, T Roca, R Griera, M Bassa and J Bosch, *J. Org. Chem.*, 2002, **67**, 6268.
31. H Miyabe, Y Fujishima and T Naito, *J. Org. Chem.*, 1999, **64**, 2174.
32. H Miyabe, M Ueda and T Naito, *Chem. Commun.*, 2000, 2059.
33. H Miyabe, M Ueda, A Nishimura and T Naito, *Org. Lett.*, 2002, **4**, 131.
34. S Fukuzawa, A Nakanishi, T Fujinami and S Sakai, *Chem. Commun.*, 1986, 624.
35. K Otsubo, J Inanaga and M Yamaguchi, *Tetrahedron Lett.*, 1986, **27**, 5763.
36. O Ujikawa, J Inanaga and M Yamaguchi, *Tetrahedron Lett.*, 1989, **30**, 2837.
37. L Forti, F Ghelfi and UM Pagnoni, *Tetrahedron Lett.*, 1995, **36**, 2509.
38. K Nozaki, K Oshima and K Utimoto, *Tetrahedron Lett.*, 1988, **29**, 6125 and 6127.
39. K Miura, M Taniguchi, K Nozaki, K Oshima and K Utimoto, *Tetrahedron Lett.*, 1990, **31**, 6391.
40. X Lu, Z Wang and J Ji, *Tetrahedron Lett.*, 1994, **35**, 613.

41. S Bergeron, T Brigaud, G Foulard, RP Royon and C Portella, *Tetrahedron Lett.*, 1994, **35**, 1985.
42. E Beyou, P Babin, B Bennetau, J Dunogues, D Teyssie and S Boileau, *Tetrahedron Lett.*, 1995, **36**, 1843.
43. BA Trofimov, NK Gusarova and SF Malysheva, *Synthesis*, 2002, 2207.
44. C Lopin, A Gautier, G Gouhier and SR Piettre, *Tetrahedron Lett.*, 2000, **41**, 10195.
45. A Gautier, G Garipova, O Dubert, H Oulyadi and SR Piettre, *Tetrahedron Lett.*, 2001, **42**, 5673.
46. S Deprele and J-L Montchamp, *J. Org. Chem.*, 2001, **66**, 6745.
47. O Dubert, A Gautier, E Condamine and SR Piettre, *Org. Lett.*, 2002, **4**, 359.
48. BA Trofimov and SF Malysheva, *Tetrahedron Lett.*, 2003, **44**, 2629.
49. T Nakano, M Kayama, H Matsumoto and Y Nagai, *Chem. Lett.*, 1981, 415.
50. M Masnyk, *Tetrahedron Lett.*, 1991, **32**, 3259.
51. CL Bumgardner and JP Burgess, *Tetrahedron Lett.*, 1992, **33**, 1683.
52. S-K Kang, B-S Ko and Y-H Ha, *J. Org. Chem.*, 2001, **66**, 3630.
53. CP Chuang, *Synlett*, 1991, 859.
54. JY Nedelec and K Nohair, *Synlett*, 1991, 660.
55. GL Edwards and KA Walker, *Tetrahedron Lett.*, 1992, **33**, 1779.
56. CP Chuang, *Tetrahedron Lett.*, 1992, **33**, 6311.
57. T Mochizuki, S Hayakawa and K Narasaka, *Bull. Chem. Soc. Jpn*, 1996, **69**, 2317.
58. V Nair, A Augustine and TD Suja, *Synthesis*, 2002, 2259.
59. G Bar, AF Parsons and CB Thomas, *Synlett*, 2002, 1069.
60. E Baciocchi, G Civitarese and R Ruzziconi, *Tetrahedron Lett.*, 1987, **28**, 5357.
61. T Varea, MEG Nunez and JR Chiner, *Tetrahedron Lett.*, 1989, **30**, 4709.
62. A Clerici and O Porta, *Tetrahedron Lett.*, 1990, **31**, 2069.
63. Y Kohno and K Narasaka, *Chem. Lett.*, 1993, 1689.
64. JH Byers and GC Lane, *Tetrahedron Lett.*, 1990, **31**, 5697.
65. JH Byers, JG Thissell and MA Thomas, *Tetrahedron Lett.*, 1995, **36**, 6403.
66. JH Byers, CC Whitehead and ME Duff, *Tetrahedron Lett.*, 1996, **37**, 2743.
67. ZY Long and QY Chen, *J. Org. Chem.*, 1999, **64**, 4775.
68. M Yoshida, Y Morinaga, M Iyoda, K Kikuchi, I Ikemoto and Y Achiba, *Tetrahedron Lett.*, 1993, **34**, 7629.
69. K Mikami and S Matsumoto, *Synlett*, 1995, 229.
70. BF Reid, RC Anderson, DR Hicks and DL Walker, *Can. J. Chem.*, 1977, **55**, 3986.
71. JE Baldwin, DR Kelly and CB Ziegler, *Chem. Commun.*, 1984, 133.
72. JE Baldwin and DR Kelly, *Chem. Commun.*, 1985, 682.
73. S Kim, IY Lee, JY Yoon and DH Oh, *J. Am. Chem. Soc.*, 1996, **118**, 5138.
74. S Kim and IY Lee, *Tetrahedron Lett.*, 1998, **39**, 1587.
75. S Kim, N Kim, JY Yoon and DH Oh, *Synlett*, 2001, 1148.
76. S Kim, HJ Song, TL Choi and JY Yoon, *Angew. Chem. Int. Ed.*, 2001, **40**, 2524.
77. S Kim and R Kavali, *Tetrahedron Lett.*, 2002, **43**, 7189.
78. J Xiang and PL Fuchs, *Tetrahedron Lett.*, 1998, **39**, 8597.
79. F Bertrand, BQ Sire and SZ Zard, *Angew. Chem. Int. Ed. Eng*, 1999, **38**, 203.
80. J Liu and C Yao, *Tetrahedron Lett.*, 2001, **42**, 6147.
81. J Liu, Y Jang, Y Shih, S Hu, C Chu and C Yao, *J. Org. Chem.*, 2001, **66**, 6021.
82. J Liu, J Liu and C Yao, *Tetrahedron Lett.*, 2001, **42**, 3613.
83. RR Webb, II and S Danishefsky, *Tetrahedron Lett.*, 1983, **24**, 1357.
84. GE Keck and JB Yates, *J. Am. Chem. Soc.*, 1982, **104**, 5829.
85. N Ono, K Zinsmeister and A Kaji, *Bull. Chem. Soc. Jpn*, 1985, **58**, 1069.
86. JE Baldwin, RM Adlington, DJ Birch, JA Crawford and JB Sweeney, *Chem. Commun.*, 1986, 1339.
87. T Toru, Y Yamada, T Ueno, E Maekawa and Y Ueno, *J. Am. Chem. Soc.*, 1988, **110**, 4815.
88. K Mizuno, M Ikeda, S Toda and Y Otsuji, *J. Am. Chem. Soc.*, 1988, **110**, 1288.
89. CK Chu, B Doboszewski, W Schmidt and GV Ullas, *J. Org. Chem.*, 1989, **54**, 2767.
90. DP Curran, W Shen, J Zhang and TA Heffner, *J. Am. Chem. Soc.*, 1990, **112**, 6738.

91. A Giardina, R Giovannini and M Petrini, *Tetrahedron Lett.*, 1997, **38**, 1995.
92. EJ Enholm, ME Gallagher, KM Moran, JS Lombardi and JP Schulte, II, *Org. Lett.*, 1999, **1**, 689.
93. EJ Enholm, ME Gallagher, S Jiang and WA Batson, *Org. Lett.*, 2000, **2**, 3355.
94. I Ryu, S Kreimerman, T Niguma, S Minakata, M Komatsu, Z Luo and DP Curran, *Tetrahedron Lett.*, 2001, **42**, 947.
95. MK Gurjar, SV Ravindranadh and S Karmarkar, *Chem. Commun.*, 2001, 241.
96. Y Ueno, S Aoki and M Okawara, *J. Am. Chem. Soc.*, 1979, **101**, 5414.
97. GE Keck and JH Byers, *J. Org. Chem.*, 1985, **50**, 5442.
98. B Barlaam, J Boivin and SZ Zard, *Tetrahedron Lett.*, 1990, **31**, 7429.
99. CC Huval and DA Singleton, *Tetrahedron Lett.*, 1993, **34**, 3041.
100. C Chatgilialoglu, A Alberti, M Ballestri, D Macieantelli and DP Curran, *Tetrahedron Lett.*, 1996, **37**, 6391.
101. C Chatgilialoglu, C Ferreri, M Ballestri and DP Curran, *Tetrahedron Lett.*, 1996, **37**, 6387.
102. C Chatgilialoglu, M Ballestri, D Vecchi and DP Curran, *Tetrahedron Lett.*, 1996, **37**, 6383.
103. BQ Sire and SZ Zard, *J. Am. Chem. Soc.*, 1996, **118**, 209.
104. FL Guyader, BQ Sire, S Sequin and SZ Zard, *J. Am. Chem. Soc.*, 1997, **119**, 7410.
105. K Miura, H Saito, T Nakagawa, T Hondo, J Tateiwa, M Sonoda and A Hosomi, *J. Org. Chem.*, 1998, **63**, 5740.
106. P Breuilles and D Uguen, *Tetrahedron Lett.*, 1990, **31**, 357.
107. JM Mellor and S Mohammed, *Tetrahedron Lett.*, 1991, **32**, 7107.
108. S Kang, H Seo and Y Ha, *Synlett*, 2001, 1321.
109. AG Angoh and DLJ Clive, *Chem. Commun.*, 1985, 980.
110. Z Cekovic and R Saicic, *Tetrahedron Lett.*, 1986, **27**, 5893.
111. DP Curran and MH Chen, *J. Am. Chem. Soc.*, 1987, **109**, 6558.
112. K Miura, K Fugami, K Oshima and K Utimoto, *Tetrahedron Lett.*, 1988, **29**, 1543.
113. P Renaud, JPV Vionnet and P Vogel, *Tetrahedron Lett.*, 1991, **32**, 3491.
114. LD Miranda, R Cruz-Almanza, M Pavon, Y Romero and JM Muchowski, *Tetrahedron Lett.*, 2000, **41**, 10181.
115. DC Harrowven, JC Hannam, MC Lucas, NA Newman and PD Howes, *Tetrahedron Lett.*, 2000, **41**, 9345.
116. M-L Bennasar, T Roca, R Griera and J Bosch, *J. Org. Chem.*, 2001, **66**, 7547.
117. M Ohkoshi, M Yoshida, H Matsuyama and M Iyoda, *Tetrahedron Lett.*, 2001, **42**, 33.
118. H Miyabe, K Fujii, T Goto and T Naito, *Org. Lett.*, 2000, **2**, 4071.
119. O Kitagawa, H Fujiwara and T Taguchi, *J. Org. Chem.*, 2002, **67**, 922.
120. O Kitagawa, Y Yamada, H Fujiwara and T Taguchi, *Angew. Chem. Int. Ed.*, 2001, **40**, 3865.
121. JM Barks, BC Gilbert, AF Parsons and B Upeandran, *Tetrahedron Lett.*, 2001, **42**, 3137.
122. A Robertson, C Bradaric, CS Frampton, J McNulty and A Capretta, *Tetrahedron Lett.*, 2001, **42**, 2609.
123. DHR Barton, ED Silva and SZ Zard, *Chem. Commun.*, 1988, 285.
124. NJ Cornwall, S Linehan and RT Weavers, *Tetrahedron Lett.*, 1997, **38**, 2919.
125. WS Murphy, D Neville and G Ferguson, *Tetrahedron Lett.*, 1996, **42**, 7615.
126. CP Chuang and SF Wang, *Synlett*, 1996, 829.
127. G Bar, AF Parsons and CB Thomas, *Tetrahedron Lett.*, 2000, **41**, 7751.
128. MC Jiang and CP Chuang, *J. Org. Chem.*, 2000, **65**, 5409.
129. R Ruzziconi, H Couthon-Gourves, J Gourves and B Corbel, *Synlett*, 2001, 703.
130. K Hirase, T Iwahama, S Sakaguchi and Y Ishii, *J. Org. Chem.*, 2002, **67**, 970.
131. V Gyollai, D Schanzenbach, L Somsak and T Linker, *Chem. Commun.*, 2002, 1294.
132. C-T Tseng, Y-L Wu and C-P Chuang, *Tetrahedron*, 2002, **58**, 7625.
133. M Hein and R Miethchen, *Eur. J. Org. Chem.*, 1999, 2429.
134. I Ryu, K Kusano, A Ogawa, N Kambe and N Sonoda, *J. Am. Chem. Soc.*, 1990, **112**, 1295.
135. I Ryu, K Kusano, N Masumi, H Yamazaki, A Ogawa and N Sonoda, *Tetrahedron Lett.*, 1990, **31**, 6887.
136. I Ryu, *Yuki Gosei Kagaku Kyokaishi*, 1992, **50**, 737.
137. I Ryu, K Kusano, H Yamazaki and N Sonoda, *J. Org. Chem.*, 1991, **56**, 5003.

138. I Ryu, H Yamazaki, K Kusano, A Ogawa and N Sonoda, *J. Am. Chem. Soc.*, 1991, **113**, 8558.
139. I Ryu, K Kusano, M Hasegawa, N Kambe and N Sonoda, *Chem. Commun.*, 1991, 1018.
140. I Ryu, K Nagahara, H Yamazaki, S Tsunoi and N Sonoda, *Synlett*, 1994, 643.
141. S Berlin, C Ericsson and L Engman, *Org. Lett.*, 2002, **4**, 3.
142. S Tsunoi, I Ryu and N Sonoda, *J. Am. Chem. Soc.*, 1994, **116**, 5473.
143. S Tsunoi, I Ryu, Y Tamura, S Yamasaki and N Sonoda, *Synlett*, 1994, 1007.
144. S Tsunoi, I Ryu, T Okuda, M Tanaka, M Komatsu and N Sonoda, *J. Am. Chem. Soc.*, 1998, **120**, 8692.
145. S Kreimerman, I Ryu, S Minakata and M Komatsu, *Org. Lett.*, 2000, **2**, 389.
146. Y Kishimoto and T Ikariya, *J. Org. Chem.*, 2000, **65**, 7656.
147. I Ryu, K Matsu, S Minakata and M Komatsu, *J. Am. Chem. Soc.*, 1998, **120**, 5838.
148. I Ryu and H Alper, *J. Am. Chem. Soc.*, 1993, **115**, 7543.
149. G Stork and R Mook, Jr., *J. Am. Chem. Soc.*, 1987, **109**, 2829.
150. R Mook, Jr. and PM Sher, *Org. Synth.*, 1987, **66**, 75.
151. C Nativi and M Taddei, *J. Org. Chem.*, 1988, **53**, 820.
152. E Lee, C-U Hur and J-H Park, *Tetrahedron Lett.*, 1989, **30**, 7219.
153. EJ Enholm, JA Burroff and LM Jaramillo, *Tetrahedron Lett.*, 1990, **31**, 3727.
154. N Moufid and Y Chapleur, *Tetrahedron Lett.*, 1991, **32**, 1799.
155. B Kopping, C Chatgilialoglu, M Zehnder and B Giese, *J. Org. Chem.*, 1992, **57**, 3994.
156. K Miura, K Oshima and K Utimoto, *Bull. Chem. Soc. Jpn*, 1993, **66**, 2356.
157. J Marko-Contelles and E de Opazo, *J. Org. Chem.*, 2002, **67**, 3705.
158. L Commeiras, M Santelli and J-L Parrain, *Tetrahedron Lett.*, 2003, **44**, 2311.
159. O Kitagawa, H Fujiwara and T Taguchi, *Tetrahedron Lett.*, 2001, **42**, 2165.
160. K Nozaki, K Oshima and K Utimoto, *J. Am. Chem. Soc.*, 1987, **109**, 2547.

– 5 –

Alkylation of Aromatics

5.1 OXIDATIVE CONDITIONS

The Friedel-Crafts reaction is polar (ionic) alkylation or acylation of electron-rich aromatics by alkyl cation or acyl cation species, derived from the reactions of alkyl halides or acyl halides with $AlCl_3$. Therefore, electron-rich aromatics such as anisole are very reactive, but electron-deficient aromatics such as pyridine are inert.

As mentioned before, alkyl radicals and acyl radicals have a nucleophilic character; therefore, radical alkylation and acylation of aromatics shows the opposite reactivity and selectivity to polar alkylation and acylation with the Friedel-Crafts reaction. Thus, alkyl radicals and acyl radicals do not react with anisole, but may react with pyridine. Eq. 5.1 shows the reaction of an alkyl radical with γ-picoline (**1**). The nucleophilic alkyl radical reacts at the 2-position of γ-picoline (**1**), where electron density is lower than that of the 3-position. So, 2-alkyl-4-methylpyridine (**2**) is obtained with complete regioselectivity. When pyridine is used instead of γ-picoline, a mixture of 2-alkylpyridine and 4-alkylpyridine is obtained. Generally, radical alkylation or radical acylation onto aromatics is not a radical chain reaction, since it is just a substitution reaction of a hydrogen atom of aromatics by an alkyl radical or an acyl radical through the addition–elimination reaction. Therefore, the intermediate adduct radical (σ complex) must be rearomatized to form a product and a hydrogen atom (or H^+ and e^-). Thus, this type of reactions proceeds effectively under oxidative conditions [1–6].

$$R^\bullet + \mathbf{1} \longrightarrow [\text{adduct radical}] \longrightarrow \mathbf{2} + H^\bullet\ (H^\oplus + e^-) \qquad (5.1)$$

Reactivities of *n*-Bu˙ (sp^3 carbon-centered radical) and Ph˙ (sp^2 carbon-centered radical) to benzene, γ-picoline, protonated γ-picoline, protonated benzothiazole are shown in Table 5.1. As can be seen here, the reactivity is greatly increased in both radicals, as the electron-density of aromatics is decreased. Therefore, the reactivity of heteroatomatic bases such as γ-picoline and benzothiazole, can be increased by their

Table 5.1

Rate constants for reactions of *n*-Bu˙ and Ph˙ to aromatics ($M^{-1} \cdot s^{-1}$)

	k_1(*n*-Bu•)	k_2(Ph•)
 Benzene	3.8×10^2	1.0×10^6
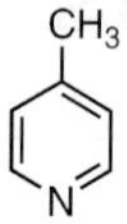 4-Methylpyridine	1.3×10^3	3.2×10^5
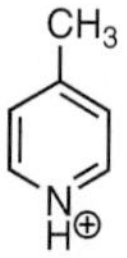 4-Methylpyridinium (N–H⊕)	1.1×10^5	1.8×10^6
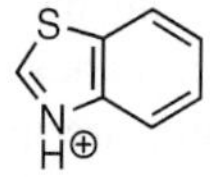 Benzothiazolium (N–H⊕)	9.9×10^5	7×10^5

protonation. Moreover, the sp^2 carbon-centered radical is much more reactive than the sp^3 carbon-centered radical. This characteristic reflects the fact that the bond dissociation energy of $C(sp^2)$-H is ~15 kcal/mol higher than that of $C(sp^3)$-H.

Reactivity of nucleophilic alkyl radicals is increased in the following order,

$$CH_3^{\cdot} < prim\text{-}R^{\cdot} < CH_3C(O)^{\cdot} < PhC(O)^{\cdot} < sec - R^{\cdot} < tert - R^{\cdot}$$

Since radical alkylation of heteroaromatics is consequently the substitution reaction of a hydrogen atom by an alkyl radical as shown in eq. 5.1, oxidative conditions are generally preferable. Thus, in the alkylation of heteroaromatics, $(RCO_2)_2$/SET metal agent (such as Cu^+, Fe^{2+}, Ag^+) system is the typical method. The Fenton system with Fe^{2+} and H_2O_2 generates $HO^{\cdot}$, HO^-, and Fe^{3+} through the SET from Fe^{2+} to H_2O_2. Once $HO^{\cdot}$ is formed, extremely reactive $HO^{\cdot}$ rapidly abstracts a hydrogen atom from the substrate. For example, treatment of α-hydroxycarboxylic acid with the Fenton system gives α-keto acid through the abstraction of an α-hydrogen atom from α-hydroxycarboxylic acid. Here, Fe^{2+} functions as a SET catalyst.

Eq. 5.2 shows the preparation of 3-chloro-4-acetyl-6-methylpyridazine (**4**) by the treatment of 3-chloro-6-methylpyridazine (**3**) and acetaldehyde in an aq. acidic solution with the Fenton system. $HO^{\cdot}$ formed from the Fenton system, abstracts a formyl hydrogen atom from acetaldehyde to form a nucleophilic acetyl radical. This radical reacts at the electrophilic 4-position of 3-chloro-6-methylpyridazinium salt formed in acidic solution, to give 3-chloro-4-acetyl-6-methylpyridazine as shown in eq. 5.2 [7–9].

$$Fe^{2\oplus} + H_2O_2 \longrightarrow Fe^{3\oplus} + [HO^{\bullet}] + HO^{\ominus}$$

$$CH_3CHO + [HO^{\bullet}] \longrightarrow [CH_3\text{-}\dot{C}{=}O] + H_2O$$

$$\mathbf{3} + [CH_3\dot{C}{=}O] \longrightarrow [\text{radical adduct}] \xrightarrow[(-Fe^{2\oplus})\,(-H^{\oplus})]{Fe^{3\oplus}} \mathbf{4} \tag{5.2}$$

Experimental procedure 1 (eq. 5.2).

To a mixture of 3-chloro-6-methylpyridazine (1 mmol), H_2SO_4 (3%, 6 ml), acetone (3 ~ 5 ml), and acetaldehyde (20 mmol) were added hydrogen peroxide (34%, 20 mmol) and sat. aq. $FeSO_4$ (20 mmol), simultaneously over 10 min. During the addition, the temperature of the reaction mixture must be kept in the range of 30 to 35 °C. The reaction mixture was then stirred for 5 min. After the reaction, water and 20% aq. Na_2CO_3 were added to the reaction mixture, and the organic layer was extracted with dichloromethane, and dried over Na_2SO_4. After filtration and removal of the solvent, the residue was chromatographed on silica gel to give 3-chloro-4-acetyl-6-methylpyridazine in 37% yield [8].

With diacyl peroxide and SET metal ions such as Cu^+, alkylation of heteroaromatics such as pyrazine can be carried out. This method can be also used for alkylation of a ferrocene complex. Eq. 5.3 shows alkylation of quinoxaline (**5**) with diaceyl peroxide and Cu^+.

$$(RCO_2)_2 \xrightarrow[(-Cu^{2\oplus})]{Cu^{\oplus}} [(RCO_2)_2]^{\dot{\ominus}} \xrightarrow[(-RCO_2^{\ominus})]{} [RCO_2^{\bullet}] \xrightarrow[(-CO_2)]{} [R^{\bullet}]$$

$$\mathbf{5} \longrightarrow [\text{radical adduct}] \xrightarrow[(-Cu^{\oplus})]{Cu^{2\oplus}} \mathbf{6} + H^{\oplus} \tag{5.3}$$

When carboxylic acids are used as an alkyl radical precursor ($R^{\cdot}$), the RCO_2H/$Na_2S_2O_8$/$AgNO_3$ system is often used. Moreover, radical acylation and alkoxycarbonylation can be also carried out by using the $RCOCO_2H$ (α-keto acid)/$Na_2S_2O_8$/$AgNO_3$ system and RO_2CCO_2H (half ester of oxalic acid)/$Na_2S_2O_8$/$AgNO_3$ system, respectively.

Eq. 5.4 shows radical alkylation, acylation, and alkoxycarbonylation of γ-picoline (**1**) with the $RCO_2H/Na_2S_2O_8/AgNO_3$ system, $RCOCO_2H/Na_2S_2O_8/AgNO_3$ system, and $RO_2CCO_2H/Na_2S_2O_8/AgNO_3$ system, respectively. These reactions comprise of the initial oxidation of Ag^+ to Ag^{2+} by peroxide, oxidation of RCO_2H, $RCOCO_2H$, and RO_2CCO_2H to $RCO_2^{\cdot}$, $RCOCO_2^{\cdot}$ and $RO_2CCO_2^{\cdot}$ with Ag^{2+}, their rapid decarboxylation to $R^{\cdot}$, $RC(O)^{\cdot}$, and $RO_2C^{\cdot}$, then reaction at the electron-deficient α-position of γ-picoline (**1**), and finally oxidation of the adducts to form 2-alkyl-4-methylpyridine (**2**), 2-acyl-4-methylpyridine (**7**), and 2-alkoxycarbonyl-4-methylpyridine (**8**), respectively [10–13].

$$2Ag^{\oplus} + S_2O_8^{2\ominus} \longrightarrow 2Ag^{2\oplus} + 2SO_4^{2\ominus}$$
$$RCO_2^{\ominus} + Ag^{2\oplus} \longrightarrow RCO_2^{\cdot} + Ag^{\oplus}$$
$$RCO_2^{\cdot} \longrightarrow R^{\cdot} + CO_2 \qquad (5.4)$$

Reagents: $Na_2S_2O_8$, H_2SO_4, RCO_2H, $AgNO_3$, Δ (**1** → **2**); $RC(O)-CO_2H$, H_2SO_4, $Na_2S_2O_8$, $AgNO_3$, Δ (**1** → **7**); $ROC(O)-CO_2H$, H_2SO_4, $Na_2S_2O_8$, $AgNO_3$, Δ (**1** → **8**); + $R^{\cdot}$; + $Ag^{2\oplus}$ → + $H^{\oplus}$ + $Ag^{\oplus}$

Experimental procedure 2.

A flask equipped with an addition funnel and mechanical stirrer was charged with 1-*N*-methyl-1,2,4-triazole (10 g, 0.12 mol), water (10 ml), and trifluoroacetic acid (27.8 ml, 0.36 mol). The mixture was heated to 70 °C. Then, homogeneous solution of $(NH_4)_2S_2O_8$ (68.66 g, 0.30 mol), water (69 ml), cyclopropanecarboxylic acid (0.24 mol) and 5 M NaOH (48 ml, 0.24 mol) was charged to the additional funnel. A solution of $AgNO_3$, prepared by dissolving $AgNO_3$ (2 g, 0.012 mol) in water (6 ml), was charged to the reaction flask. Dropwise addition of a mixture of $(NH_4)_2S_2O_8$ and carboxylic acid solution was initiated and in the space of 5 to 10% of the addition, the temperature began to rise rapidly and gas evolution was observed. The temperature was maintained between 80 ~ 90 °C by controlling the rate of addition. After the addition, the reaction mixture was stirred for one hour as it cooled to room temperature. Conc. aq. NH_4OH was slowly added until the pH was adjusted to about 12 at 0 °C. The mixture was extracted with ethyl acetate, and the organic layer was dried over $MgSO_4$. After filtration, the solvent was removed and the residue was chromatographed on silica gel (eluent: ethyl acetate/hexane = 1/1) to provide 1-*N*-methyl-5-cyclopropyl-1,2,4-triazole in 60% yield [13].

With the Fenton system, treatment of excess crown ethers (**9**) and 2-methylquinolinium salt in the presence of $FeSO_4$ and *t*-butyl hydrogen peroxide generates crown-substituted 2-methylquinolines (**10**) in good yields. Here, the crown ether is selectively introduced into the electron-deficient 4-position (eq. 5.5) [14].

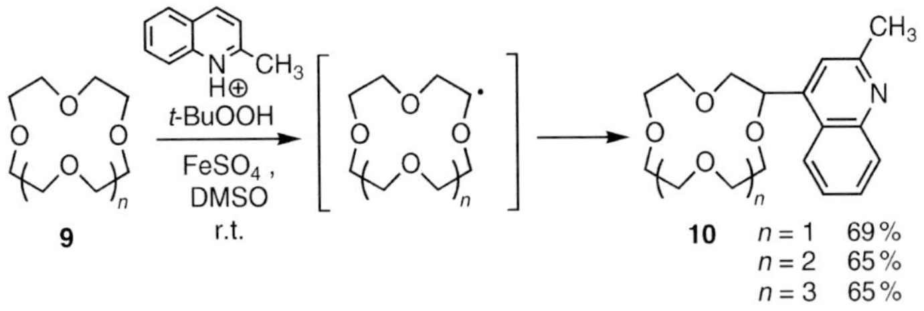

(5.5)

As a unique reaction with the Fenton system, the alkylation of heteroaromatics with alkyl iodide, hydrogen peroxide, and dimethyl sulfoxide in the presence of $FeSO_4$ can be carried out. This reaction comprises of the initial formation of reactive $HO^{\cdot}$ by the reaction of $FeSO_4$ and hydrogen peroxide, reaction of $HO^{\cdot}$ on the sulfur atom of dimethyl sulfoxide to form $CH_3^{\cdot}$ and methanesulfinic acid (S_H2 reaction), reaction of $CH_3^{\cdot}$ on the iodine atom of alkyl iodide via S_H2 pathway to form more stable $R^{\cdot}$ and methyl iodide, and then addition of $R^{\cdot}$ to the α-position of γ-picoline (**1**) to form an addition-intermediate radical which is rearomatized under oxidative conditions to 2-alkyl-4-methylpyridine (**2**) [15, 16].

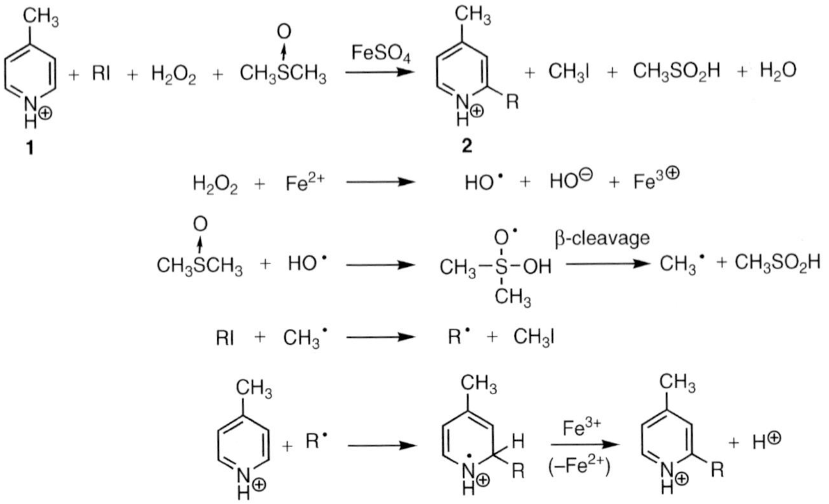

(5.6)

(Diacyloxyiodo)arenes (**11**) generate the corresponding alkyl radical through single decarboxylation under heating or irradiation conditions, and it can be used for the alkylation of heteroaromatics, as shown in eq. 5.7. However, one of two acyloxy groups in (diacyloxyiodo)arene generates an alkyl radical and the other acyloxy group must be recovered as a carboxylic acid [17–19]. The same alkyl radicals can be generated from alcohols through double decarboxylation by changing the acyloxy groups in (diacyloxyiodo)arenes, with half esters of oxalic acid prepared from alcohols and oxalyl chloride.

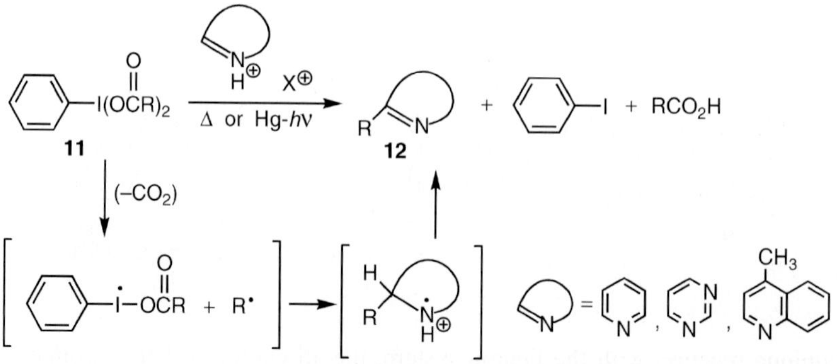

(5.7)

Treatment of carboxylic acids with $Pb(OAc)_4$ can also be used for the alkylation of heteroaromatics, but it is toxic. Eq. 5.8 shows the introduction of a phenyl group to the cubane by the treatment of 1,4-cubanedicarboxylic acid monoester (**13**) with $Pb(OAc)_4$ in benzene [20].

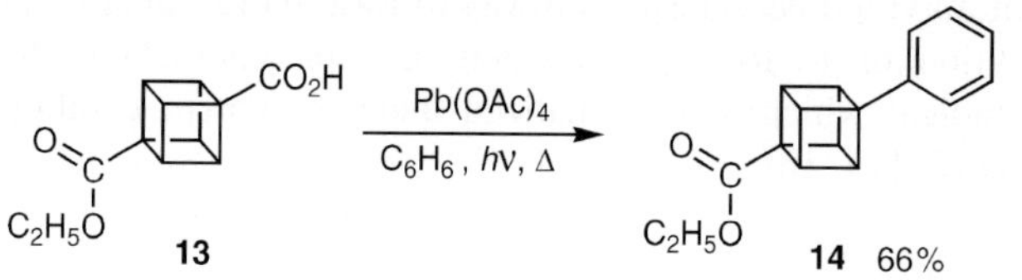

(5.8)

Ce^{4+} and Mn^{3+} can generate an electrophilic malonyl radical from malonate diester. Thus, the introduction of a malonyl group to an uracil derivative is effective by treatment of the uracil derivatives (**15**) with $CH_2(CO_2R)_2$ and $Mn(OAc)_3$ in acetic acid under heating conditions as shown in eq. 5.9 [21, 22].

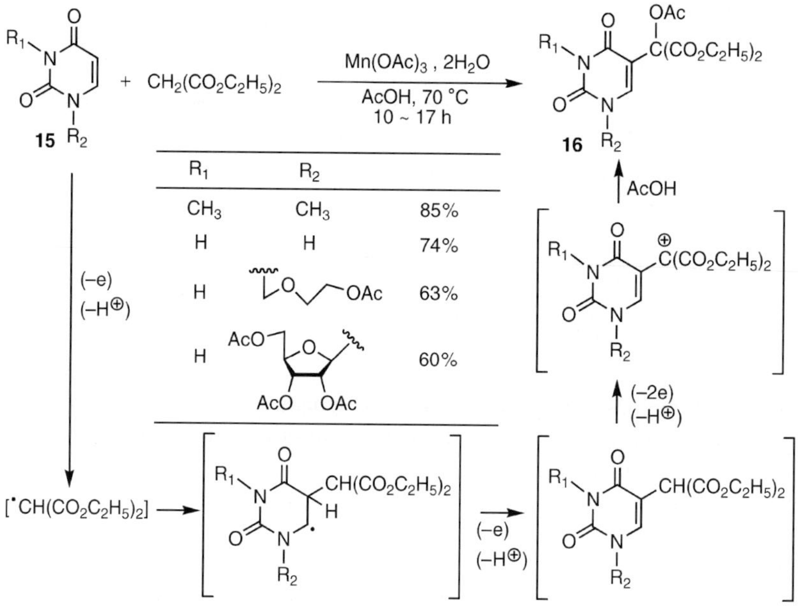

R_1	R_2	
CH_3	CH_3	85%
H	H	74%
H	OAc	63%
H	AcO AcO OAc	60%

(5.9)

Experimental procedure 3 (eq. 5.9).

Diethyl malonate (481 mg) was added to acetic acid (4 ml) solution of *N,N'*-dimethyluracil (70 mg, 0.5 mmol), $Mn(OAc)_3 \cdot 2H_2O$ (804 mg, 3 mmol) at 70 °C under a nitrogen atmosphere, and the mixture was stirred for 12 h at the same temperature. Then, the reaction mixture was filtered. Ethyl acetate was added to the filtrate and the organic layer was washed with sat. aq. $NaHCO_3$, and then water. The organic layer was dried over Na_2SO_4 and filtered. After removal of the solvent, the residue was chromatographed on silica gel (eluent: ethyl acetate/hexane = 1/1) to give diethyl α-acetoxy-α-(*N,N'*-dimethyluracil)malonate in 85% yield (151 mg) [21].

5.2 NON-OXIDATIVE CONDITIONS

From the practical point of view, alkylation of heteroaromatics with Barton decarboxylation of *N*-acyloxy-2-thiopyridones (**17**), prepared from carboxylic acids and *N*-hydroxy-2-thiopyridone, is very useful, since it can be used for various kinds of carboxylic acids such as sugars and nucleosides [23–26]. This reaction comprises of the initial homolytic cleavage of the N–O bond in *N*-acyloxy-2-thiopyridone to form an acyloxyl radical and PyS˙, β-cleavage of the acyloxyl radical to generate an alkyl radical and CO_2, addition to the electron-deficient position of heteroaromatics by the alkyl radical to form the adduct, and finally, abstraction of a hydrogen atom from the adduct by PyS˙, as shown in eq. 5.10.

W-hν, 1 h; CH_2Cl_2, r.t.; + PySH
17 **12**
84% (α : γ = 43 : 41) 64% 76% 66%
(5.10)

Experimental procedure 4 (eq. 5.10).

A solution of *N*-adamantanecarbonyloxy-2-thiopyridone (145 mg, 0.5 mmol), camphorsulfonate salt of lepidine (1.23 g, 3.27 mmol), and dichloromethane (6 ml) was irradiated with a tungsten lamp (200 W, W-hν) under nitrogen atmosphere at the range of 20 ~ 25 °C for 1 h. After the reaction, the reaction mixture was poured into sat. aq. $NaHCO_3$ and extracted with ether. The organic layer was dried over $MgSO_4$ and filtered. After removal of the solvent, the residue was chromatographed on silica gel (eluent: dichloromethane) to give 2-adamantyl-4-methylquinoline in 97% (135 mg) yield [23].

Here, biologically interesting heteroaromatics such as thiazole, caffeine, adenine, etc., are easily alkylated. In nature, biologically active *C*-nucleosides such as showdomycin, pyrazomycin, formycin, oxetanocin-A that have potent antiviral and antitumour activities, are known. The present method can be used for the preparation of these *C*-nucleosides through radical coupling reaction of the sugar moiety and the base moiety, via sugar anomeric radicals, as shown in eq. 5.11 (Figure 5.1).

18 → (DCC, THF) → **19** → (W-$h\nu$, CH_2Cl_2, r.t.) → **20**

70% (α : β = 13 : 87)

(5.11)

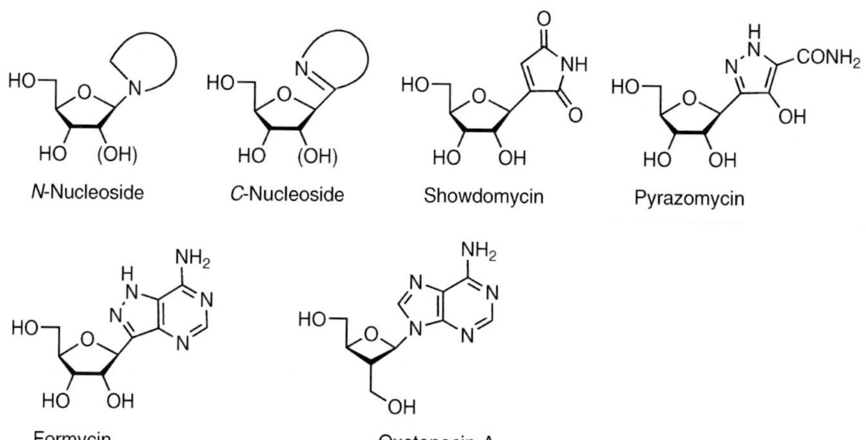

Figure 5.1

Experimental procedure 5 (eq. 5.11).

Carboxylic acid (0.5 mmol) derived from 2-deoxy-D-ribose was dissolved in dry THF (3 ml), and then *N*-hydroxy-2-thiopyridone (0.53 mmol) and DCC (0.6 mmol) were added to the solution at 0 °C. After being stirred for 1.5 h at room temperature in the dark, the reaction mixture was filtered under an argon atmosphere. To the filtrate, 4-methylquinolinium camphorsulfonate (3.5 mmol) and dry THF (4 ml) were added. The yellow solution obtained was stirred under an argon atmosphere and irradiated with a 500 W tungsten lamp for 2.5 h at 30 °C. The reaction mixture was quenched with sat. aq. $NaHCO_3$ solution, extracted with CH_2Cl_2. The organic layer was dried over Na_2SO_4. After filtration and removal of the solvent, the residue was chromatographed on silica gel (eluent: ethyl acetate/hexane = 1/3 ~ 1/1) to give *C*-nucleoside in 70% yield ($\alpha/\beta = 13/87$) [26].

When the introduction of a carbocylic acid group to the sugar anomeric position is difficult, sugar anisyl tellurides can be used as the precursor of sugar anomeric radicals, through S_H2 reaction on the tellurium atom of sugar anisyl tellurides by the ethyl radical derived from triethylborane (eq. 5.12) [27–30].

PhCH$_2$O, PhCH$_2$O, (OCH$_2$Ph), Te, OCH$_3$, **21** — Et_3B, Air / r.t. → [PhCH$_2$O, PhCH$_2$O, (OCH$_2$Ph) radical] + Et-Te–C$_6$H$_4$–OCH$_3$ (5.12)

Since SmI_2 is an effective SET agent, coupling reaction of 1,10-phenanthroline and ketone is promoted by SmI_2 via a ketyl radical, to produce 2-(1-hydroxyalkyl)-1,10-phenanthroline (**23**). Thus, various kinds of α-hydroxyalkyl-substituted 1,10-phenanthrolines, which can be used as a metal ligand, are prepared (eq. 5.13) [31].

22 — 1,10-phenanthroline, SmI_2 / THF, r.t. → [ketyl radical OSmI$_2$] → **23** 88% (OH) (5.13)

Bu_3SnH is an effective reducing agent, so generally it cannot be used for the alkylation (substitution) of heteroaromatics due to the rapid reduction of the alkyl radical formed. However, $(Me_3Si)_3SiH$ can be used for the alkylation of heteroaromatics in the presence of excess AIBN, as shown in eq. 5.14 [32–35].

$$\mathrm{RX}\ \mathbf{24} \xrightarrow[\mathrm{C_6H_6},\ \Delta]{\text{(heteroarene)-}\mathrm{H^{\oplus}\,X^{\ominus}},\ \mathrm{(Me_3Si)_3SiH,\ AIBN}} \mathrm{R\text{-}(heteroarene)}\ \mathbf{12}\ \ 45 \sim 90\% \tag{5.14}$$

R = 1°-, 2°-, 3°-alkyl; heteroarene = pyridine, thiazole, pyrimidine, etc

---

Experimental procedure 6 (eq. 5.14).

To a mixture of adamantyl bromide (0.5 mmol) and trifluoroacetate salt of lepidine (2.5 mmol) in benzene (6 ml) were added *tris*(trimethylsilyl)silane (1.0 mmol) and AIBN (1.0 mmol). The mixture was heated at 80 °C for 14 h under an argon atmosphere. After the reaction, the reaction mixture was washed with sat. aq. $NaHCO_3$. The organic layer was dried over Na_2SO_4. After removal of the solvent, the residue was chromatographed on silica gel to give 2-(1-adamantyl)-4-methylquinoline in 90% (125 mg) yield [33].

1,1,2,2-Tetraphenyldisilane (TPDS) is a new stable radical reagent (crystalline solid); no decomposition was observed for 3 months under air at room temperature, and it is sufficiently stable in ethanol. Treatment of alkyl bromides, TPDS, and heteroaromatics, which were activated by protonation with trifluoroacetic acid, in ethanol at refluxing temperature gives good to moderate yields of the corresponding alkylated hetroaromatics, as shown in eq. 5.15 [36–38].

$$\mathbf{25} + \mathbf{26} \xrightarrow[\mathrm{EtOH}\ \Delta]{\mathrm{AIBN},\ \mathrm{Ph_4Si_2H_2}} \mathbf{27}\ \ 55\% \tag{5.15}$$

Experimental procedure 7 (eq. 5.15).

AIBN (0.45 mmol) was added 5 times over 8 h (2 h intervals) to a solution of trifluoroacetate salt of heteroaromatics (1.5 mmol), alkyl halide (0.3 mmol), and 1,1,2,2-tetraphenyldisilane (0.45 mmol) in ethanol (4 ml) at refluxing temperature. 1,1,2,2-Tetraphenyldisilane (0.45 mmol) was added again after 4 h. The solution was stirred for 22 h at the same temperature. The resulting solution was quenched with sat. aq. $NaHCO_3$. The organic layer was extracted with ethyl acetate and dried over Na_2SO_4. After removal of the solvent, the residue was chromatographed on silica gel to give the product [37].

Intramolecular radical cyclization of *N*-haloalkyl pyridinium salts (**28**) with Bu_3SnH readily takes place to generate cyclic pyridinium salts (**29**) (eq. 5.16), although intermolecular radical alkylation with Bu_3SnH does not work due to the direct reduction of the alkyl radical [39]. Since it is well known that natural quinolidine and indolidine alkaloids bear potent biologically activity, the present reaction becomes a ring-construction method of these alkaloids, by the reduction of the formed cyclic pyridinium salts. Here, the α-isobutyronitrile radical formed from AIBN play an important role as a hydrogen acceptor to form the cyclization adduct.

Bu$_3$SnH, AIBN
CH$_3$CN / THF
Δ

28 → **29**

$n = 1$ 65%
$n = 2$ 60%
$n = 3$ 58%

natural indole alkaloid

(5.16)

Eq. 5.17 shows *ipso* substitution of imidazoles (**30**) at the 2-position by an sp^3 carbon-centered radical. Here, the tosyl group plays an important role in both activation of the *ipso* 2-position and as an effective leaving group, Ts$^\cdot$ [40–46]. The phosphoryl group also shows the same leaving ability as the tosyl group. Preparation of [1.2-b]-fused bicyclic pyrazoles (**33**) from 1-(phenylselenoalkyl)pyrazoles (**32**) also works well, as shown in eq. 5.18.

Bu$_3$SnH, AIBN
Δ

30 → **31**

$n = 1$ 52%
$n = 2$ 48%
$n = 3$ 63%

Ts$^\cdot$ $\xrightarrow{Bu_3SnH}$ TsH

(5.17)

ACCN,
Bu$_3$SnH
C$_6$H$_5$CH$_3$ Δ

32 → **33**

$n = 1$ 38%
$n = 2$ 63%
$n = 3$ 37%

(5.18)

5.3 PHOTOCHEMICAL CONDITIONS

It is well known that benzophenone generates a biradical through $n-\pi^*$ electronic transition under irradiation (~340 nm). Irradiation of a mixture of 1,4-benzoquinone (**34**) and aromatic aldehydes in the presence of benzophenone generates 2-aroyl-1,4-dihydroxybenzene (**35**) [47–49]. This reaction comprises of the abstraction of a formyl hydrogen atom of an aromatic aldehyde by the oxygen-centered radical of the benzophenone biradical to form an aroyl radical and a 1,1-diphenylhydroxymethyl radical, and addition of the nucleophilic aroyl radical to 1,4-benzoquinone (**34**) to form a phenoxyl radical derivative, which finally abstracts a hydrogen atom from an aromatic aldehyde to give 2-aroyl-1,4-dihydroxybenzene (**35**) as shown in eq. 5.19. As an environmentally benign method, this reaction also proceeds in supercritical CO_2 under the same conditions without toxic benzene.

hν (~340 nm)

pinacol coupling

X-C6H4-CHO Hg-hν

C_6H_6, Ph_2CO, r.t.

34 **35**

$X = OCH_3$ 72%
$= CH_3$ 65%
$= CO_2CH_3$ 58%

(5.19)

Experimental procedure 8 (eq. 5.19).

A benzene (240 ml) solution of 1,4-benzoquinone (4.8 g, 44.4 mmol), *o*-chlorobenzaldehyde (20 ml, 346.7 mmol), and benzophenone (2.5 mmol) was bubbled with nitrogen gas for 15 min. Then the mixture was irradiated with a high pressure mercury lamp under a nitrogen atmosphere for 5 days. After the reaction, the solvent was removed and the residue was chromatographed on silica gel to give *o*-chlorobenzoyl-1,4-dihydroxybenzene in 78% yield [47].

5.4 OTHERS

In the previous sections, the reactions of nucleophilic alkyl and acyl radicals with electron-deficient aromatics via SOMO–LUMO interaction have been described. At this point, we introduce the reactions of electrophilic alkyl radicals and electron-rich aromatics via SOMO–HOMO interaction, though the study is quite limited. Treatment of ethyl iodoacetate with triethylborane in the presence of electron-rich aromatics (**36**) such as pyrrole, thiophene, furan, etc. produces the corresponding ethyl arylacetates (**37**) [50–54].

This reaction comprises firstly of S_H2 reaction on the iodine atom of ethyl iodoacetate by an ethyl radical, formed from triethylborane and molecular oxygen, to form a more stable α-ester radical and ethyl iodide. Electrophilic addition of the α-ester radical to electron-rich aromatics (**36**) forms an adduct radical, and finally abstraction of a hydrogen atom from the adduct by the ethyl radical or oxidation by molecular oxygen generates ethyl arylacetate (**37**), as shown in eq. 5.20. Here, a nucleophilic ethyl radical does not react with electron-rich aromatics (**36**), while only an electrophilic α-ester radical reacts with electron-rich aromatics via SOMO–HOMO interaction.

$ICH_2CO_2C_2H_5$, Et_3B, DMSO

36 → **37** ($CH_2CO_2C_2H_5$) (5.20)

X = NH 47%
= O 60%
= S 56%

Experimental procedure 9 (eq. 5.20).

To a mixture of pyrrole (10 mmol), ethyl iodoacetate (1 mmol), and DMSO (5 ml) was added Et_3B (1 ml, 1 M hexane solution) under aerobic conditions. After 45 min, Et_3B (1 ml) was added again. After the reaction, water was added to the mixture, and the organic layer was extracted with ether, and dried over Na_2SO_4. After filtration and removal of the solvent, the residue was chromatographed on silica gel to obtain the product in 47% yield [51].

Irradiation of ethyl iodoacetate in the presence of pyrrole (**38**) and $Na_2S_2O_3$ provides ethyl pyrrolylacetate (**39**) as shown in eq. 5.21.

38 + $ICH_2CO_2C_2H_5$ —(Hg-$h\nu$, $Na_2S_2O_3$)→ **39** ($CH_2CO_2C_2H_5$) 90% (5.21)

40 ($NHNH_2$·HCl) —($Mn(OAc)_3$, furan/thiophene, Δ)→ **41** (5.22)

X = O 60%
X = S 70%

Arylation of furan and thiophene with arylhydrazine (**40**) and $Mn(OAc)_3$ gives 2-aryl substituted furan and thiophene (**41**) as shown in eq. 5.22.

REFERENCES

1. F Minisci, *Synthesis*, 1973, 1.
2. F Minisci, E Vismara and F Fontana, *Heterocycles*, 1989, **28**, 489.
3. F Coppa, F Fontana, F Minisci and E Vismara, *Tetrahedron*, 1991, **47**, 7343.
4. T Caronna, G Fronza, F Minisci, O Porta and GP Gardini, *J. Chem. Soc., Perkin Trans. 2*, 1972, 1477.

5. F Minisci, C Giordano, E Vismara, S Levi and V Tortelli, *J. Am. Chem. Soc.*, 1984, **106**, 7146.
6. M Tada and S Totoki, *J. Heterocyclic Chem.*, 1988, **25**, 1295.
7. S Araneo, F Fontana, F Minisci, F Recupero and A Serri, *Tetrahedron Lett.*, 1995, **36**, 4307.
8. VD Piaz, MP Giovannoni and G Ciciani, *Tetrahedron Lett.*, 1993, **34**, 3903.
9. C Lampard, JA Murphy and N Lewis, *Tetrahedron Lett.*, 1991, **32**, 4993.
10. JJ Rao and WC Agosta, *Tetrahedron Lett.*, 1992, **33**, 4133.
11. F Fontana, F Minisci, MCN Barbosa and E Vismara, *J. Org. Chem.*, 1991, **56**, 2866.
12. F Coppa, F Fontana, E Lazzarini, F Minisci, G Pianese and L Zhao, *Tetrahedron Lett.*, 1992, **33**, 3057.
13. KB Hansen, SA Springfield, R Desmond, PN Devine, EJJ Grabowski and PJ Reider, *Tetrahedron Lett.*, 2001, **42**, 7353.
14. YB Zeletchonok, VV Zorin and MC Klyavlin, *Tetrahedron Lett.*, 1988, **29**, 5037.
15. F Fontana, F Minisci and E Vismara, *Tetrahedron Lett.*, 1988, **29**, 1975.
16. E Baciocchi, E Muraglia and G Sleiter, *J. Org. Chem.*, 1992, **57**, 6817.
17. H Togo, M Aoki and M Yokoyama, *Chem. Lett.*, 1991, 1691.
18. H Togo, M Aoki, T Kuramochi and M Yokoyama, *J. Chem. Soc., Perkin Trans. 1*, 1993, 2417.
19. H Togo, R Taguchi, K Yamaguchi and M Yokoyama, *J. Chem. Soc., Perkin Trans. 1*, 1995, 2135.
20. RM Moriarty, JS Khosrowshahi, RM Miller, JF Andersen and R Gilardi, *J. Am. Chem. Soc.*, 1989, **111**, 8953.
21. YH Kim, DH Lee and SG Yang, *Tetrahedron Lett.*, 1995, **36**, 5027.
22. N Arai and K Narasaka, *Bull. Chem. Soc. Jpn*, 1995, **68**, 1707.
23. DHR Barton, B Garcia, H Togo and SZ Zard, *Tetrahedron Lett.*, 1986, **27**, 1327.
24. E Castagino, S Corsano, DHR Barton and SZ Zard, *Tetrahedron Lett.*, 1986, **27**, 6337.
25. H Togo, M Fujii and T Ikuma, *Tetrahedron Lett.*, 1991, **32**, 3377.
26. H Togo, S Ishigami, M Fujii, T Ikuma and M Yokoyama, *J. Chem. Soc., Perkin Trans. 1*, 1994, 2931 see also 2407.
27. W He, H Togo, H Ogawa and M Yokoyama, *Heteroatom Chem.*, 1997, **8**, 411.
28. W He and H Togo, *Tetrahedron Lett.*, 1997, **38**, 5541.
29. W He, H Togo, Y Waki and M Yokoyama, *J. Chem. Soc., Perkin Trans. 1*, 1998, 2425.
30. H Togo, W He, Y Waki and M Yokoyama, *Synlett*, 1998, 700.
31. DJ O'Neill and P Helquist, *Org. Lett.*, 1999, **1**, 1659.
32. H Togo, K Hayashi and M Yokoyama, *Chem. Lett.*, 1991, 2063.
33. H Togo, K Hayashi and M Yokoyama, *Chem. Lett.*, 1993, 641.
34. H Togo, K Hayashi and M Yokoyama, *Bull. Chem. Soc. Jpn*, 1994, **67**, 2522.
35. VM Barrasa, AG Viedma, C Burgos and JA Builla, *Org. Lett.*, 2000, **2**, 3933.
36. O Yamazaki, H Togo, S Matsubayashi and M Yokoyama, *Tetrahedron Lett.*, 1998, **39**, 1921.
37. O Yamazaki, H Togo, S Matsubayashi and M Yokoyama, *Tetrahedron*, 1999, **55**, 3735.
38. H Togo, M Sugi and K Toyama, *C. R. Acad. Sci. Paris, Chim.*, 2001, **4**, 539.
39. JA Murphy and MS Sherburm, *Tetrahedron Lett.*, 1990, **31**, 1625 see also 3495.
40. F Aldabbagh and WR Bowman, *Tetrahedron Lett.*, 1997, **38**, 3793.
41. MLEN Da Mata, WB Motherwell and F Ujjainwalla, *Tetrahedron Lett.*, 1997, **38**, 137 see also 141.
42. CJ Moody and CL Norton, *J. Chem. Soc., Perkin Trans. 1*, 1997, 2639.
43. B Alcaide and AR Vicente, *Tetrahedron Lett.*, 1998, **39**, 6589.
44. F Aldabbagh and WR Bowman, *Tetrahedron*, 1999, **55**, 4109.
45. DLJ Clive and S Kang, *Tetrahedron Lett.*, 2000, **41**, 1315.
46. SM Allin, WRS Barton, WR Bowman and T McInally, *Tetrahedron Lett.*, 2002, **43**, 4191.
47. GA Kraus and M Kirihara, *J. Org. Chem.*, 1992, **57**, 3256.
48. GA Kraus and P Liu, *Tetrahedron Lett.*, 1994, **35**, 7723.
49. R Pacut, ML Grimm, GA Kraus and JM Tanko, *Tetrahedron Lett.*, 2001, **42**, 1415.
50. E Baciocchi, E Muraglis and C Villani, *Synlett*, 1994, 821.
51. E Baciocchi and E Muraglia, *Tetrahedron Lett.*, 1993, **34**, 5015.
52. JH Byers, JE Campbell, FH Knapp and JG Thissell, *Tetrahedron Lett.*, 1999, **40**, 2677.
53. JH Byers, JE Campbell, FH Knapp and JG Thissell, *Tetrahedron Lett.*, 1999, **40**, 2677.
54. AS Demir, O Reis and M Emrullahoglu, *Tetrahedron*, 2002, **58**, 8055.

– 6 –

Intramolecular Hydrogen-Atom Abstraction

6.1 HYDROGEN-ATOM ABSTRACTION BY OXYGEN-CENTERED RADICALS: BARTON REACTION

The Barton reaction is as follows. Irradiation of alkyl nitrite, RO–NO, with a mercury lamp generates the corresponding alkoxyl radical, RO˙ through homolytic cleavage of the weak O–N bond. The formed reactive alkoxyl radical abstracts a hydrogen atom to form a carbon-centered radical and a strong O–H bond, via 6-membered transition state (1,5-H shift) or 7-membered transition state (1,6-H shift). The formed carbon-centered radical reacts on the nitrogen atom of alkyl nitrite to give δ-nitrosoalcohol or ε-nitrosoalcohol, and alkoxyl radical again. This reaction takes place through a chain reaction. Generally, nitroso compounds tautomerize to oximes, when they have an α-hydrogen atom. The general scheme of the Barton reaction and a typical example are shown in eqs. 6.1 and 6.2, respectively [1–4]. The 1,5-H shift is approximately 10 times faster than the 1,6-H shift.

Barton reaction

(6.1)

(6.2)

In eq. 6.2, the initially formed axial oxygen-centered radical at the 6-position of the steroid abstracts a hydrogen atom of the axial methyl group at the 10-position through the 6-membered transition state with chair form. The driving force of this hydrogen atom abstraction from the inert methyl group is the formation of a stronger O–H bond ($\sim$103 kcal/mol) than C–H bond (89 $\sim$ 96 kcal/mol) in methyl or alkyl groups. The Barton reaction is a typically radical specific reaction, and is very useful for the introduction of a functional group to the inert alkyl side chain. Barton discovered this reaction mainly with steroids, and applied it to develop steroidal pharmaceutical compounds. In recognition of the major contribution these reactions made to the stereochemistry of organic molecules, Barton, along with Hassel, received the Nobel Prize in 1969. I greatly respected him as a chemist and a gentleman.

When compound (**3**) was treated under an oxygen atmosphere, the formed carbon-centered radical via 1,5-H shift reacts with molecular oxygen to form ultimately alkyl nitrate (**4**) via the formation of peroxy nitrite (eq. 6.3).

(6.3)

Instead of alkyl nitrite, other alkoxyl radical precursors such as ROOH, ROOR′, ROI, ROCl, etc. can also be used for the same type of reaction. The high reactivity of these compounds comes from the weak bond dissociation energies in O–O, O–I, and O–Cl bonds. Another simple method is as follows. Photolytical treatment of alcohol (**5**) with NIS (*N*-iodosuccinimide) provides the tetrahydrofuran skeleton (**6**), through the formation of alkyl hypoiodite (ROI), homolytic cleavage of the O–I bond to form an alkoxyl radical, 1,5-H shift to form a carbon-centered radical, reaction with ROI to form δ-iodoalcohol, and finally ionic cyclization to form a tetrahydrofuran skeleton, together with HI. Eq. 6.4 shows the formation of TBS-protected γ-lactol (**6**) from TBS-protected alcohol (**5**) under photolytic conditions [5–8].

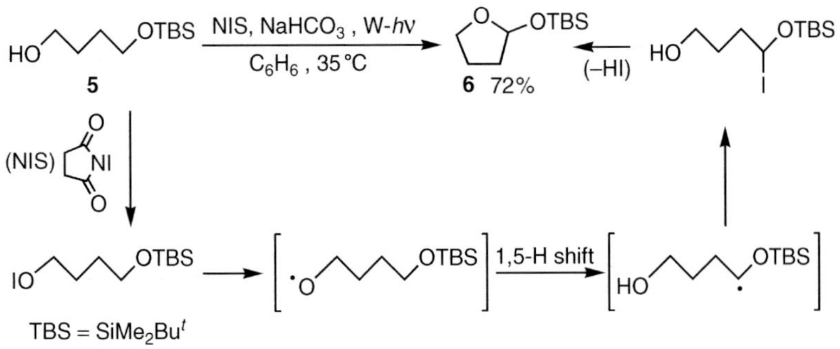

(6.4)

TBS-Protected γ-lactol is obtained through the formation of hypoiodite, followed by the formation of an alkoxyl radical, 1,5-H shift, and then formation of δ-iodoalcohol, and finally polar cyclization. The Fenton system with ROOH and Fe^{2+} also generates alkoxyl radical, $RO^{\cdot}$. These reactions indicate that the Barton (-type) reactions are remote functionalization of non-activated C–H bonds at the δ- or ε-position.

Sulfenate ester (RO-SAr) (**7**) has a weak S–O bond, so the corresponding alkoxyl radical is formed under photolytic conditions with a mercury lamp, followed by 1,5-H shift where the formed carbon-centered radical reacts with the starting sulfenate ester to give δ-hydroxy sulfide (**8**) via a chain pathway, as shown in eq. 6.5 [9, 10].

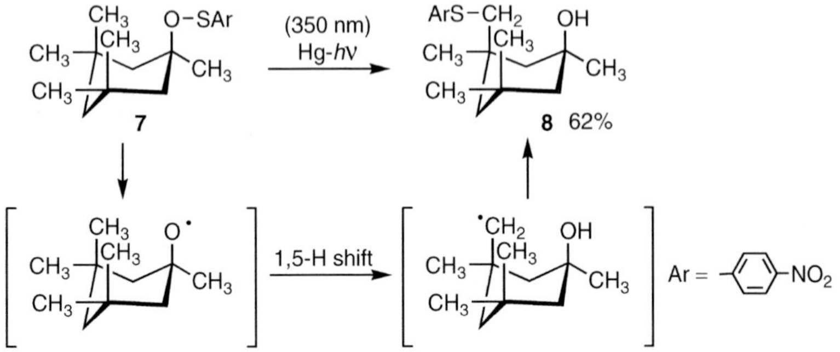

(6.5)

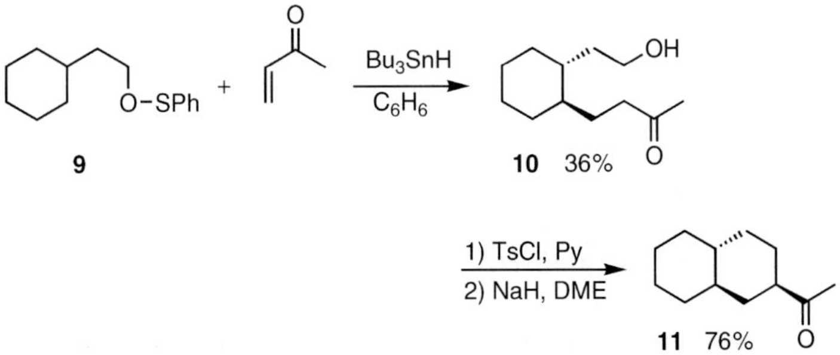

(6.6)

Under the same photolytic conditions together with Bu_3SnH, annulation of the cyclohexane ring can be carried out through the generation of an alkoxyl radical from sulfenate ester (**9**), 1,5-H shift, nucleophilic addition of the formed carbon-centered radical to methyl vinyl ketone, and finally, treatment with a polar sequence method as shown in eq. 6.6.

Since the 1960s, $Pb(OAc)_4$ has been used for the generation of alkoxyl radicals from the alcohols. However, it is a toxic reagent, so recently less-toxic hypervalent iodine reagents, such as (diacetoxyiodo)benzene along with molecular iodine, have been used for the same types of reaction. Treatment of alcohol (**12**) with (diacetoxyiodo)benzene in the presence of molecular iodine under irradiation with a tungsten lamp generates tetrahydrofuran derivative (**13**), by means of the formation of hypoiodite, generation of the alkoxyl radical, 1,5-H shift, formation of a carbon cation by the oxidation with (diacetoxyiodo)benzene, and finally, polar cyclization [11–20]. Eq. 6.7 shows the formation of spiro-ketal (**13**). In this case, ultrasonic conditions can be also used instead of photolytic conditions.

W-$h\nu$, $PhI(OAc)_2$, I_2, c-C_6H_{12}, 24 h; **12** → **13** 53%; (−e) (6.7)

In nature, there are many kinds of polycyclic ether antibiotics, so this method can be used for the construction of polycyclic ethers. Eq. 6.8 shows the preparation of spiro sugar ketals (**15**) at anomeric position [21–23]. Biologically attractive spiro-nucleosides at the anomeric position may be prepared by this procedure.

W-$h\nu$, $PhI(OAc)_2$, I_2, c-C_6H_{12}, 40 °C, 1 h; **14** → **15a** 51% + **15b** 17%; AcOH (6.8)

Since aromatic carboxylic acids ($ArCO_2H$) are more acidic than acetic acid, (diacetoxyiodo)benzene is not effective for the formation of carboxyl radicals, $ArCO_2^{\bullet}$. However, treatment of aromatic carboxylic acid (**16**) with [*bis*(trifluoroacetoxy)iodo]-benzene in the presence of molecular iodine under irradiation with a tungsten lamp effectively generates benzo-γ-lactone (**17**) [24]. Decarboxylation of the aromatic carboxyl radical is not so fast ($\sim 10^5\ s^{-1}$), so the 1,5-H shift readily occurs.

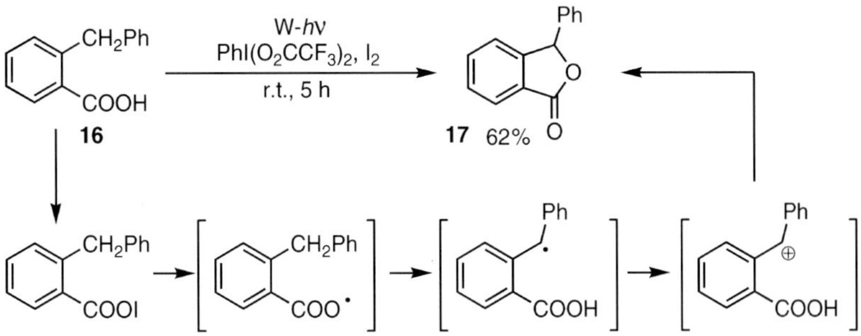

(6.9)

The 1,5-H shift or 1,6-H shift of alkoxyl radicals generated from *prim*-alcohols or *sec*-alcohols usually occurs easily, and so β-cleavage of these alkoxyl radicals does not proceed. However, β-cleavage reaction ($\sim 10^5\ s^{-1}$, 60 °C) of the alkoxyl radicals derived from *tert*-alcohols becomes faster than that of the 1,5-H shift reaction. Part of the reason behind this β-cleavage reaction is the steric hindrance in the radical center of the *tert*-alkoxyl radical. Reduction of steric hindrance happens readily via the β-cleavage to generate a carbon-centered radical and a stable ketone. Eq. 6.10 shows the treatment of unsaturated 8α-decanol (**18**) with (diacetoxyiodo)benzene in the presence of molecular iodine under irradiation with a tungsten lamp to generate 1-iodomethyl bicyclo[5.3.0]-deca-2-one (**19**), through β-cleavage of the formed *tert*-alkoxyl radical and 5-*exo-trig* cyclization [25–35]. This method can be used for the preparation of ring-expanded steroidal analogues from steroidal alcohols.

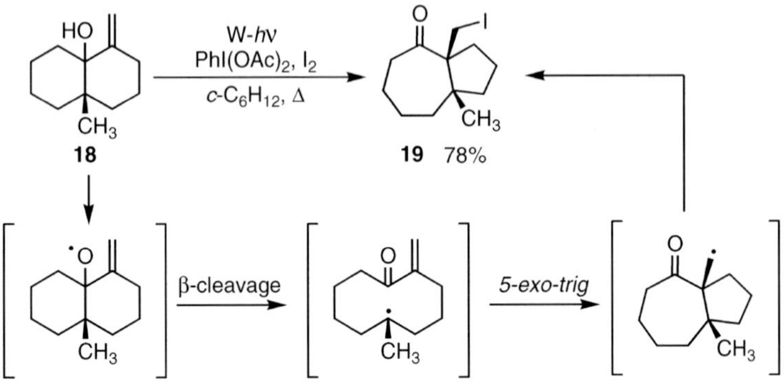

(6.10)

Synthetic application of the β-cleavage reaction to sugars is useful for the preparation of new types of sugars. For example, as shown in eq. 6.11, the same treatment of protected glucose (**20**) bearing Bn group at 2, 3, and 4 positions, with iodosylbenzene and molecular iodine generates an arabinose derivative (**21**), through the formation of an anomeric alkoxyl radical, followed by β-cleavage reaction of the C_1-C_2 bond, and then polar cyclization by the 6-OH group. These new types of sugars can be used as chiral building blocks for organic synthesis [36–43].

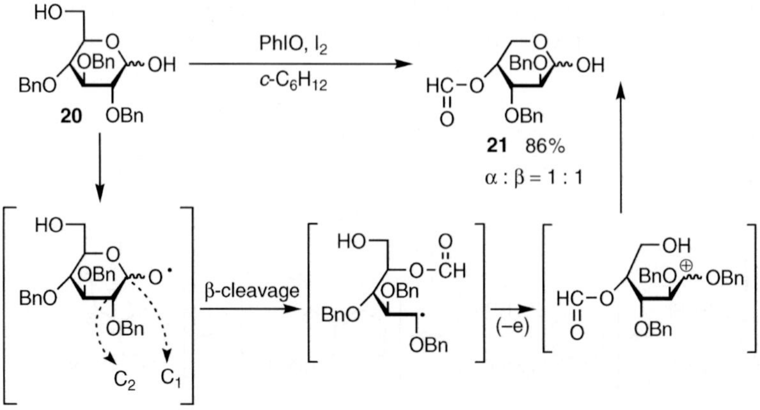

(6.11)

Experimental procedure 1 (eq. 6.11).

To a mixture of glucopyranose (0.22 mmol) in dry cyclohexane (24 ml) under an argon atmosphere were added iodosylbenzene (0.43 mmol) and iodine (0.22 mmol) at room temperature. The mixture was stirred for 20 h at the same temperature. After the reaction, the reaction mixture was poured into water and extracted with ether. The organic layer was washed with aq. sodium thiosulfate solution, water, and dried over Na_2SO_4. After removal of the solvent, the residue was chromatographed on silica gel (eluent: hexane/ethyl acetate = 65/35) to give D-arabinose derivative in 86% yield [68].

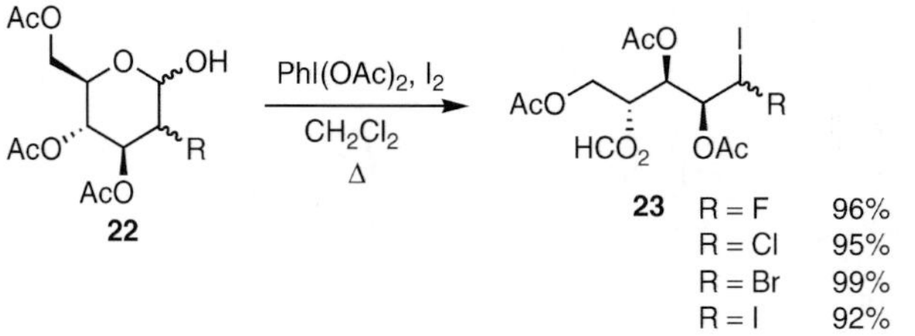

(6.12)

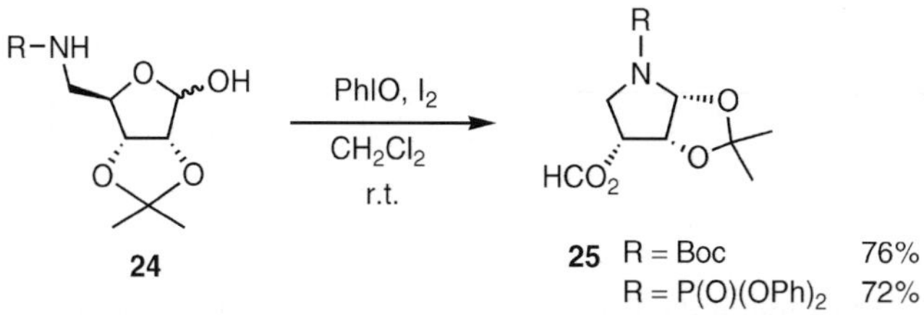

(6.13)

Treatment of protected glucose halohydrin (**22**) with (diacetoxyiodo)benzene and molecular iodine in refluxing dichloromethane under irradiation with a tungsten lamp generates 1-halo-1-iodo-arabinitol derivative (**23**) (eq. 6.12). Amino sugars (aza sugars) of the piperidine and pyrrolidine types (**25**) can also be obtained by the same treatment of 5-amino-5-deoxyfuranopentose (**24**) and 6-amino-6-deoxyhexose respectively, by means of β-cleavage and subsequent polar cyclization as shown in eq. 6.13 [44, 45]. Since it is well known that nojirimycin, a piperidine-type sugar, has potent biological activity as glycosidase inhibitor, the present method for the preparation of new aza sugars is attractive. One-pot synthesis of aryl glycine from the protected serine (**26**) with (diacetoxyiodo)benzene and molecular iodine in the presence of arene proceeds effectively as shown in eq. 6.14.

PhCONH, CO_2CH_3, OH, **26** — W-$h\nu$, $PhI(OAc)_2$, I_2, naphthalene, CH_2Cl_2, r.t. → [PhCONH, CO_2CH_3, OAc] — $BF_3{\cdot}Et_2O$ → PhCONH, CO_2CH_3, **27** 63% (6.14)

H, H, O, HO, H, O, **28** — Py, $Pb(OAc)_4$, $Cu(OAc)_2$, C_6H_6 Δ → HCO_2, H, O, **29** 82% (6.15)

Treatment of lactol (**28**) with $Pb(OAc)_4$ and $Cu(OAc)_2$ in refluxing benzene provides the corresponding δ-vinyl carbonyl product, 5-vinyl[4.3.0]nonane as shown in eq. 6.15.

6.2 HYDROGEN-ATOM ABSTRACTION BY NITROGEN-CENTERED RADICALS: THE HOFMANN–LÖFFLER–FREYTAG REACTION

Heating or photolytic treatment of *N,N*-dialkyl-*N*-haloamine in sulfuric acid or trifluoroacetic acid, followed by neutralization with a base, generates a pyrrolidine or piperidine skeleton. This is the Hofmann–Löffler–Freytag reaction, and the reaction comprises of the formation of an electrophilic aminium radical, 1,5-H shift (6-membered transition state) or 1,6-H shift (7-membered transition state), formation of 4-haloalkyl ammonium or 5-haloalkyl ammonium, and its polar cyclization by neutralization with a base. Eq. 6.16 shows the formation of *N*-alkyl pyrrolidine (**31**) from *N*-chloro-*N*-alkyl-*N*-butylamine (**30**) in sulfuric acid [46, 47].

$R{-}\overset{\oplus}{N}H(CH_2)_3CH_3$ (N–Cl) **30** — Δ or $h\nu$, H_2SO_4 → N–R **31**

30 ↓ [$R{-}\overset{\oplus}{\dot{N}}H(CH_2)_3CH_3$] — 1,5-H shift → [$R{-}\overset{\oplus}{N}H_2{-}(CH_2)_3{-}CH_2^{\bullet}$] — chain pathway → $R{-}\overset{\oplus}{N}H_2{-}(CH_2)_3{-}CH_2Cl$ — neutralization → **31** (6.16)

The driving force in this reaction is the weak N–Cl bond in the starting haloamine and the formation of a strong N–H bond in the product. Moreover, the N–H bond dissociation energy in amine is about 103 kcal/mol, while that in ammonium is approximately 112 kcal/mol. So, under strong acidic conditions, an electrophilic and reactive aminium radical abstracts a hydrogen atom to form ammonium, through 5- or 6-membered transition state (1,5-H shift and 1,6-H shift). However, this reaction requires quite acidic conditions, so it is not very useful. Recently, hypervalent iodine reagents such as (diaceoxyiodo)benzene, have become popular, as the reactions can be carried out under neutral conditions. Thus, photolytic treatment of *N*-nitroamine (**32**) or *N*-cyanoamine with (diacetoxyiodo)benzene in the presence of molecular iodine generates a pyrrolidine skeleton (**33**) as shown in eq. 6.17, through the formation of the N–I bond and its homolytic cleavage to form an aminyl radical, 1,5-H shift, formation of δ-iodoamino compound, and finally polar cyclization [48–53]. It is important to introduce a cyano or nitro group at the amino group, in order to increase the acidity and electrophilicity of the formed nitrogen-centered radical. The homochiral 7-oxa-2-azabicyclo[2.2.1]heptane (**35**) ring can be constructed by the same photolytic treatment of a phosphoramidate derivative (**34**) of carbohydrate with iodosylbenzene and iodine, as shown in eq. 6.18.

CH_3 C_8H_{17} CH_3 NH–NO_2 **32** — W-*hν*, $PhI(OAc)_2$, I_2, 90 °C, *c*-C_6H_{12}, 1 h → CH_3 C_8H_{17} N–NO_2 **33**

CH_3 I–N–NO_2 → [CH_3 $^{\bullet}$N–NO_2] — 1,5-H shift → [$^{\bullet}CH_2$ HN–NO_2] → ICH_2 HN–NO_2 → **33** (6.17)

$(PhO)_2(O)P$ NH O H O O OMe O **34** — PhIO, I_2, CH_2Cl_2 → $P(O)(OPh)_2$ N O OMe O O O **35** 75% (6.18)

Experimental procedure 2 (eq. 6.18).

A solution of amide (0.053 mmol) in a mixture of CH_2Cl_2 (1.5 ml) and cyclohexane (1.5 ml) containing iodosylbenzene (0.106 mmol) and iodine (0.064 mmol) was irradiated with a tungsten lamp for 80 min at room temperature. After the reaction, the mixture was poured into chloroform, and washed with aq. Na_2SO_3 solution, and dried over Na_2SO_4. After removal of the solvent, treatment of the residue by column chromatography (eluent: hexane/ethyl acetate = 85/15) produces cyclized sugar in 75% yield [53].

When this system is applied to sulfonamide (**36**) bearing an aromatic ring at the δ-position, the corresponding *N*-sulfonyl α-arylpyrrolidine (**37**) is obtained in good yield [54]. Thus, this method can be used for the preparation of aza-*C*-nucleosides.

36 (NHZ) → [W-$h\nu$, $PhI(OAc)_2$, I_2; $ClCH_2CH_2Cl$, 60 °C, 2 h] → **37** 84% (N–Z)

$Z = SO_2–C_6H_4–NO_2$ (6.19)

Experimental procedure 3 (eq. 6.19).

To a mixture of sulfonamide (0.5 mmol) in 1,2-dichloromethane (7 mL) were added (diacetoxyiodo)benzene (0.8 mmol) and iodine (0.5 mmol) under an argon atmosphere. The mixture was warmed in the range of 60 to 70 °C under irradiation with a tungsten lamp for 2 h. After the reaction, the mixture was poured into chloroform, and washed with aq. Na_2SO_3 solution, dried over Na_2SO_4. After removal of the solvent, the residue was chromatographed on silica gel with preparative TLC to obtain a pyrrolidine derivative in 84% yield [54].

The same treatment of *N*-methyl(*o*-methyl)arenesulfonamide (**38**) generates the corresponding 1,2-benzisothiazoline-3-one 1,1,-dioxide (*N*-methylsaccharins) (**39**), through triple 1,5-H shift and hydrolysis of the formed *N*-methyl(*o*-triiodomethyl)arenesulfonamides [55]. This method is very useful, since biologically attractive 1,2-benzisothiazoline-3-one 1,1-dioxides bearing various kinds of substituents on the aromatic ring can be prepared under neutral conditions.

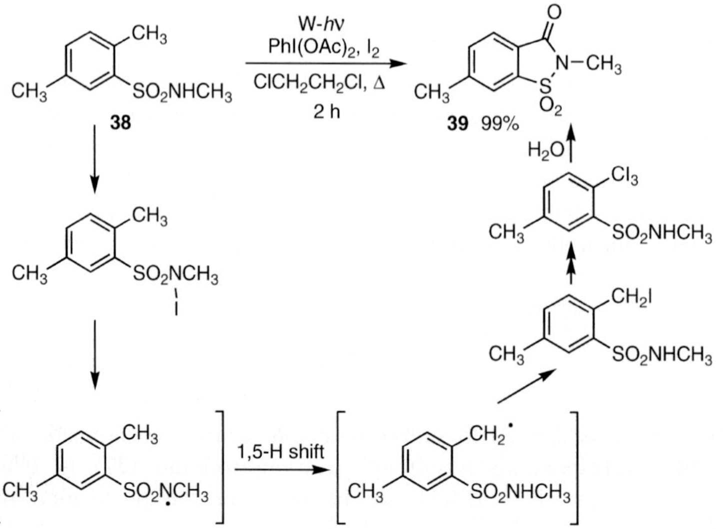

(6.20)

Experimental procedure 4 (eq. 6.20).

To a mixture of sulfonamide (0.5 mmol) in 1,2-dichloromethane (7 mL) were added (diacetoxyiodo)benzene (1.5 mmol) and iodine (0.5 mmol) under an argon atmosphere. The mixture was warmed to refluxing conditions under irradiation with a tungsten lamp for 2 h. After the reaction, the mixture was poured into chloroform, and washed with aq. Na_2SO_3 solution, dried over Na_2SO_4. After removal of the solvent, the residue was chromatographed on silica gel with preparative TLC to obtain *N*-methylsaccharin in 99% yield [55].

6.3 HYDROGEN-ATOM ABSTRACTION BY CARBON-CENTERED RADICALS

Since the Barton reaction and the Hofmann–Löffler–Freytag reaction generate very reactive oxygen-centered and nitrogen-centered radicals respectively, the next 1,5- and 1,6-hydrogen atom abstraction reaction readily happens. However, 1,5-H shift does not proceed effectively by carbon-centered radicals, because there is not so much energy difference between the C–H bond before and after 1,5-H shift. So the reactions are quite limited. Eq. 6.21 shows iodine transfer from reactive 1-iodoheptyl phenyl sulfone (**40**) to a mixture of 5-iodoheptyl phenyl sulfone (**41a**) and 6-iodoheptyl phenyl sulfone (**41b**) initiated by benzoyl peroxide, through 1,5-H shift by an sp^3 carbon-centered radical [56–58].

(6.21)

However, the sp^3 carbon-centered radical does not generally give rise to 1,5-H shift, although the sp^2 carbon-centered radical can be used for 1,5-H shift, since it is more reactive than the sp^3 carbon-centered radical and it can form a strong $C(sp^2)$-H bond (10 ~ 15 kcal/mol stronger than that of $C(sp^3)$-H). Eq. 6.22 shows the product distribution of a direct reduction product (**43a**) and an indirect reduction product (**43b**) via 1,5-H shift, in the reaction of arylbromide (**42**) with Bu_3SnD. It suggests that more than half of the amount of reduction product is formed via 1,5-H shift [59].

X	43a	43b
O	21 %	32 %
CH_2	11 %	67 %
S	20 %	30 %
NH	55 %	45 %

(6.22)

(6.23)

As a synthetic use of 1,5-H shift by an sp^2 carbon-centered radical, treatment of *o*-bromobenzyl 4-*tert*-butylcyclohexyl ether (**44**) with Bu_3SnH in the presence of AIBN generates 4-*tert*-butylcyclohexanone (**45**) by means of 1,5-H shift by a phenyl radical, followed by β-cleavage, as shown in eq. 6.23 [60]. The reaction looks like oxidation.

Treatment of the following indole amides (**46**) and (**48**) with Bu_3SnH in the presence of AIBN provides spiroindolenine (**47**) and hexahydropyrrolo[3,4-b]indole (**49**), through

1,5-H shift followed by 5-*exo-trig* and 5-*endo-trig* cyclization, respectively (eqs. 6.24 and 6.25) [61, 62].

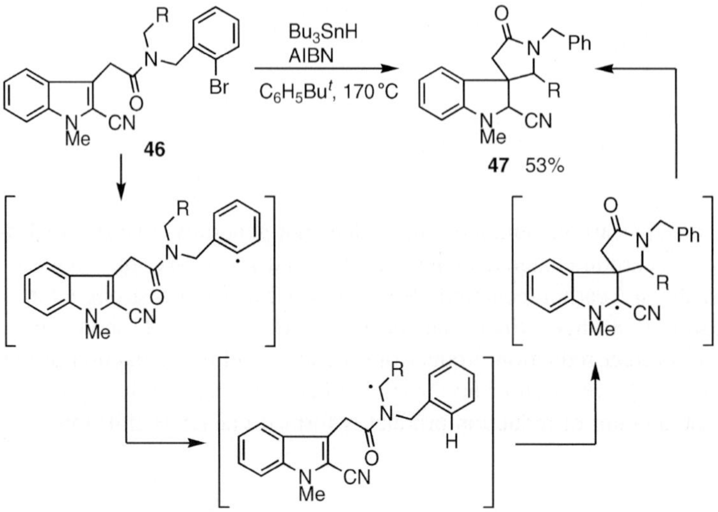

(6.24)

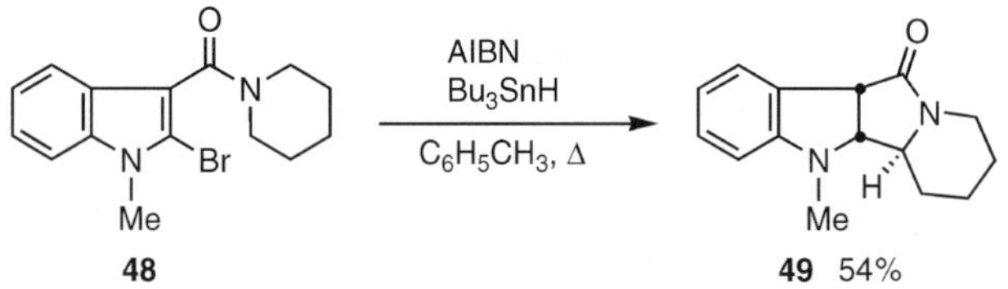

(6.25)

Instead of a hydrogen atom transfer, 1,5-Ar transfer by an sp^3 carbon-centered radical (eq. 6.26) and 1,4-Ar transfer by an sp^2 carbon-centered radical (eq. 6.27) proceeds via radical *ipso*-substitution [63–67]. Thus, treatment of *N*-methyl-*N*-(2-bromophenyl)-1-naphthalenesulfonamide (**52**) with Bu_3SnH/AIBM, $(Me_3Si)_3SiH$/AIBN, and $Ph_4Si_2H_2$/AIBN generates good yields of biaryl (**53**) with high *ipso*-selectivity, especially with $(Me_3Si)_3SiH$/AIBN and $Ph_4Si_2H_2$/AIBN, respectively.

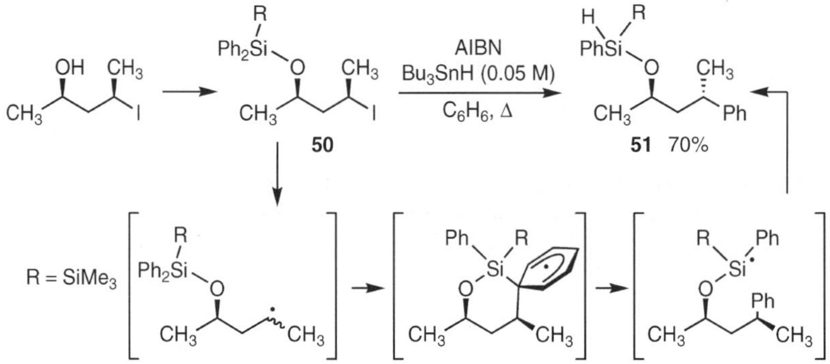

(6.26)

AIBN
MH
m-xylene, Δ

MH	53	54
Bu_3SnH	41%	37%
$(Me_3Si)_3SiH$	56%	22%
$Ph_4Si_2H_2$	60%	31%

(6.27)

Experimental procedure 5 (eq. 6.27).

A solution of AIBN (1.0 mmol) in *m*-xylene (10 ml) was added dropwise over 22 h to a refluxing solution of *N*-methyl-*N*-(2-bromophenyl)-1-naphthalenesulfonamide (0.5 mmol) and 1,1,2,2-tetraphenyldisilane (1.0 mmol) in *m*-xylene (2 ml). After a few hours, 1,1,2,2-tetraphenyldisilane (1.0 mmol) was added again. After the reaction, the solvent was removed and the residue was purified by preparative TLC (eluent: hexane/ethyl acetate = 4/1 ~ 10/1) to provide the biaryl [67].

REFERENCES

1. DHR Barton, JM Beaton, LE Geller and MM Pechet, *J. Am. Chem. Soc.*, 1960, **82**, 2640.
2. J Allen, RB Boar, JF McGhie and DHR Barton, *J. Chem. Soc., Perkin Trans. 1*, 1973, 2402.
3. J Kalvoda and K Heusler, *Synthesis*, 1971, 501.
4. ML Mihailovic and Z Cekovic, *Synthesis*, 1970, 209.
5. CE McDonald, TR Beebe, M Beard, D McMillen and D Selski, *Tetrahedron Lett.*, 1989, **30**, 4791.
6. Z Cekovic and D Ilijev, *Tetrahedron Lett.*, 1988, **29**, 1441.
7. Z Cekovic and MM Green, *J. Am. Chem. Soc.*, 1974, **96**, 3000.
8. Z Cekovic, L Dimitrijevic, G Djokic and T Srnic, *Tetrahedron*, 1979, **35**, 2021.
9. DJ Pasto and F Cottard, *Tetrahedron Lett.*, 1994, **35**, 4303.
10. G Petrovic and Z Cekovic, *Org. Lett.*, 2000, **2**, 3769.
11. H Togo, G Nogami and Y Hoshina, *Yuki Gosei Kagaku Kyokaishi*, 1997, **55**, 90.
12. H Togo and M Katohgi, *Synlett*, 2001, 565.
13. R Baker and MA Brimble, *Chem. Commun.*, 1985, 78.

14. R Baker, MA Brimble and JA Robinson, *Tetrahedron Lett.*, 1985, **26**, 2115.
15. MA Brimble and GM Williams, *Tetrahedron Lett.*, 1990, **31**, 3043.
16. K Furuta, T Nagata and H Yamamoto, *Tetrahedron Lett.*, 1988, **29**, 2215.
17. PD Armas, JI Concepcion, CG Francisco, R Hernandez, JA Salazar and E Suarez, *J. Chem. Soc., Perkin Trans. 1*, 1989, 405.
18. JI Concepcion, CG Francisco, R Hernandez, JA Salazar and E Suarez, *Tetrahedron Lett.*, 1984, **25**, 1953.
19. SCP Costa, MJSM Moreno, MLS Melo and ASC Neves, *Tetrahedron Lett.*, 1999, **40**, 8711.
20. LA Paquette, LQ Sun, D Friedrich and PB Savage, *Tetrahedron Lett.*, 1997, **38**, 195.
21. A Martin, JA Salazar and E Suarez, *Tetrahedron Lett.*, 1995, **36**, 4489.
22. A Martin, JA Salazar and E Suarez, *J. Org. Chem.*, 1996, **61**, 3999.
23. RL Dorta, A Martin, JA Salazar and E Suarez, *Tetrahedron Lett.*, 1996, **37**, 6021.
24. H Togo, T Muraki and M Yokoyama, *Tetrahedron Lett.*, 1995, **36**, 7089.
25. CG Francisco, R Freire, MS Rodriguez and E Suarez, *Tetrahedron Lett.*, 1987, **28**, 3397.
26. R Hernandez, SM Velazqueg and E Suarez, *J. Org. Chem.*, 1994, **59**, 6395.
27. CW Ellwood and G Pattenden, *Tetrahedron Lett.*, 1991, **32**, 1591.
28. CE Mowbray and G Pattenden, *Tetrahedron Lett.*, 1993, **34**, 127.
29. T Arencible, JA Salazar and E Suarez, *Tetrahedron Lett.*, 1994, **35**, 7463.
30. A Boto, C Betancor and E Suarez, *Tetrahedron Lett.*, 1994, **35**, 5509.
31. A Boto, C Betancor, R Hernandez, MS Rodriguez and E Suarez, *Tetrahedron Lett.*, 1993, **34**, 4865.
32. R Freire, R Hernandez, MS Rodriguez, E Suarez and A Perales, *Tetrahedron Lett.*, 1987, **28**, 981.
33. R Freire, JI Marrero, MS Rodriguez and E Suarez, *Tetrahedron Lett.*, 1986, **27**, 383.
34. R Hernandez, JJ Marrero and E Suarez, *Tetrahedron Lett.*, 1989, **30**, 5501.
35. R Hernandez and E Suarez, *J. Org. Chem.*, 1994, **59**, 2766.
36. PD Armas, CG Francisco and E Suarez, *J. Am. Chem. Soc.*, 1993, **115**, 8865.
37. PD Armas, CG Francisco and E Suarez, *Angew. Chem. Int. Ed. Engl.*, 1992, **31**, 772.
38. RL Dorta, A Martin, JA Salazar and E Suarez, *J. Org. Chem.*, 1998, **63**, 2251 see also 2099.
39. PD Armas, FG Tellado, JJM Tellado and J Robles, *Tetrahedron Lett.*, 1997, **38**, 8081.
40. CC Gonzalez, AR Kennedy, EI Leon, C Riesco-Fagundo and E Suarez, *Angew. Chem. Int. Ed.*, 2001, **40**, 2326.
41. JH Rigby, A Payen and N Warshakonn, *Tetrahedron Lett.*, 2001, **42**, 2047.
42. A Boto, R Hernandez, A Montoya and E Suarez, *Tetrahedron Lett.*, 2002, **43**, 8269.
43. CG Francisco, R Freire, AJ Herrera, I Perez-Martin and E Suarez, *Org. Lett.*, 2002, **4**, 1959.
44. CG Francisco, R Freire, CC Gonzalez and E Suarez, *Tetrahedron: Asymmetry*, 1997, **8**, 1971.
45. CG Francisco, R Freire, CC Gonzalez, EI Leon, C Riesco-Fagundo and E Suarez, *J. Org. Chem.*, 2001, **66**, 1861.
46. RS Neale, *Synthesis*, 1971, 1.
47. ME Wolff, *Chem. Rev.*, 1963, **63**, 55.
48. PD Armas, R Carrau, JI Concepcion, CG Francisco, R Hernandez and E Suarez, *Tetrahedron Lett.*, 1985, **26**, 2493.
49. PD Armas, CG Francisco, R Hernandez, JA Salazar and E Suarez, *J. Chem. Soc., Perkin Trans. 1*, 1988, 3255.
50. R Carrau, R Hernandez and E Suarez, *J. Chem. Soc., Perkin Trans. 1*, 1987, 937.
51. RL Dorta, CG Francisco and E Suarez, *Chem. Commun.*, 1989, 1168.
52. CG Francisco, AJ Herrera and E Suarez, *Tetrahedron: Asymmetry*, 2000, **11**, 3879.
53. CG Francisco, AJ Herrera and E Suarez, *J. Org. Chem.*, 2003, **68**, 1012.
54. H Togo, Y Hoshina, T Muraki and H Nakayama, *J. Org. Chem.*, 1998, **63**, 5193.
55. M Katohgi, H Togo, K Yamaguchi and M Yokoyama, *Tetrahedron*, 1999, **55**, 14885.
56. M Masnyk, *Tetrahedron Lett.*, 1997, **38**, 879.
57. J Brunckova, D Crich and Q Yao, *Tetrahedron Lett.*, 1994, **35**, 6619.
58. JC Baldwin, D Brown, PH Scudder and ME Wood, *Tetrahedron Lett.*, 1995, **36**, 2105.
59. D Denenmark, P Hoffmann, T Winkler, A Waldner and AD Mesmaeker, *Synlett*, 1991, 621.
60. DP Curran and H Yu, *Synthesis*, 1992, 123.

61. ST Hilton, TCT Ho, G Pljevaljcic and K Jones, *Org. Lett.*, 2000, **2**, 2639.
62. GW Gribble, HL Fraser and JC Badenock, *Chem.Commun.*, 2001, 805.
63. R Loven and WN Speckamp, *Tetrahedron Lett.*, 1972, 1567.
64. A Studer, M Bossart and H Steen, *Tetrahedron Lett.*, 1998, **39**, 8829.
65. JJ Kohler and WN Speckamp, *Tetrahedron Lett.*, 1977, 635.
66. A Studer, M Bossart and T Vasella, *Org. Lett.*, 2000, **2**, 985.
67. A Ryokawa and H Togo, *Tetrahedron*, 2001, **57**, 5915.
68. CG Francisco, EI Leon, P Moreno and E Suarez, *Tetrahedron: Asymmetry*, 1998, **9**, 2975.

– 7 –

Synthetic Uses of Free Radicals for Nucleosides and Sugars: Barton–McCombie Reaction

Deoxygenative reduction of alcohols via xanthates (**1**), with Bu_3SnH in the presence of AIBN, is known as the Barton–McCombie reaction [1–3]. The reaction mechanism is shown in eq. 7.1. The driving force of this reaction is the formation of a strong C=O bond from the C=S bond, and the former bond is approximately 10 kcal/mol stronger than the latter bond. Generally, alkyl (RX) or aryl (ArX) halides are used as radical precursors for $R^{\cdot}$ or $Ar^{\cdot}$; however, halogenation of sugars and nucleosides, which have many OH groups and other delicate functional groups, is rather difficult. So the Barton–McCombie reaction is very useful for the radical reactions in sugars, nucleosides, and peptides. Other thiocarbonyl derivatives formed from alcohols (down side in eq. 7.1) with phenoxythiocarbonyl chloride (PhOC=SCl), *N,N'*-thiocarbonyl diimidazole, etc. can also be used instead of methyl xanthate. The deoxygenative reduction of *sec*-alcohols was studied to begin with, then later, it was found that the same deoxygenative reduction of *tert*-alcohols and *prim*-alcohols could be carried out.

R–OH → (1) NaOH, CS_2; 2) CH_3I) → R–OC(=S)SCH₃ **1** (methyl xanthate) → (Bu_3SnH, AIBN) → RH **2**

1 + [$Bu_3Sn^{\cdot}$] → [R–O–Ċ(SSnBu₃)–SCH₃] → [$R^{\cdot}$] + $Bu_3SnS–C(=O)–SCH_3$; [$R^{\cdot}$] + Bu_3SnH → RH **2** (7.1)

R–O–C(=S)–X

X = $-SCH_3$, $-O-C_6H_5$, $-C_6H_5$, –N-imidazolyl, $-O-C_6H_4-F$, $-O-C_6H_4-OCH_3$, $-NH-C_6H_5$

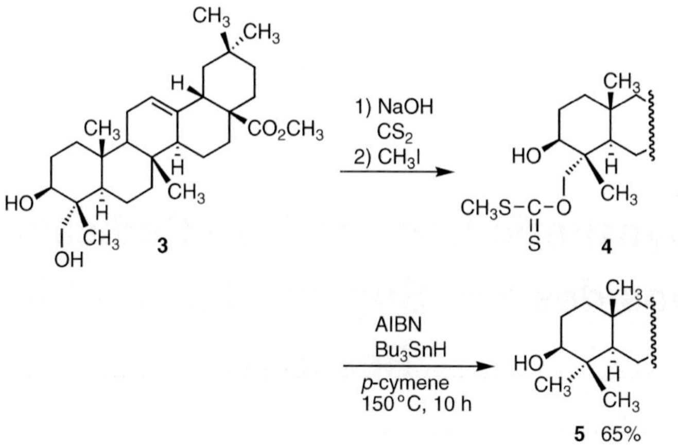

(7.2)

Eq. 7.2 shows a typical deoxygenative reaction of hederagenin (**3**) via methyl xanthate (**4**) with Bu_3SnH in the presence of AIBN under benzene refluxing conditions.

Experimental procedure 1 (eq. 7.2).

A solution of Bu_3SnH (200 mg, 0.69 mmol) in *p*-cymene (3 ml) was added dropwise over 2 h to a solution of methyl xanthate of hederagenin (50 mg, 0.087 mmol) in *p*-cymene (3 ml) at 150 °C. After the addition, the mixture was stirred at the same temperature for 10 h. After the reaction, CCl_4 was added and the mixture was refluxed for 3 h. After removal of the solvent, molecular iodine in ether was added until the iodine color remains. Then, the solution was washed with 10% aq. KF solution (5 ml), and the organic layer was dried over Na_2SO_4. After removal of the solvent, the residue was recrystallized to give the deoxygenated product [2].

In detailed reaction mechanism, two pathways are proposed, as shown in eq. 7.3. Path *a* is the formation of an alkoxy thiocarbonyl radical [**I**], together with Bu_3SnSCH_3, while path *b* is the formation of an adduct radical [**II**] onto methyl xanthate (**1**) by $Bu_3Sn^{\cdot}$. Based on the study with ^{119}Sn-NMR, it was found that the deoxygenation reaction proceeds via path *b* [4, 5]. Other examples are shown in Tables 7.1 and 7.2.

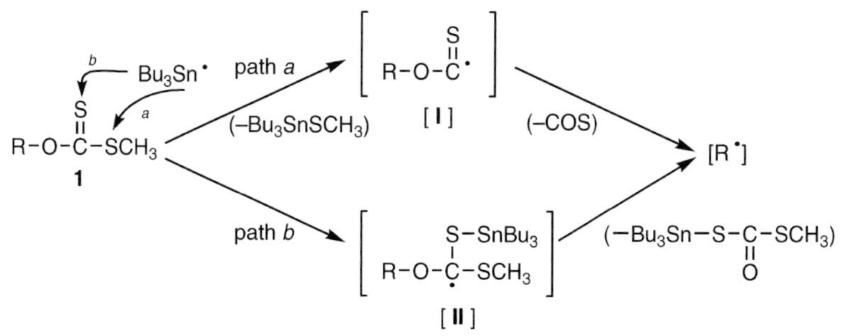

(7.3)

Table 7.1

Barton–McCombie reaction of imidazole thiocarbonates with Bu_3SnH

R–OH → (thiocarbonyldiimidazole) → R–O–C(=S)–N(imidazolyl) **6** → (AIBN, Bu_3SnH, $C_6H_5CH_3$, Δ) → R–H **2**

	6	2
BzO, HO, OCH$_3$, BzO, OBz (Bz = PhCO)	94%	92%
Ph, O, OBn, HO, NHAc (Bn = PhCH$_2$)	59%	57%

One great advantage of the Barton–McCombie reaction is that acetal, ketone, ester, hydroxy, and amino groups are not affected during the reaction [6–18]. Polystyrene-supported di-*n*-butylstannane can reduce a thiocarbonyl derivative under the same

Table 7.2

Barton–McCombie reaction with Bu_2POH and $(PhCO_2)_2$

R–OH → (1) NaOH, CS_2; 2) CH_3I) → ($(PhCO_2)_2$, Bu_2POH, dioxane, Δ, 24 h) → R–H

Substrate	Yield
(diacetone glucose, OH)	93%
(1-adamantanol, OH)	90%
$CH_3(CH_2)_{10}CH_2$—OH	87%

Reaction Mechanism

R–O–C(=S)–SCH_3; O=PBu_2–S–C(R–O)–SCH_3; $Bu_2\dot{P}$=O; R$^\bullet$; RH; H–Bu_2P=O; Bu_2P(=O)–S–C(=O)–SCH_3

conditions, and here the reduction product can be obtained by the simple filtration of the reaction mixture (eq. 7.4).

7

A
AIBN
$C_6H_5CH_3$, Δ

8 98%

(7.4)

Polystyrene-supported di-*n*-butylstannane **A**

n–Bu–Sn–Bu–*n*
H

1) PhOCSCl, Py
CH_2Cl_2 95%
2) $(Me_3Si)_3SiH$
AIBN
$C_6H_5CH_3$, Δ 94%

9

10

(7.5)

The $(Me_3Si)_3SiH$/AIBN system, instead of the Bu_3SnH/AIBN system can also be used for the same reduction (eq. 7.5).

Experimental procedure 2 (eq. 7.5).

To a solution of cholesterol (0.77 g, 2 mmol) in dichloromethane (10 ml) were added pyridine (0.6 ml, 8 mmol) and phenoxy thiocarbonyl chloride (0.4 ml, 2.2 mmol). After 2 h, methanol (1 ml) was added, and the mixture obtained was washed with 1 M hydrochloric acid solution twice, and dried over Na_2SO_4. After removal of the solvent, the residue was recrystallized from acetone to give phenyl thiocarbonate in 95% yield. Then, $(Me_3Si)_3SiH$ (0.7 ml, 2.2 mmol) and AIBN (50 mg, 0.5 mmol) were added to the solution of the phenyl thiocarbonate (0.75 g, 1.5 mmol) in toluene (20 ml) and the obtained mixture was heated at 80 °C for 2 h under nitrogen atmosphere. After the reaction, the solvent was removed and the residue was recrystallized from acetone to give cholest-5-ene in 94% yield [14].

11

AIBN
$(Me_3Si)_3SiH$
$C_6H_5CH_3$, Δ

12 87%

(7.6)

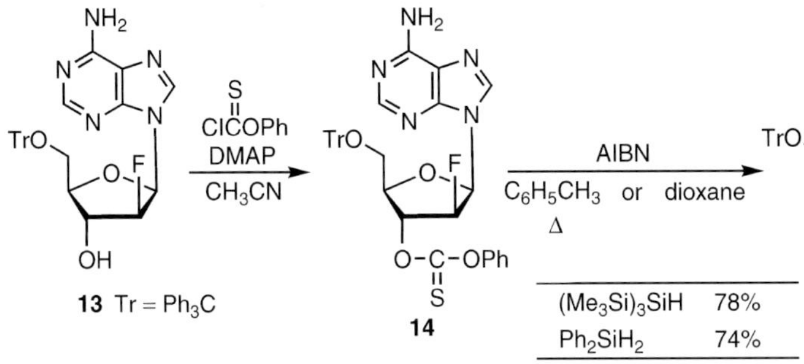

(7.7)

2,4-*Bis*-dithiocarbonate (**11**) is reduced to 2,4-dideoxy-1,6-anhydro-D-glucose (**12**) (eq. 7.6) and 3′-OH of nucleoside (**13**) is deoxygenated to *anti*-HIV nucleoside (**15**) (eq. 7.7).

N-Phenyl thiocarbamate (**16**) is also reduced with the Bu_3SnD/AIBN system to provide 2′-deoxy-2′-*d*-3′,5′-O-TIPDS-uridine (**17**) with 100% of *d*-content (eq. 7.8). The less toxic 1,1,2,2-tetraphenyldisilane ($Ph_4Si_2H_2$)/AIBN or Et_3B system also reduces methyl xanthate (**18**) in good yield (eq. 7.9).

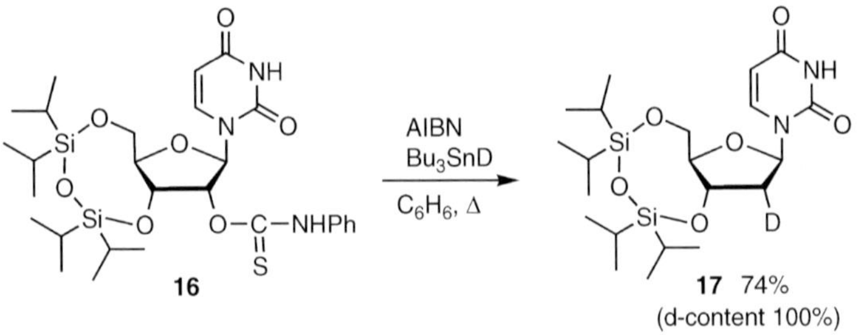

(7.8)

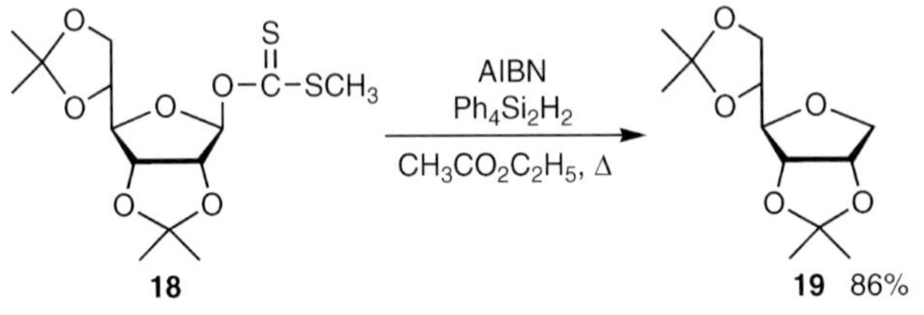

(7.9)

Experimental procedure 3 (eq. 7.8).

A mixture of substrate (0.2 mmol), Bu_3SnD (0.4 mmol), and AIBN (0.04 mmol) in benzene was refluxed for 4 h under an argon atmosphere. After the reaction, the solvent was removed and the residue was chromatographed on silica gel with column (eluent: hexane/ethyl acetate = 7/3) to give 2′-deoxy-2′-*d*-3′,5′-O-TIPDS-uridine in 74% yield [17].

Experimental procedure 4 (eq. 7.9).

A mixture of methyl xanthate (0.2 mmol), $Ph_4Si_2H_2$ (0.22 mmol), and AIBN (0.06 mmol) in ethyl acetate (1.5 ml) was refluxed for 16 h under an argon atmosphere. After the reaction, the solvent was removed and the residue was chromatographed on silica gel with column to obtain the reduction product [18].

Experimental procedure 5 (Table 7.2).

A mixture of methyl xanthate (0.4 mmol) and Bu_2POH (dibutylphosphine oxide, 2.0 mmol) in dry dioxane (3 ml) was refluxed under an argon atmosphere, and then a solution of benzoyl peroxide (0.4 mmol) in dioxane (3 ml) was added dropwise to the mixture. After 90 min, the solvent was removed and the residue was chromatographed on silica gel to give the reduction product [22].

Deoxygenative reduction can be carried out at room temperature with Et_3B as an initiator [19,20]. However, this reaction is limited to thiocarbonates derived from *sec*- and *tert*-alcohols, since the C–O bond cleavage of thiocarbonates derived from *prim*-alcohols does not occur easily at room temperature.

$$\mathbf{20}\ (\text{cyclododecyl–O–C(=S)–SCH}_3) \xrightarrow[C_6H_6,\ \Delta]{Et_3B,\ \text{Air},\ Bu_3SnH} \mathbf{21}\ 93\% \tag{7.10}$$

Experimental procedure 6 (eq. 7.10).

To a solution of *O*-cyclododecyl *S*-methyl dithiocarbonate (1.0 mmol), and Bu_3SnH (320 mg, 1.1 mmol) in benzene (5 ml) was added Et_3B (1 *M* hexane solution, 1.1 ml, 1.1 mmol). The mixture was stirred for 20 min at 20 °C. After the reaction, the solvent was removed and the residue was chromatographed on silica gel with column to provide cyclododecane in 93% yield [19].

Instead of Bu_3SnH, other less toxic silanes such as $(Me_3Si)_3SiH$, $Ph_4Si_2H_2$, 5,10-dihydro-5,10-disilanthracene (**B**), $PhSiH_3$, or Ph_2SiH_2, or trivalent phosphorus compounds such as $(EtO)_2POH$, Bu_2POH, or H_3PO_2, in the presence of AIBN or peroxide can be used for the deoxygenative reactions, as shown below (eqs. 7.11–7.14) [21–33]. However, monosilanes are rather less reactive, so it is recommended to use benzoyl peroxide or Et_3B as an initiator to carry out the chain reduction effectively.

22 → **23** (Et_3B, H_3PO_2, AIBN, $CH_3OCH_2CH_2OCH_3$, Δ) (7.11)

24 → **25** 96% (**B**) (7.12a)

26 → **27** 85% (**B**) (7.12b)

28 → **29** 95% ($(PhCO_2)_2$, $PhSiH_3$, $C_6H_5CH_3$, Δ) (7.13)

30 → **31** 87% (Et_3B, O_2, Ph_2SiH_2, C_6H_6, Δ) (7.14)

Treatment of 1,2-dixanthates (**32**) derived from 1,2-diols, with Bu_3SnH in the presence of AIBN under refluxing conditions generates the corresponding olefins (**33**) via radical β-elimination, as shown in eq. 7.15 [34–38].

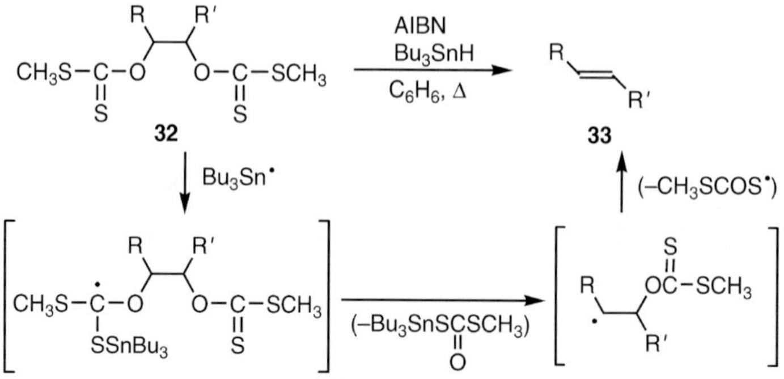

(7.15)

In general, thermodynamically stable *trans* olefins are formed, while *cis* olefins are obtained in the case of 1,2-dixanthates derived from cycloalkan-1,2-diols. Here, instead of the Bu_3SnH/AIBN system, environmentally benign $Ph_4Si_2H_2$/AIBN, Ph_2SiH_2/AIBN, and *N*-ethylpiperidine hypophosphite/AIBN systems can also be used. As a typically useful method, preparation of D4T (2′,3′-didehydro-3′-deoxythymidine) analogue, which has the same potent *anti*-HIV activity as AZT, is shown in eq. 7.16. In practice, D2C (2′,3′-dideoxycytidine) and D4C (2′,3′-didehydro-2′,3′-dideoxycytidine) were also prepared by this method [39, 40].

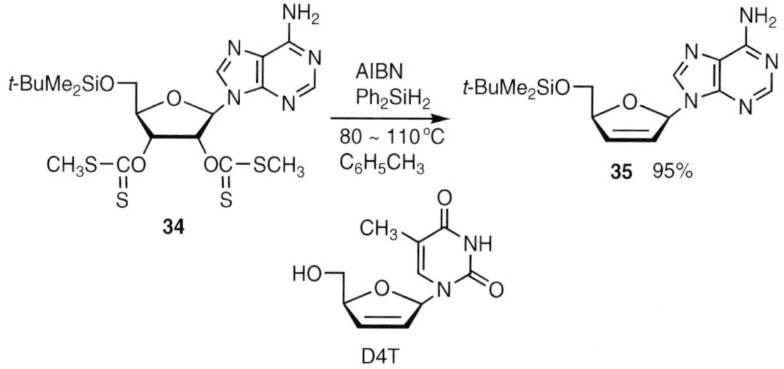

(7.16)

Experimental procedure 7 (eq. 7.16).

A mixture of AIBN (262 mg) in toluene (6 ml) was added over 30 min intervals to a refluxing solution of 2′,3′-dixanthate of adenosine (0.4 mmol) and Ph_2SiH_2 (0.8 mmol) in toluene (3 ml) under an argon atmosphere. After 90 min, the solvent was removed and the residue was chromatographed on silica gel to give 2′,3′-dideoxy-2′,3′-didehydroadenosine in 95% yield [36].

Treatment of cyclic thionocarbonate (**36**) derived from 1,3-diol, with the Bu_3SnH/AIBN system generates the corresponding monodeoxygenated product (**37**), as shown in eq. 7.17. In this case, selective O–C bond cleavage of the *O-sec*-alkyl group occurs, not that of the *O-prim*-alkyl group [41].

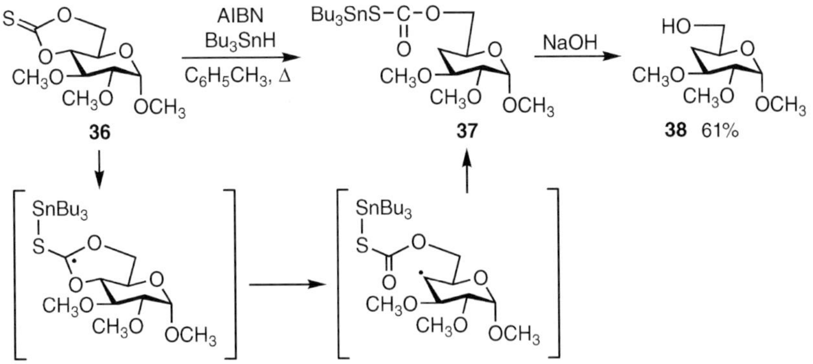
AIBN
Bu_3SnH
$C_6H_5CH_3$, Δ
NaOH
36
37
38 61%

(7.17)

As an oxidative condition, refluxing treatment of methyl xanthate (**39**) with $[CH_3(CH_2)_{10}CO_2]_2$ (lauroyl peroxide) in isopropanol generates the corresponding reduction product (**40**). Here, the formed $CH_3(CH_2)_{10}^{\cdot}$ plays the same role as $Bu_3Sn^{\cdot}$, and isopropanol donates an α-H hydrogen atom, as shown in eq. 7.18 [42, 43].

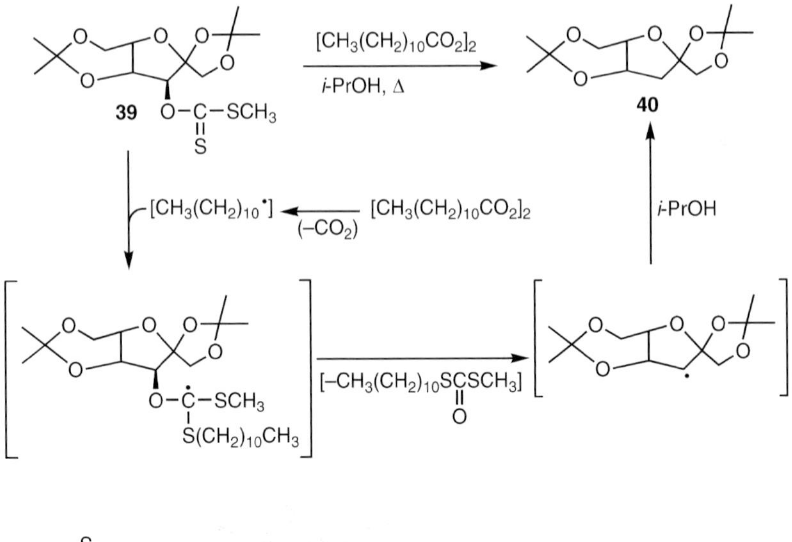
$[CH_3(CH_2)_{10}CO_2]_2$
i-PrOH, Δ
39
40
$[CH_3(CH_2)_{10}^{\cdot}]$
$(-CO_2)$
i-PrOH
$[-CH_3(CH_2)_{10}SCSCH_3]$

(7.18)

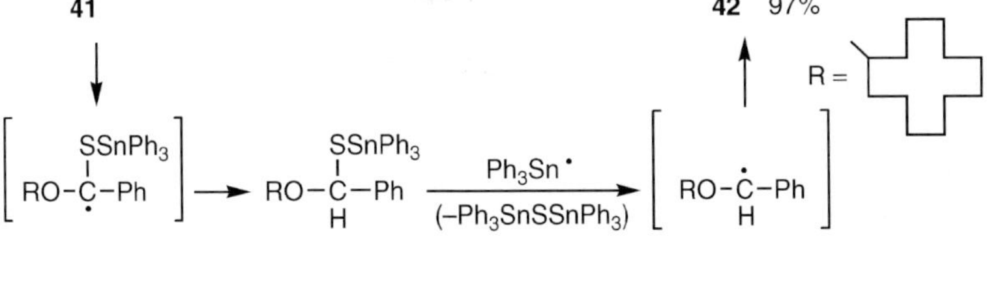
Ph_3SnH, Et_3B
r.t.
41
$ROCH_2Ph$
42 97%
R =
$Ph_3Sn^{\cdot}$
$(-Ph_3SnSSnPh_3)$

(7.19)

As a unique reaction, treatment of thionoesters (**41**) or thionolactones with Ph_3SnH in the presence of Et_3B at room temperature provides the reduction products (**42**) of thiocarbonyl groups to methylene groups (eq. 7.19) [44, 45]. Thus, this is the indirect reduction of esters to ethers, and lactones to cyclic ethers, since thionoesters and thionolactones are prepared from the corresponding esters or lactones with Lawesson's reagent.

Finally, acetate ester (**43**) and trifluoroacetate ester can be reduced with silane in the presence of peroxide under sealed conditions (eq. 7.20) [46, 47].

O–C(=O)–CH$_3$ (t-BuO)$_2$ Ph$_2$Si(H)–C$_6$H$_4$–SiPh$_2$(H), 140 °C, sealed tube → **43** → **44** 89% (7.20)

REFERENCES

1. DHR Barton and SW McCombie, *J. Chem. Soc., Perkin Trans 1*, 1975, 1574.
2. DHR Barton, WB Motherwell and A Stange, *Synthesis*, 1981, 743.
3. DHR Barton, W Hartwig, RS Motherwell and WB Motherwell, *Tetrahedron Lett.*, 1982, **23**, 2019.
4. PJ Barker and ALJ Beckwith, *Chem. Commun.*, 1984, 683.
5. DHR Barton, DO Jang and JC Jaszberenyi, *Tetrahedron Lett.*, 1990, **31**, 3991.
6. MJ Robins, JS Wilson and F Hansske, *J. Am. Chem. Soc.*, 1983, **105**, 4059.
7. JR Rasmussen, CJ Slinger, RJ Kordish and DD Newman-Evans, *J. Org. Chem.*, 1981, **46**, 4843.
8. DHR Barton, DO Jang and JC Jaszberenyi, *Tetrahedron Lett.*, 1990, **31**, 4681.
9. DHR Barton, D Crich, A Lobberding and SZ Zard, *Chem. Commun.*, 1985, 646.
10. DHR Barton, D Crich, A Lobberding and SZ Zard, *Tetrahedron*, 1986, **42**, 2329.
11. DHR Barton and JC Jaszberenyi, *Tetrahedron Lett.*, 1989, **30**, 2619.
12. M Gerlach, F Jördens, H Kuhn, WP Neumann and M Peterseim, *J. Org. Chem.*, 1991, **56**, 5972.
13. WP Neumann and M Peterseim, *Synlett*, 1992, 801.
14. D Schummer and G Höfle, *Synlett*, 1990, 705.
15. DHR Barton, DO Jang and JC Jaszberenyi, *Tetrahedron Lett.*, 1992, **33**, 6629.
16. S Takamatsu, T Maruyama, S Katayama, N Hirose and K Izawa, *Tetrahedron Lett.*, 2001, **42**, 2321.
17. M Oba and K Nishiyama, *Tetrahedron*, 1994, **50**, 10193.
18. H Togo, S Matsubayashi, O Yamazaki and M Yokoyama, *J. Org. Chem.*, 2000, **65**, 2816.
19. K Nozaki, K Oshima and K Utimoto, *Tetrahedron Lett.*, 1988, **29**, 6125.
20. DHR Barton, SI Parekh and CL Tse, *Tetrahedron Lett.*, 1993, **34**, 2733.
21. DHR Barton, DO Jang and JC Jaszberenyi, *Synlett*, 1991, 435.
22. DO Jang, DH Cho and DHR Barton, *Synlett*, 1998, 39.
23. DHR Barton, DO Jang and JC Jaszberenyi, *Tetrahedron Lett.*, 1992, **33**, 2311.
24. DHR Barton, DO Jang and JC Jaszberenyi, *Tetrahedron Lett.*, 1992, **33**, 5709.
25. DHR Barton, DO Jang and JC Jaszberenyi, *Tetrahedron*, 1993, **49**, 2793.
26. T Gimisis, M Ballestri, C Ferreri and C Chatgilialoglu, *Tetrahedron Lett.*, 1995, **36**, 3897.
27. DHR Barton, DO Jang and JC Jaszberenyi, *Synlett*, 1991, 435.
28. DHR Barton, DO Jang and JC Jaszberenyi, *Tetrahedron Lett.*, 1993, **49**, 2793.
29. DHR Barton, DO Jang and JC Jaszberenyi, *Tetrahedron Lett.*, 1990, **31**, 4681.

30. O Jimenez, MP Bosch and A Guerrero, *Synthesis*, 2000, 1917.
31. S Takamatsu, S Katayama, N Hirose, M Naito and K Izawa, *Tetrahedron Lett.*, 2001, **42**, 7605.
32. S Takamatsu, T Maruyama, S Katayama, N Hirose and K Izawa, *Tetrahedron Lett.*, 2001, **42**, 2321.
33. DO Jang, J Kim, DH Cho and C-M Chung, *Tetrahedron Lett.*, 2001, **42**, 1073.
34. AGM Barrett, DHR Barton, R Bielski and SW McCombie, *Chem. Commun.*, 1977, 866.
35. DHR Barton, DO Jang and JC Jaszberenyi, *Tetrahedron Lett.*, 1991, **32**, 7187.
36. DHR Barton, DO Jang and JC Jaszberenyi, *Tetrahedron Lett.*, 1991, **32**, 2569.
37. H Togo, M Sugi and K Toyama, *C.R. Acad, Sci. Paris, Chim.*, 2001, **4**, 539.
38. DO Jang and DH Cho, *Tetrahedron Lett.*, 2002, **43**, 5921.
39. B Lythgoe and I Waterhouse, *Tetrahedron Lett.*, 1977, 4223.
40. TS Lin, JH Yang, MC Liu and JL Zhu, *Tetrahedron Lett.*, 1990, **31**, 3829.
41. DHR Barton and R Subramanian, *Chem. Commun.*, 1976, 867.
42. BQ Sire and SZ Zard, *Tetrahedron Lett.*, 1998, **39**, 9435.
43. JE Forbes and SZ Zard, *Tetrahedron Lett.*, 1989, **30**, 4367.
44. DO Jang, SH Song and DH Cho, *Tetrahedron*, 1999, **55**, 3479.
45. DO Jang and SH Song, *Synlett*, 2000, 811.
46. H Sano, T Takeda and T Migita, *Chem. Lett.*, 1988, 119.
47. DO Jang, J Kim, DH Cho and C-M Chung, *Tetrahedron Lett.*, 2001, **42**, 1073.

– 8 –

Barton Decarboxylation Reaction with *N*-Hydroxy-2-thiopyridone

The Barton decarboxylation reaction is very useful for organic synthesis. In the early 1980s, radical decarboxylation of *O*-acyl esters prepared from carboxylic acids and *N*-hydroxy-2-thiopyridone, was first reported by Barton. This reaction does not require any toxic radical reagents such as Bu_3SnH, and the decarboxylatively formed alkyl radicals, derived from the corresponding carboxylic acids, can be used for functional group conversion and C–C bond formations [1–5]. Sodium salt of *N*-hydroxy-2-thiopyridine is commercially available, and its aqueous 40% solution is also available as sodium omadine for keeping microorganisms active. Free *N*-hydroxy-2-thiopyridone is a stable crystals or solids under protection from light. *O*-Acyl esters of *N*-hydroxy-2-thiopyridone can be purified by flash column chromatography on silica gel. The esters are light-sensitive yellow solids or oil and have typically strong absorption in the range of 1800 cm^{-1} on the IR spectra.

Under refluxing conditions in benzene or toluene, or by irradiation with a tungsten lamp (100–300 W) at room temperature, N–O bond cleavage of the *O*-acyl ester (**2**) readily occurs to generate a carboxyl radical and a pyridine 2-thiyl radical. β-Cleavage of the formed carboxyl radical rapidly occurs to form an alkyl radical and CO_2 (the rate constant for β-cleavage of $RCO_2^{\bullet}$ is approximately $10^9\ s^{-1}$). The formed alkyl radical then reacts on the thiocarbonyl sulfur atom of *O*-acyl ester (**2**) to generate alkyl 2-pyridyl sulfide and carboxyl radical again. So this reaction proceeds via a chain pathway (eq. 8.1).

R–COOH (**1**) —[1) $(COCl)_2$; 2) HO–N(2-thiopyridone)]→ R–C(=O)–O–N(2-thiopyridone) (**2**, yellow solids or oil) —[W-*hv* or Δ]→ $[R^{\bullet}]$ + CO_2 + [2-pyridylthiyl radical] (8.1)

Half-life of the decay of *O*-acyl esters (**2**) at 80 °C is shown below (eq. 8.2) and it depends on the used primary-, secondary-, and tertiary-alkyl chained carboxylic acids.

Thus, the reactivity is increased as follows: *prim*-R < *sec*-R < *tert*-R.

$R = CH_3(CH_2)_{14}: t_{\frac{1}{2}} = 31$ min
$R = (CH_3)_2CH : t_{\frac{1}{2}} = 26$ min
$R = (CH_3)_3C : t_{\frac{1}{2}} = 14$ min (8.2)

When an *O*-acyl ester (**2**) derived from phenylacetic acid is irradiated, a beautiful hyperfine structure of a benzyl radical can be observed by ESR [6]. This can be explained by the fact that the rate constant for the reaction of the formed R$^{\cdot}$ and *O*-acyl esters (**2**) is about 10^6 s^{-1} and it is rather slower than that of decarboxylation of $RCO_2^{\cdot}$ (eq. 8.3) [7, 8]. Generally, under photolytic conditions, approximately 100% of the chain reaction occurs, and under benzene refluxing conditions, approximately 80% of the chain reaction and approximately 20% of the caged reaction takes place.

$R = CH_3(CH_2)_{14}-$
4°C $k = 0.87 \times 10^6$ M^{-1} s^{-1}
80°C $k = 4.2 \times 10^6$ M^{-1} s^{-1} (8.3)

8.1 REDUCTION

Refluxing or photolytic treatment of *O*-acyl esters (**2**) in the presence of a hydrogen donor such as Bu_3SnH or *tert*-BuSH, provides the corresponding reduction products via alkyl radicals. This reaction can be applied to primary-, secondary-, and tertiary-alkyl chained carboxylic acids, and can also be used for steroids, sugars, and peptides as shown in eq. 8.4 [9–14]. Racemization does not occur at other chiral centers.

(8.4)

Experimental procedure 1 (eq. 8.4).

To a solution of *N*-hydroxy-2-thiopyridone (140 mg, 1.1 mmol) and pyridine (0.1 ml) in toluene (10 ml) was added a solution of 3β-acetoxy-11-ketobisnorallocholanic acid chloride (1 mmol) in toluene (1 ml) under a nitrogen atmosphere at room temperature. During the reaction, the reaction vessel was protected from light with aluminum-foil. After 10 min, the reaction mixture was filtered quickly, and the filtrate was added dropwise over 30 min to a refluxing solution of *tert*-BuSH (0.5 ml) in toluene (20 ml). During this addition, the dropping vessel was again protected from light with aluminum-foil. After 1 h, the solvent was removed and the residue was purified by column chromatography on silica gel to give 3β-acetoxy-5αH-pregnan-11-one in 82% yield.

The same reaction can be carried out by the dropwise addition of 3β-acetoxy-11-ketobisnorallocholanic acid chloride (1 mmol) in toluene (5 ml) to a refluxing solution of dried sodium salt of *N*-hydroxy-2-thiopyridone (1.2 mmol), 4-(dimethylamino)pyridine (0.1 mmol), and *tert*-BuSH (4.5 mmol) in toluene (20 ml), over 15 min [1].

With this reduction method, β-stereoselective preparation of *O*-glycosides and *C*-glycosides can be carried out as shown in eq. 8.5 [15–19].

1) NaOH
2) [N-(chlorocarbonyloxy)-2-thiopyridinium chloride]
3) *t*-BuSH, 8 °C, W-*hν*

7 → **8** 76% (8.5)

equatorial radical [**I**] ⇌ axial radical [**II**] → β-*O*-glycoside (8.6)

The formed anomeric radical adopts both α form (equatorial radical) [**I**] and β form (axial radical) [**II**], and these are in equilibrium state. However, the β form radical is more reactive and nucleophilic, because of its orbital interaction between the singly occupied radical orbital and the axial lone-pair on the neighboring ring-oxygen atom (like *anti*-periplanar effect) [20–22]. Therefore, the β form radical (axial radical) [**II**] predominantly reacted with a hydrogen donor to generate β-*O*-glycoside.

8.2 CONVERSION TO HALIDES

Photolytic treatment of *O*-acyl esters (**2**), derived from primary-, secondary-, and tertiary-alkyl chained carboxylic acids, in CCl_4, $BrCCl_3$, and CHI_3 with a tungsten lamp generates good yields of the corresponding alkyl chlorides, alkyl bromides, and alkyl iodides (**9**), respectively, via Hunsdiecker-type reaction (Table 8.1). Here, the by-products are 2-Py-S-CCl_3 for CCl_4 and $BrCCl_3$, and Py-S-CHI_2 for CHI_3, respectively.

Table 8.1

Conversion of carboxylic acids to alkyl halides

$RCO_2H \rightarrow RX + PySCCl_3$ (9)

R–C(=O)–O–N (2) $\xrightarrow[20\,°C,\ XCCl_3\ (X = Cl, Br)\ \text{or}\ CHI_3\ (X = I)]{W\text{-}h\nu}$ $R^{\bullet} + PyS^{\bullet} + CO_2$

RCO_2H	RX	Yield
$CH_3(CH_2)_{14}–CO_2H$	$CH_3(CH_2)_{14}Cl$	70%
$CH_3(CH_2)_{14}–CO_2H$	$CH_3(CH_2)_{14}Br$	95%
$CH_3(CH_2)_{14}–CO_2H$	$CH_3(CH_2)_{14}I$	74%
Adamantyl–CO_2H	Adamantyl–Br	98%
Steroid (AcO, CH_3, CH_3, H)–CO_2H	Steroid–Br	72%

The same treatment of *O*-acyl ester of *N*-hydroxy-2-thiopyridone, derived from cubanecarboxylic acid in CF_3CH_2I gives iodocubane in good yield. As an interesting result, refluxing treatment of the *O*-acyl ester (**10**) derived from optically active α-fluorocyclopropanecarboxylic acid, in $BrCCl_3$, provides the corresponding α-bromo-α-fluorocyclopropane (**11**) with retention [23]. Generally, radical inversion at the cyclopropane ring rapidly occurs ($10^{11}–10^{12}\ s^{-1}$). However, radical inversion of the α-fluorocyclopropyl radical is rather slow relatively ($\sim 10^6\ s^{-1}$), as shown in eq. 8.7.

10 $\xrightarrow[BrCCl_3]{\Delta}$ [radical] → **11** 78% complete retention (8.7)

$k = 10^{11} \sim 10^{12}\ s^{-1}$ (71 °C)

$k' = 10^6\ s^{-1}$ (–108 °C)

activation energy 10.5 kcal/mol

One great advantage of the decarboxylative halogenation with *O*-acyl esters of *N*-hydroxy-2-thiopyridone is that the reaction does not require any heavy metal such as Ag or Hg, unlike the Hunsdiecker reaction [24, 25]. Moreover, decarboxylative bromination of *p*-methoxybenzoic acid can be also carried out in good yield, while it does not proceed with the Hunsdiecker reaction; instead, electrophilic bromination on the aromatic ring occurs.

8.3 CONVERSION TO CHALCOGENIDES

As mentioned before, thermal or photolytic treatment of *O*-acyl esters (**2**) generates the corresponding alkyl 2-pyridyl sulfides. Oxidation of the sulfides with *m*CPBA provides the corresponding sulfoxides, which can be further derived to olefins under heating conditions via Ei mechanism. One example is shown below (eq. 8.8) [26].

OAc COOH —1) $(COCl)_2$; 2) NaO–N(2-thiopyridone), Δ→ OAc SPy **12** —1) *m*CPBA; 2) Δ→ OAc **13** (8.8)

The same photolytic treatment of *O*-acyl esters (**2**) in the presence of disulfides, diselenides, and ditellurides effectively produces the corresponding alkyl sulfides, alkyl selenides, and alkyl tellurides respectively, through S_Hi reaction on the chalcogen atoms by alkyl radicals, as shown in eq. 8.9. The reactivities somewhat depend on the kind of chalcogenides. Thus, the effective formation of alkyl sulfides requires 30 eq. of disulfides, that of alkyl selenides requires 10 eq. of diselenides, and that of alkyl tellurides requires 2 eq. of ditellurides [27, 28].

$$\underset{\mathbf{2}}{R{-}C(=O){-}O{-}N(\text{2-thiopyridone})} \xrightarrow[0^\circ C]{W\text{-}h\nu,\ ArXXAr} \underset{\mathbf{14}}{RXAr} + PySXAr \quad (X = S, Se, Te) \qquad (8.9)$$

It is well known that trace amounts of seleno-amino acids like selenocysteine play an important role in living bodies, probably for oxidation and reduction. Seleno-amino acids too can be prepared by this method. In this way, seleno-methionine (**17**) can be prepared from glutamic acid under mild conditions, without racemization (eq. 8.10) [29].

Vinyl-glycine with 100% *e.e.* can also be prepared from low-temperature elimination of the selenoxide, derived from seleno-amino acid.

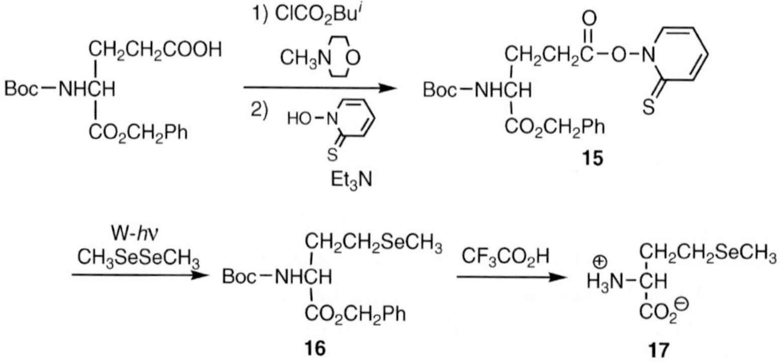

(8.10)

8.4 CONVERSION TO OTHER FUNCTIONAL GROUPS

It is rather difficult to convert carboxylic acids to decarboxylative alcohols. However, treatment of *O*-acyl esters (**2**) in the presence of $Sb(SPh)_3$ and molecular oxygen, followed by hydrolysis, generates the corresponding decarboxylative alcohols. Eq. 8.11 shows the preparation of a sex pheromone of the citrus mealybug from (+)-*cis*-pinonic acid [30–32]. When $^{18}O_2$ instead of $^{16}O_2$, is used in this reaction, ^{18}O-alcohols can be obtained.

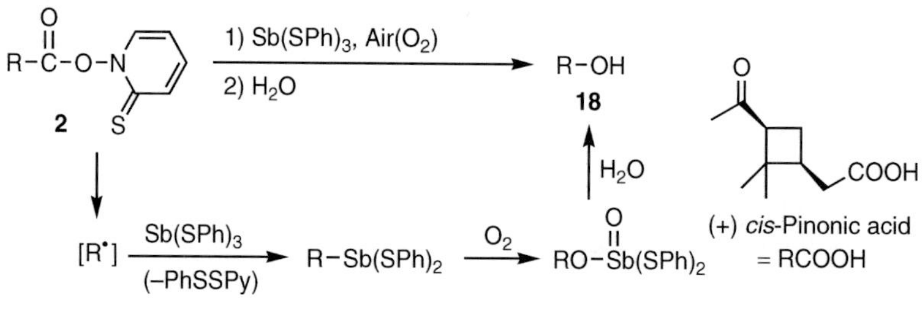

(8.11)

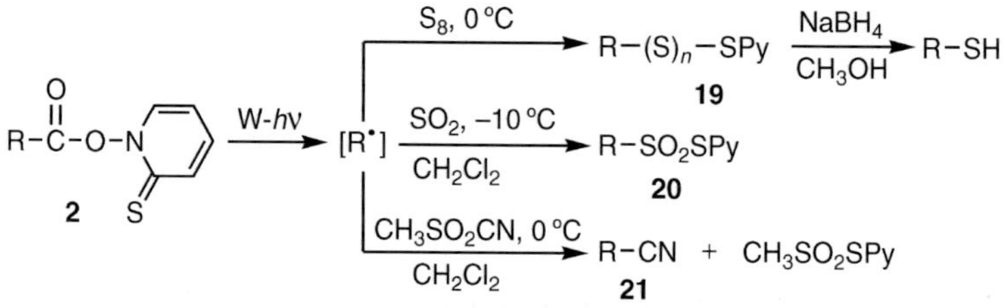

(8.12)

Photolytic treatment of *O*-acyl esters (**2**) with sulfur (S_8), SO_2, and CH_3SO_2CN provides the corresponding thiols (**19**), 2-pyridyl thiolsulfonates (**20**), and nitriles (**21**) in good yields, as shown in eq. 8.12 [33–35]. Here, the rate constant for decomposition (β-cleavage) of $RSO_2^{\cdot}$. is not as rapid ($4 \times 10^5 M^{-1} s^{-1}$ at 25 °C), as with that of $RCO_2^{\cdot}$. The formed $RSO_2^{\cdot}$ readily reacts with *O*-acyl esters (**2**) via a chain process, before the reverse β-cleavage reaction.

Experimental procedure 2 (eq. 8.12).

A solution of sulfur (S_8, 74 mg, 0.29 mmol) and adamantanecarboxylate ester of *N*-hydroxy-2-thiopyridone (290 mg, 1 mmol) in dichloromethane (15 ml) was irradiated with a tungsten lamp for 1 h under a nitrogen atmosphere at 0 °C. After the reaction, a solution of $NaBH_4$ (20 mmol) in methanol (10 ml) was added dropwise over 15 min to the reaction mixture at room temperature, and the mixture was stirred for one hour. After the reaction, 1 M sulfuric acid (60 ml) was added to the mixture, and then the solution was extracted with dichloromethane (4 × 20 ml), and dried over Na_2SO_4. After removal of the solvent, the residue was purified by column chromatography on silica gel (eluent: hexane/ethyl acetate = 97/3) to provide adamantanethiol in 88% yield [33].

8.5 CARBON–CARBON BOND FORMATION

8.5.1 Intramolecular radical cyclization

Preparation of cyclic compounds via 5-*exo-trig* or 6-*exo-trig* manner with *O*-acyl esters (**2**) proceeds effectively [36–40]. Eq. 8.13 shows the preparation of biologically active perhydroindole-2-carboxylic acids (**22a**) and (**22b**) from an aspartic acid derivative via 5-*exo-trig* manner.

1) HO–N(2-thiopyridone), *i*-BuOCOCl, CH_3–N(morpholine)
2) W-*hν*, r.t. 1 h

22a + **22b** 69% (8.13)

8.5.2 Intermolecular coupling reaction

Photolytic treatment of *O*-acyl esters (**2**) at −64 °C gives the corresponding radical coupling products, R–R via $R^{\cdot}$, since the radical chain reaction is retarded due to the low temperature [41].

8.5.3 Intermolecular addition reaction

Intermolecular addition reaction is important. Generally, alkyl radicals derived from *O*-acyl esters (**2**) are nucleophilic, so treatment with electron-deficient olefins such as nitroethylene, vinyl phosphonium, or vinyl sulfone generates the corresponding addition products effectively. Eq. 8.14 shows the photolytic reaction of *O*-acyl esters (**2**) with nitroethylene and 2-nitropropene to form the corresponding addition products (**23**) and (**25**) respectively. The addition products (**23**) with nitroethylene are further converted to one-carbon homologenized carboxylic acids (**24**) by treatment with H_2O_2/K_2CO_3 (pK_a of nitromethane is approximately 10), as in the Arndt-Eistert reaction, and the addition products (**25**) with 2-nitropropene are further treated with $TiCl_3$ to give methyl ketones (**26**) [42–51].

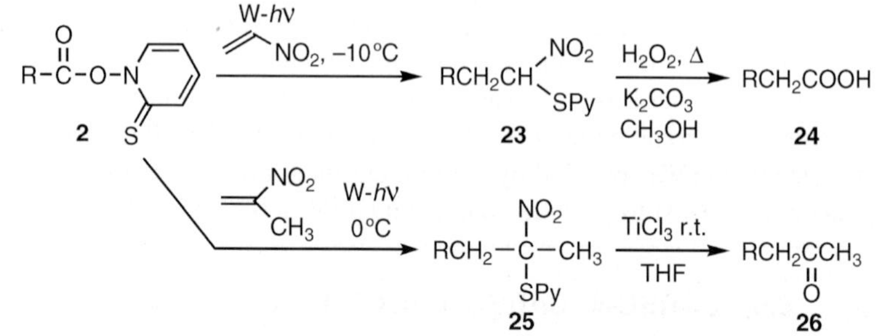

(8.14)

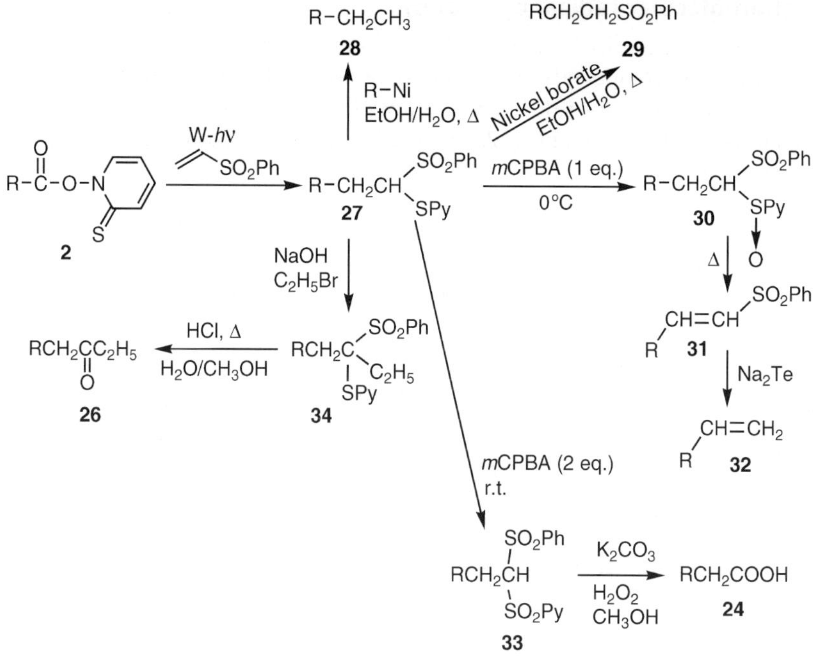

(8.15)

Experimental procedure 3 (eq. 8.15).

To a solution of β,β-diphenylpropionic acid (2 mmol) in $CHCl_3$ (8 ml) were added oxalyl chloride (1.5 g) and DMF (one drop). After stirring for 1 h, the solvent and excess oxalyl chloride were removed. The residue was dissolved in $CHCl_3$ (6 ml) and then *N*-hydroxy-2-thiopyridone (2.2 mmol) was added to the mixture under a nitrogen atmosphere at 0 °C.

Then a mixture of pyridine (5 mmol) in $CHCl_3$ (2 ml) was slowly added dropwise. During this reaction, the vessel was protected from light by aluminum-foil. After 30 min, the reaction mixture was filtered quickly. $CHCl_3$ (5 ml) and phenyl vinyl sulfone (10 mmol) were added to the yellowish filtrate, and the mixture was irradiated with a tungsten lamp (200 W) for 20 min under a nitrogen atmosphere at 20 °C. After the reaction, the solvent was removed, and the residue was dissolved in THF (15 ml). $NH_2NH_2–H_2O$ (1.6 g) was added to the solution and the mixture was stirred for 15 min. After the reaction, the solvent was removed and the residue was chromatographed on silica gel to give 4,4-diphenyl-1-benzenesulfonyl-1-thiopyridylbutane in 74% yield [43].

The addition products (**27**) formed from the same photolytic treatment of *O*-acyl esters (**2**) with phenyl vinyl sulfone, can be converted to various kinds of compounds (**24**), (**26**), and (**28**) to (**34**), as shown in eq. 8.15, and this method really becomes valuable chemistry.

Generally, treatment with electron-deficient olefins such as nitroethylene or vinyl sulfone is effective for radical addition reactions, since alkyl radicals derived from *O*-acyl esters (**2**) are nucleophilic and take SOMO–LUMO interaction. However, treatment of *O*-acyl esters (**2′**) derived from perfluoroalkyl carboxylic acids (R_fCO_2H) generates electrophilic radicals, $R_f^{\cdot}$, which react preferably with electron-rich olefins such as vinyl ether, as shown in eq. 8.16 [52].

$(R_f–CO)_2O$ —(NaO–N(pyridine-2-thione))→ $R_f–C(=O)–O–N$(pyridine-2-thione) **2′** (unstable) —(=/OEt, W-$h\nu$)→ $R_fCH_2CH(OEt)SPy$ **35** (8.16)

2′ → $[R_f\cdot]$ → $[R_fCH_2\dot{C}HR]$ → **35**

$R–C(=O)–O–N$(pyridine-2-thione) **2** —($F_2C{=}CCl_2$, W-$h\nu$, r.t.)→ $RCF_2CCl_2–SPy$ **36** —($AgNO_3$, THF–H_2O)→ RCF_2COOH **37** (8.17)

Treatment of *O*-acyl esters (**2**) with 1,1-dichloro-2,2-difluoroethylene provides α,α-difluorocarboxylic acids (**37**) through photolysis, followed by the hydrolysis of the adducts (**36**) with $AgNO_3$ (eq. 8.17) [53]. Eq. 8.18 shows the preparation of α-keto carboxylic acids (**40**) from carboxylic acids, by means of the radical addition to ethyl acrylate, oxidation to the sulfoxides by *m*CPBA, the Pummerer reaction with

$(CF_3CO)_2O$, and finally their hydrolysis under basic conditions [54].

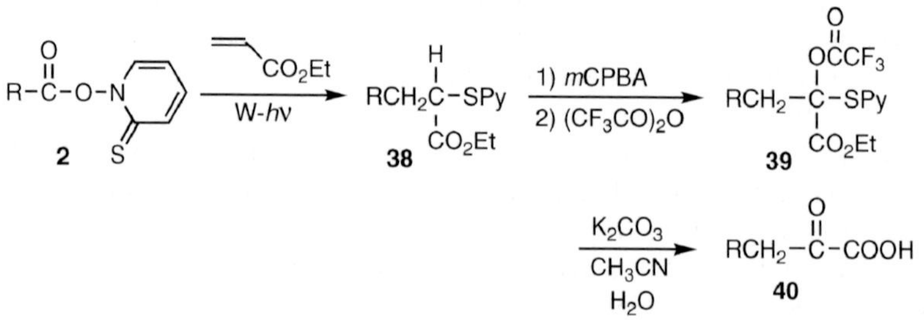

(8.18)

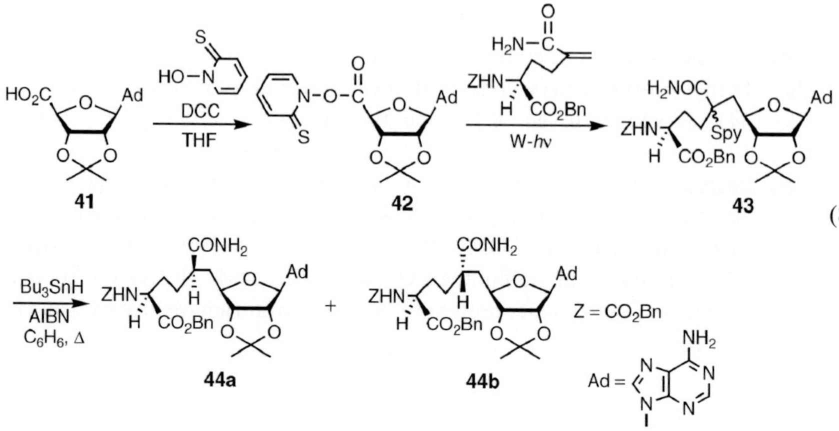

(8.19a)

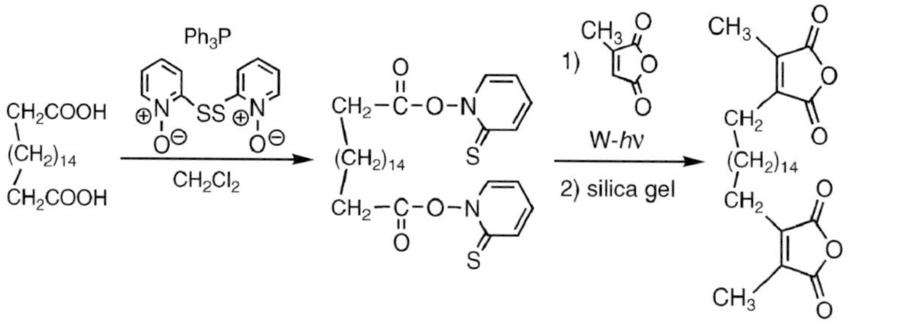

(8.19b)

As a synthetic application to biologically active compounds, eq. 8.19a shows the preparation of nucleoside antibiotics (**44a**) and (**44b**). Tyromycin (**45**) inhibits the leucine and cysteine aminopeptidases, and it can be prepared in good yield from the photolytic treatment of the *N*-acyloxy diester with citraconic anhydride, followed by silica gel treatment, as shown in eq. 8.19b [55–60]. Other synthetic applications using intra- and intermolecular tandem reactions were also studied [61, 62].

8.5.4 Alkylation of aromatics

Bu_3SnH cannot generally be used for radical alkylation of heteroaromatics, since it is a reductive radical reagent. However, *O*-acyl esters (**2**) are not a reductive reagent.

Thus, photolytic treatment of *O*-acyl esters (**2**) with electron-deficient heteroaromatics salts prepared from the reaction of heteroaromatics with camphorsulfonic acid or trifluoroacetic acid, effectively produces the corresponding alkylated heteroaromatics. Details on the alkylation of heteroaromatics with *O*-acyl esters (**2**) are given in Chapter 5.

Benzoquinones are not aromatics, but electron-deficient compounds. Photolytic treatment of *O*-acyl esters (**2**) with benzoquinones generates the corresponding 1,2-adducts. Practically, 1,2-disubstituted benzoquinones (**46**) bearing both alkyl and 2-thiopyridyl groups can be obtained. Since the initially formed product is 1,2-disubstituted dihydrobenzoquinone, it is smoothly oxidized to 1,2-disubstituted benzoquinone by excess benzoquinone, due to the difference in their oxidation potential, as shown in eq. 8.20 [63, 64].

(8.20)

(8.21)

As a synthetic application, stereoselective synthesis of potent *anti*-HIV (−)-ilimaquinone, is shown in eq. 8.21.

Experimental procedure 4 (eq. 8.20).

A mixture of *O*-acyl ester (1 mmol) of *N*-hydroxy-2-thiopyridone and benzoquinone (5–7 mmol) in degassed dichloromethane (20 ml) was irradiated with a 300 W tungsten lamp for 30 min. under a nitrogen atmosphere at 0 °C. After the reaction, the solvent was removed and the residue was chromatographed on silica gel to give 1,2-disubstituted benzoquinone [63].

8.5.5 Others

The tellurium atom has high radicophilicity. Thus, photolytic treatment of 2,3,4,6-tetra-*O*-acetyl-D-glucosyl *p*-anisyl telluride (**50**) with *N*-hydroxy-2-thiopyridone-*O*-acetate in the presence of methyl acrylate provides the *C*-glycoside (**51**), through S_Hi reaction on the tellurium atom by a methyl radical, to form 2,3,4,6-tetra-*O*-acetyl-D-glucosyl radical and methyl *p*-anisyl telluride [65–69]. The driving force of this reaction is the high radicophilicity of the tellurium atom and the stability of the 2,3,4,6-tetra-*O*-acetyl-D-glucosyl radical as compared with the methyl radical. Here, nucleophilicity of the methyl radical is much lower than that of the 2,3,4,6-tetra-*O*-acetyl-D-glucosyl radical, so the addition of the methyl radical to methyl acrylate does not proceed.

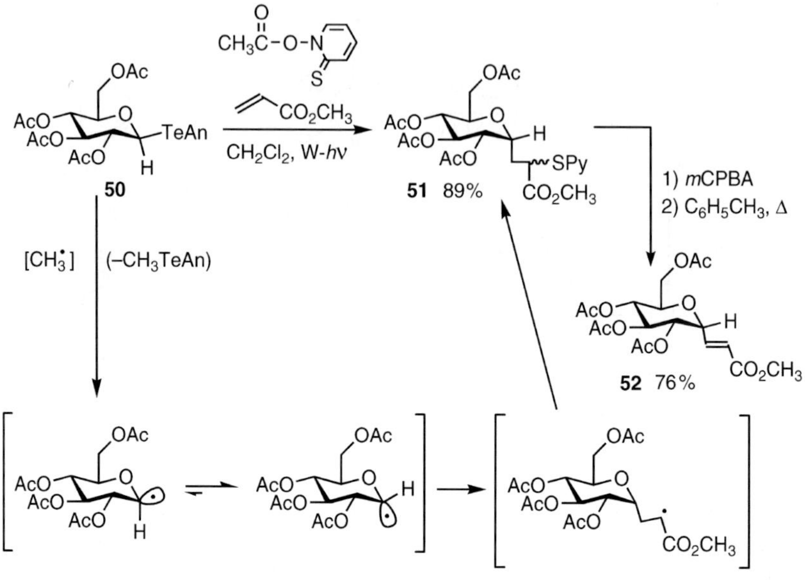

$CH_3C(=O)-O-N$(pyridine-2-thione), $CH_2=CHCO_2CH_3$, CH_2Cl_2, W-$h\nu$: **50** → **51** 89%; [$CH_3^{\bullet}$] (–CH_3TeAn); 1) *m*CPBA 2) $C_6H_5CH_3$, Δ → **52** 76% (8.22)

Unlike *N*-hydroxy-2-thiopyridone, *N*-hydroxy-2-selenopyridone is rather unstable and is easily oxidized. However, once photolytic treatment of *O*-acyl ester (**53**) of *N*-hydroxy-2-selenopyridone is carried out, the formed 2-pyridyl selenide is very useful. Thus, elimination of the selenoxides, formed from the oxidation of the selenide with *m*CPBA or ozone, proceeds effectively at low temperature. For example, eq. 8.23 shows the preparation of (L)-vinylglycine (**56**) from (L)-glutamic acid without racemization at all [70].

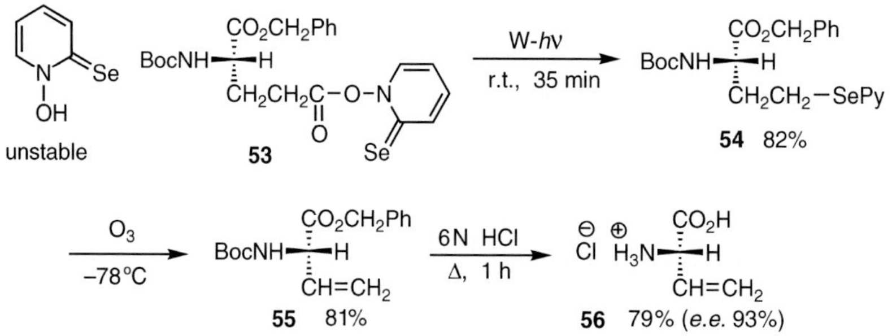

unstable; **53** → (W-$h\nu$, r.t., 35 min) **54** 82% → (O_3, –78°C) **55** 81% → (6N HCl, Δ, 1 h) **56** 79% (*e.e.* 93%) (8.23)

Carbonate and carbamate derivatives, prepared from the reaction of alcohols and amines with phosgene and *N*-hydroxy-2-thiopyridone, provide the corresponding alkoxyl radicals and aminyl radicals respectively, via radical decarboxylation [71–81]. For example, photolytic treatment of carbamate (**57**) derived from 4-pentenylamine generates the corresponding 4-pentenylaminyl radical, as shown in eq. 8.24. Under neutral conditions in the presence of *tert*-BuSH, a direct reduction product (**A**) of 4-pentenylaminyl radical is formed, while, in acidic conditions, cyclized product (**C**), pyrrolidine, via 5-*exo-trig* manner from 4-pentenylaminyl radical is formed. Under the latter conditions, the real reactive species is an electrophilic 4-pentenylaminium radical which rapidly cyclized via 5-*exo-trig* manner.

Additive / C_6H_6, 25 °C

Additive	Yield(%)	**A**	**B**	**C**
t-BuSH	80	100	0	0
CF_3COOH	90	0	0	100

R = $-(CH_2)_3CH_3$ (8.24)

Refluxing treatment of oxalate esters (**59**), prepared from the reaction of alcohols, oxalyl chloride and *N*-hydroxy-2-thiopyridone, generates the corresponding alkyl radicals via double decarboxylation. Thus, when these reactions are carried out in the presence of *tert*-BuSH and CCl_4, the corresponding reduction products RH and alkyl chlorides respectively, are provided [82–85]. However, this double decarboxylation reaction is limited to secondary and tertiary alcohols. Double decarboxylation of the oxalate ester derived from primary alcohol does not take place; instead, 2-pyridyl alkyl thiocarbonate is obtained through single decarboxylation, because the bond dissociation energy of the primary alkyl C–OH bond is strong (eq. 8.25) (Figure 8.1).

ROH $\xrightarrow[C_6H_6,\ r.t.]{(COCl)_2}$ ROCOCOCl (**58**) $\xrightarrow[DMAP,\ C_6H_6\ \Delta]{t\text{-BuSH, NaO–N(2-thiopyridone)}}$ R–H (**60**)

R = 2°-, 3°-alkyl

59 oxalate ester $\xrightarrow{(-PyS^\bullet)}$ [RO–C(O)–C(O)–O•] $\xrightarrow{(-CO_2)}$ [RO–C(O)•] $\xrightarrow{(-CO_2)}$ [R•] $\xrightarrow[(-t\text{-BuSPy})]{t\text{-BuSH}}$ R–H (8.25)

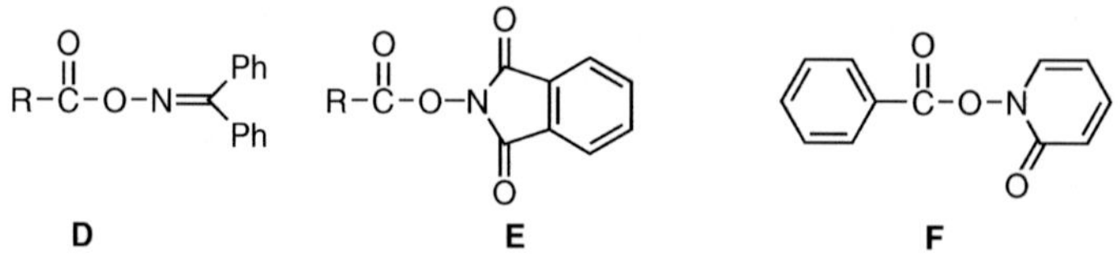

D E F

Figure 8.1

Finally, as examples of similar types of reactions, photolytic treatment of *O*-acyl ester (**D**) of benzophenone oxime, *N*-acyloxy-phthalimide (**E**), and *O*-acyl ester (**F**) of *N*-hydroxy-2-pyridone with a mercury lamp generates the corresponding alkyl radicals via decarboxylation. However, these reactions can be used only for the alkylation of aromatics (solvents such as benzene) and reduction [86–89], so their synthetic utility is extremely limited.

REFERENCES

1. DHR Barton, D Crich and WB Motherwell, *Tetrahedron*, 1985, **41**, 3901.
2. DHR Barton, D Crich and G Kretzschmar, *J. Chem. Soc., Perkin Trans. 1*, 1986, 39.
3. DHR Barton and WB Motherwell, *Heterocycles*, 1984, **21**, 1.
4. DHR Barton, D Crich and P Potier, *Tetrahedron Lett.*, 1985, **26**, 5943.
5. H Togo, G Nogami and Y Hoshina, *Yuki Gosei Kagaku Kyokaishi*, 1990, **48**, 641.
6. KU Ingold, J Lusztyk, B Maillard and JC Walton, *Tetrahedron Lett.*, 1988, **29**, 917.
7. M Newcomb and SU Park, *J. Am. Chem. Soc.*, 1986, **108**, 4132.
8. M Newcomb and J Kaplan, *Tetrahedron Lett.*, 1987, **28**, 1615.
9. DHR Barton, D Crich and WB Motherwell, *Chem. Commun.*, 1983, 939.
10. DHR Barton, D Crich and WB Motherwell, *Tetrahedron*, 1985, **41**, 3901.
11. P Garner, JT Anderson and S Dey, *J. Org. Chem.*, 1998, **63**, 5732.
12. DHR Barton, Y Herve, P Potier and J Thierry, *Chem. Commun.*, 1984, 1298.
13. DHR Barton, Y Herve, P Potier and J Thierry, *Tetrahedron*, 1988, **44**, 5479.
14. AV Cross, S Bory, M Gaudy and A Marquet, *Tetrahedron Lett.*, 1989, **30**, 1799.
15. D Crich and LBL Lim, *Tetrahedron Lett.*, 1991, **32**, 2565.
16. D Crich and LBL Lim, *Tetrahedron Lett.*, 1991, **31**, 1897.
17. D Crich and TJ Ritchie, *J. Chem. Soc., Perkin Trans. 1*, 1990, 945.
18. D Crich and TJ Ritchie, *Chem. Commun.*, 1988, 1461.
19. D Crich and F Hermann, *Tetrahedron Lett.*, 1993, **34**, 3385.
20. DHR Barton, D Crich and WB Motherwell, *Tetrahedron Lett.*, 1983, **24**, 4979.
21. WG Dauben, BA Kowalczyk and DP Bridon, *Tetrahedron Lett.*, 1989, **30**, 2461.
22. J Tsanaktsidis and PE Eaton, *Tetrahedron Lett.*, 1989, **30**, 6967.
23. K Gawronska, J Gawronski and HM Walborsky, *J. Org. Chem.*, 1991, **56**, 2193.
24. DHR Barton, B Lacher and SZ Zard, *Tetrahedron*, 1987, **43**, 4321.
25. FM Romero and R Ziessel, *Tetrahedron Lett.*, 1995, **36**, 6474.
26. EJ Cochrane, SW Lazer, JT Pinhey and JD Whitby, *Tetrahedron Lett.*, 1989, **30**, 7111.
27. DHR Barton, D Bridon and SZ Zard, *Tetrahedron Lett.*, 1984, **25**, 5777.
28. DHR Barton, D Bridon and SZ Zard, *Heterocycles*, 1987, **25**, 449.
29. DHR Barton, D Bridon, Y Herve, P Potier, J Thierry and SZ Zard, *Tetrahedron*, 1986, **42**, 4983.
30. DHR Barton, D Bridon and SZ Zard, *Chem. Commun.*, 1985, 1066.
31. DHR Barton, N Ozbalik and M Schmitt, *Tetrahedron Lett.*, 1989, **30**, 3263.
32. AJ Bloodworth, D Crich and T Melvin, *J. Chem. Soc., Perkin Trans. II*, 1990, 2957.
33. DHR Barton, E Castagino and JC Jaszberenyi, *Tetrahedron Lett.*, 1994, **35**, 6057.
34. DHR Barton, B Lacher, B Misterkiewicz and SZ Zard, *Tetrahedron*, 1988, **44**, 1153.

35. DHR Barton, JC Jaszberenyi and EA Theodorakis, *Tetrahedron Lett.*, 1991, **32**, 3321.
36. E Castagnino, S Corsano and DHR Barton, *Tetrahedron Lett.*, 1989, **30**, 2983.
37. AJ Walkington and DA Whiting, *Tetrahedron Lett.*, 1989, **30**, 4731.
38. AJ Bloodworth, D Crich and T Melvin, *Chem. Commun.*, 1987, 786.
39. DHR Barton, J Guilhem, Y Herve, P Potier and J Thierry, *Tetrahedron Lett.*, 1987, **28**, 1413.
40. LD Luca, G Giacomelli, G Porcu and M Taddi, *Org. Lett.*, 2001, **3**, 855.
41. DHR Barton, D Bridon, IF Picot and SZ Zard, *Tetrahedron*, 1987, **43**, 2733.
42. DHR Barton, H Togo and SZ Zard, *Tetrahedron*, 1985, **41**, 5507.
43. DHR Barton, H Togo and SZ Zard, *Tetrahedron Lett.*, 1985, **26**, 6349.
44. DHR Barton, J Boivin, E Crepon, J Sarma, H Togo and SZ Zard, *Tetrahedron*, 1991, **47**, 7091.
45. DHR Barton, D Crich and G Kretzschmer, *Tetrahedron Lett.*, 1984, **25**, 1055.
46. DHR Barton, CY Chern, JC Jaszberenyi and T Shinada, *Tetrahedron Lett.*, 1993, **34**, 6505.
47. S Poigny, M Guyot and M Samadi, *J. Chem. Soc., Perkin Trans. 1*, 1997, 2175.
48. DHR Barton, J Boivin, J Sarma, E Silva and SZ Zard, *Tetrahedron Lett.*, 1989, **30**, 4237.
49. DHR Barton and JC Sarma, *Tetrahedron Lett.*, 1990, **31**, 1965.
50. DHR Barton and W Liu, *Tetrahedron Lett.*, 1997, **38**, 2431.
51. DHR Barton, CY Chern and JC Jaszberenyi, *Tetrahedron Lett.*, 1992, **33**, 7299.
52. DHR Barton, B Lacher and SZ Zard, *Tetrahedron*, 1986, **42**, 2325.
53. T Okano, N Takakura, Y Nakano and S Eguchi, *Tetrahedron Lett.*, 1992, **33**, 3491.
54. DHR Barton, CY Chern and JC Jaszberenyi, *Tetrahedron Lett.*, 1992, **33**, 5017.
55. DHR Barton, SD Gero, BQ Sire and M Samadi, *Chem. Commun.*, 1989, 1000.
56. DHR Barton, SD Gero, BQ Sire and M Samadi, *Tetrahedron Lett.*, 1989, **30**, 4969.
57. DHR Barton, SD Gero, BQ Sire and M Samade, *Chem. Commun.*, 1988, 1373.
58. DHR Barton, SD Gero, BQ Sire and M Samadi, *J. Chem. Soc., Perkin Trans. 1*, 1991, 981.
59. DHR Barton, Y Herve, P Potier and J Thierry, *Tetrahedron*, 1987, **43**, 4297.
60. S Poigny, M Guyot and M Samadi, *J. Org. Chem.*, 1998, **63**, 1342.
61. RN Saicic and Z Cekovic, *Tetrahedron Lett.*, 1990, **31**, 4203.
62. J Boivin, E Crepon and SZ Zard, *Tetrahedron Lett.*, 1991, **32**, 199.
63. DHR Barton, D Bridon and SZ Zard, *Tetrahedron*, 1987, **43**, 5307.
64. T Ling, E Poupon, EJ Rueden and EA Theodarakis, *Org. Lett.*, 2002, **4**, 819.
65. DHR Barton, PI Dalko and SD Gero, *Tetrahedron Lett.*, 1991, **32**, 4713.
66. DHR Barton and M Ramesh, *J. Am. Chem. Soc.*, 1990, **112**, 891.
67. W He, H Togo, H Ogawa and M Yokoyama, *Heteroatom Chem.*, 1997, **8**, 411.
68. W He, H Togo and M Yokoyama, *Tetrahedron Lett.*, 1997, **38**, 5541.
69. W He, H Togo, Y Waki and M Yokoyama, *J. Chem. Soc., Perkin Trans. 1*, 1998, 2425.
70. DHR Barton, D Crich, Y Herve, P Potier and J Thierry, *Tetrahedron*, 1985, **41**, 4347.
71. ALJ Beckwith and BP Hay, *J. Am. Chem. Soc.*, 1988, **110**, 4415.
72. BP Hay and ALJ Beckwith, *J. Org. Chem.*, 1989, **54**, 4330.
73. M Newcomb and MU Kumar, *Tetrahedron Lett.*, 1991, **32**, 45.
74. ALJ Beckwith and IGE Davison, *Tetrahedron Lett.*, 1991, **32**, 49.
75. Y Togo, N Nakamura and H Iwamura, *Chem. Lett.*, 1991, 1201.
76. M Newcomb and TM Deeb, *J. Am. Chem. Soc.*, 1987, **109**, 3163.
77. M Newcomb and MU Kumar, *Tetrahedron Lett.*, 1990, **31**, 1675.
78. M Newcomb and C Ha, *Tetrahedron Lett.*, 1991, **32**, 6493.
79. M Newcomb, SU Park, J Kaplan and DJ Marquardt, *Tetrahedron Lett.*, 1985, **26**, 5651.
80. M Newcomb and JL Esker, *Tetrahedron Lett.*, 1991, **32**, 1033.
81. J Biovin, E Fouguet and SZ Zard, *Tetrahedron Lett.*, 1991, **32**, 4299.
82. DHR Barton and D Crich, *Chem. Commun.*, 1984, 774.
83. D Crich and SM Fortt, *Synthesis*, 1987, 35.
84. DHR Barton and D Crich, *Tetrahedron Lett.*, 1985, **26**, 757.
85. H Togo, M Fujii and M Yokoyama, *Bull. Chem. Soc. Jpn*, 1991, **64**, 57.
86. EC Taylor, HW Altland and F Kienzle, *J. Org. Chem.*, 1976, **41**, 24.
87. M Hasebe and T Tsuchiya, *Tetrahedron Lett.*, 1987, **28**, 6207.
88. K Okada, K Okamoto and M Oda, *J. Am. Chem. Soc.*, 1988, **110**, 8736.
89. K Okada, K Okubo and M Oda, *Tetrahedron Lett.*, 1992, **33**, 83.

35. DHR Barton, [illegible] and [illegible], *Tetrahedron Lett.*, 1991, 32, [illegible]
36. [illegible] and DHR Barton, *Tetrahedron Lett.*, 1990, 31, [illegible]
37. AJ Waldmann and DA Whiting, *Tetrahedron Lett.*, 1990, 31, [illegible]
38. AJ Blankstein, LG [illegible] and [illegible], *Chem. Commun.*, 1986, [illegible]
39. DHR Barton, J Cadilhac, Y [illegible], P [illegible] and [illegible], *Tetrahedron Lett.*, 1988, 28, [illegible]
40. ED [illegible], G [illegible], C [illegible] and M [illegible], *Org. Lett.*, 2001, 3, [illegible]
41. DHR Barton, D [illegible], [illegible] and SZ Zard, *Tetrahedron*, 1988, 44, [illegible]
42. DHR Barton, [illegible] and SZ Zard, *Tetrahedron*, 1987, 43, [illegible]
43. DHR Barton, H [illegible] and SZ Zard, *Tetrahedron Lett.*, 1985, 26, [illegible]
44. DHR Barton, [illegible], H [illegible] and SZ Zard, *Tetrahedron*, 1991, 47, [illegible]
45. DHR Barton, [illegible] and [illegible], *Tetrahedron Lett.*, 1985, 26, [illegible]
46. DHR Barton, CY Chern, JC Jaszberenyi and [illegible], *Tetrahedron Lett.*, 1993, 34, [illegible]
47. S [illegible], M [illegible] and M Samadi, *J. Chem. Soc. Perkin Trans. I*, 1997, [illegible]
48. DHR Barton, J [illegible], [illegible] and SZ Zard, *Tetrahedron Lett.*, 1989, 30, [illegible]
49. DHR Barton and [illegible], *Tetrahedron Lett.*, 1990, 31, [illegible]
50. DHR Barton and W Liu, *Tetrahedron Lett.*, 1997, 38, [illegible]
51. DHR Barton, CY Chern and JC Jaszberenyi, *Tetrahedron Lett.*, 1992, 33, [illegible]
52. DHR Barton, [illegible] and SZ Zard, *Tetrahedron*, 1998, 54, [illegible]
53. T [illegible], [illegible] and [illegible], *Tetrahedron Lett.*, 1992, 33, [illegible]
54. DHR Barton, CY Chern and JC Jaszberenyi, *Tetrahedron Lett.*, 1992, 33, [illegible]
55. DHR Barton, SD Géro, BQ Sire and M Samadi, *Chem. Commun.*, 1989, [illegible]
56. DHR Barton, SD Géro, BQ Sire and M Samadi, *Tetrahedron Lett.*, 1989, 30, [illegible]
57. DHR Barton, SD Géro, BQ Sire and M Samadi, *Chem. Commun.*, 1988, [illegible]
58. DHR Barton, SD Géro, BQ Sire and M Samadi, *J. Chem. Soc. Perkin Trans. I*, 1991, [illegible]
59. DHR Barton, [illegible], [illegible] and [illegible], *Tetrahedron*, 1987, 43, [illegible]
60. [illegible], M Géro and M Samadi, *J. Org. Chem.*, 1988, 53, [illegible]
61. [illegible] and [illegible], *Tetrahedron Lett.*, 1990, 31, [illegible]
62. [illegible] and SZ Zard, *Tetrahedron Lett.*, 1991, 32, [illegible]
63. DHR Barton, [illegible] and SZ Zard, *Tetrahedron*, 1987, 43, [illegible]
64. [illegible], [illegible], [illegible] and [illegible], *Org. Lett.*, 2001, 3, [illegible]
65. DHR Barton, [illegible] and [illegible], *Tetrahedron Lett.*, 1991, 32, [illegible]
66. DHR Barton and M Ramesh, *J. Am. Chem. Soc.*, 1990, 112, [illegible]
67. W Hu, [illegible], H [illegible] and M [illegible], *Tetrahedron*, [illegible]
68. W Hu, H [illegible] and M [illegible], *Tetrahedron Lett.*, 1997, 38, [illegible]
69. W Hu, H [illegible], [illegible] and M [illegible], *J. Chem. Soc. Perkin Trans. I*, 1998, [illegible]
70. DHR Barton, [illegible], [illegible] and [illegible], *Tetrahedron*, 1985, 41, [illegible]
71. [illegible] and [illegible], *J. Am. Chem. Soc.*, 1988, 110, [illegible]
72. [illegible] and [illegible], *J. Org. Chem.*, 1989, 54, [illegible]
73. [illegible] and [illegible], *Tetrahedron Lett.*, 1991, 32, [illegible]
74. [illegible] and [illegible], *Tetrahedron Lett.*, 1991, 32, [illegible]
75. [illegible], [illegible] and [illegible], *Chem. Lett.*, 1984, [illegible]
76. M Newcomb and [illegible], *J. Am. Chem. Soc.*, 1987, 109, [illegible]
77. M [illegible] and [illegible], *Tetrahedron Lett.*, 1990, 31, [illegible]
78. M [illegible] and [illegible], *Tetrahedron Lett.*, 1991, 32, [illegible]
79. M [illegible], [illegible], J Kaplan and [illegible], *Tetrahedron Lett.*, 1985, 26, [illegible]
80. [illegible] and [illegible], *Tetrahedron Lett.*, 1991, 32, [illegible]
81. [illegible], [illegible] and SZ Zard, *Tetrahedron Lett.*, 1991, 32, [illegible]
82. DHR Barton and [illegible], *Chem. Commun.*, 1986, [illegible]
83. [illegible] and [illegible], [illegible], 1988, [illegible]
84. DHR Barton and [illegible], *Tetrahedron Lett.*, 1984, 25, [illegible]
85. [illegible], [illegible] and [illegible], [illegible], 1976, [illegible]
86. [illegible], [illegible] and [illegible], *J. Org. Chem.*, 1976, 41, [illegible]
87. [illegible], *Tetrahedron Lett.*, 1988, 28, [illegible]
88. [illegible]

– 9 –

Free Radical Reactions with Metal Hydrides

Free radicals mentioned in this chapter are not useful for organic synthesis. However, it is important to know the radical character of metal hydride reagents, since metal hydride reagents sometimes behave not only as a polar (ionic) hydride donor species, but also a single electron donor species, depending on the substrates and reaction conditions.

Single electron transfer (SET) reactions with metals such as Fe, Cu, Mn, Ce, Sm, etc. have been described in the previous chapters. So here, SET with metal hydrides is described. It has been long believed that treatment of alkyl halides or ketones with metal hydrides such as $LiAlH_4$ or $NaBH_4$ produces the corresponding reduction products by the reaction with a hydride species ($H:^-$). However, in the 1970s some evidence for radical interventions was reported, especially radical reactions via SET, and a radical cyclization method with 5-*exo-trig* manner, was mainly used as an indirect proof for radical reactions therein, and often direct ESR observation of the radical intermediates was tried. This chemistry was well studied especially by Ashby [1–8]. Treatment of 2,2-dimethyl-5-hexenyl-1-iodide (**1**) (primaryl, but sterically hindered neopentyl-type iodide) with $LiAlH_4$ generated the corresponding cyclized 1,1,3-trimethylcyclopentane (**2**) in 96% yield, together with a direct reduction product, 5,5-dimethyl-1-hexene (**3**), in 2.5% yield. Moreover, treatment of the same

CH3 CH3 I **1** — $LiAlH_4$ or $LiAlD_4$ (SET) r.t. → [CH3 CH3 ⇌ CH3 CH3]

(9.1)

$LiAlH_4$ | $LiAlD_4$

CH3 CH3 CH3 **3** 2.5% + CH3 CH3 CH3 **2** 96%

CH2D CH3 CH3 **3′** 5.5% (d_1, 65%) + CH2D CH3 CH3 **2′** 69% (d_1, 57%)

neopentyl-type iodide with $LiAlD_4$ gave the corresponding 1,1,3-trimethylcyclopentane (**2′**) in 69% yield with 57% d_1 and 43% d_0, together with 5,5-dimethyl-1-hexene (**3′**) in 5.5% yield with 65% d_1 and 35% d_0, as shown in eq. 9.1. Here d_1 means the abstraction of a deuterium atom from $LiAlD_4$, and d_0 means the abstraction of a hydrogen atom from the solvent, by the radical species. These results indicate that most reactions of 2,2-dimethyl-5-hexenyl-1-iodide (**1**) with $LiAlH_4$ proceeds via SET, not a polar pathway, and the reaction proceeds through cyclization of the 2,2-dimethyl-5-hexen-1-yl radical derived from the 2,2-dimethyl-5-hexenyl-1-iodide anion radical formed by SET reduction with $LiAlH_4$ (eq. 9.1).

When steric hindrance in substrates is increased, and when the leaving anion group in substrates is iodide, SET reaction is much induced (Cl < Br < I). This reason comes from the fact that steric hindrance retards the direct nucleophilic reduction of substrates by a hydride species, and the σ^* energy level of C–I bond in substrates is lower than that of C–Br or C–Cl bond. Therefore, metal hydride reduction of alkyl chlorides, bromides, and tosylates generally proceeds mainly via a polar pathway, i.e. S_N2. Since LUMO energy level in aromatic halides is lower than that of aliphatic halides, SET reaction in aromatic halides is induced not only in aromatic iodides but also in aromatic bromides. Eq. 9.2 shows reductive cyclization of *o*-bromophenyl allyl ether (**4**) via an sp^2 carbon-centered radical with $LiAlH_4$.

LiAlH$_4$, Et$_2$O r.t.; **4** → **5**, 90%; (–Br⊖) (9.2)

LiAlD$_4$, Et$_2$O r.t.; **6** → **7a** : **7b** = 1 : 1; $d_1 / d_0 = 63 / 37$ (9.3)

Though reduction of aromatic halides with $NaBH_4$ does not proceed at all, photolytic treatment of the same aromatic halides and $NaBH_4$ with a mercury lamp provides the reduction products via SET pathway. Eq. 9.3 shows the reduction of *trans*-β-bromostyrene (**6**) with $LiAlD_4$ via SET pathway. The vinyl radical rapidly inverses on the sp^2 carbon-centered atom, so a mixture of *cis*- and *trans*-styrenes (**7a**) and (**7b**) with $d_1/d_0 = 63/37$ is formed. d_0-Styrene is again formed through the abstraction of a hydrogen atom from the solvent by the vinyl radical. Since the activation energy of *cis/trans* inversion in the vinyl radical is ~2 kcal/mol, this isomerization occurs even at lower temperature. As an interesting study, the results on the treatment of bicyclic cyclopropane-1,1-dibromide (**8**) with $LiAlH_4$, Bu_3SnH, CH_3MgBr, and $NaBH_4$ are shown in eq. 9.4. Isomeric ratios with $LiAlH_4$, CH_3MgBr, and $NaBH_4$ are close to that with a typical radical reagent, Bu_3SnH. Moreover, all these reactions have an induction period of about 5 ~ 10 min, and molecular oxygen retards these reactions. Based on these results, these reactions proceed via SET pathway and a bicyclic 1-bromocyclopropyl radical is the key intermediate [9–13].

8 —MH, DMF→ **9a** *cis* + **9b** *trans* (9.4)

MH	Yield	*cis / trans*
$LiAlH_4$	73%	3.0
Bn_3SnH	82%	2.5
CH_3MgBr	72%	2.5
$NaBH_4$	79%	1.8

10 —$LiAlH_4$, THF-d_8, r.t.→ [radical] → **11a** < 0.5% + **11b** 95% (9.5)

When 2,2-dimethylhexyl-1-iodide (**10**), a saturated neopentyl-type halide, was treated with $LiAlH_4$ in THF-d_8, the corresponding reduction products (**11a**) and (**11b**) were obtained quantitatively, and 95% of the reduction product contains d_1, as shown in eq. 9.5 [14]. This result again indicates a 2,2,-dimethylhexyl radical is the real intermediate for the reduction product.

Treatment of perylene (**12**) with $LiAlH_4$ at room temperature induces the appearance of deep blue color. This color corresponds to the perylene anion radical formed by the SET from $LiAlH_4$ to LUMO of perylene. ESR measurement indicates that ~80% yield of the perylene anion radical is formed [15]. Properly, formation of the anion radical depends on the kind of metal hydride reagent as shown in eq. 9.6.

perylene **12** —MH, THF, r.t.→ anion radical blue (9.6)

MH	Yield of anion radical
$LiAlH_4$	80 %
$NaAlH_4$	80 %
AlH_3	31 %
MgH_2	35 %
HMgCl	20 %

Treatment of aromatic ketones with $LiAlH_4$ or AlH_3 induces the formation of ketyl radicals via SET. As mentioned above, SET is induced when the substrates have steric hindrance at the reaction position, and when LUMO energy level is lower, such as in polycondensed aromatics and ketones.

These types of SET reactions are not useful for organic synthesis; however, it is important to know about the SET character of metal hydride reagents.

REFERENCES

1. S-K Chung and F-F Chung, *Tetrahedron Lett.*, 1979, 2473.
2. PR Singh, A Nigam and JM Khurana, *Tetrahedron Lett.*, 1980, **21**, 4753.
3. EC Ashby, AB Goel and RN Depriest, *J. Am. Chem. Soc.*, 1980, **102**, 7780.
4. EC Ashby, RN Depriest and AB Goel, *Tetrahedron Lett.*, 1981, **22**, 1763.
5. EC Ashby, RN Depriest and TN Pham, *Tetrahedron Lett.*, 1983, **24**, 2825.
6. EC Ashby, B Wenderoth, TN Pham and WS Park, *J. Org. Chem.*, 1984, **49**, 4505.
7. EC Ashby and TN Pham, *Tetraheron Lett.*, 1987, **28**, 3197.
8. CO Welder and EC Ashby, *J. Org. Chem.*, 1997, **62**, 4829.
9. JT Groves and KW Ma, *J. Am. Chem. Soc.*, 1974, **96**, 6527.
10. K Tsujimoto, S Tasaka and M Ohashi, *Chem. Commun.*, 1975, 758.
11. SK Chung, *J. Org. Chem.*, 1980, **45**, 3513.
12. EC Ashby, RN Depriest, AB Goel, B Wenderoth and TN Pham, *J. Org. Chem.*, 1984, **49**, 3545.
13. M Kropp and GB Schuster, *Tetrahedron Lett.*, 1987, **28**, 5295.
14. EC Ashby and CO Welder, *Tetrahedron Lett.*, 1995, **36**, 7171.
15. EC Ashby, AB Goel, RN Depriest and HS Prasad, *J. Am. Chem. Soc.*, 1981, **103**, 973.

– 10 –

Stereochemistry in Free Radical Reactions

Stereochemistry in radical reactions for organic synthesis has not been studied very extensively, because mild or low temperature-promoted radical reaction methods are extremely limited and the stereoselectivity in radical reactions is generally rather poor. Recently, however, stereoselective organic synthesis with radical reactions has become popular, since mild radical reaction methods such as the Barton decarboxylation, Et_3B-initiated Bu_3SnH reaction, etc. have been developed. Normally, low temperature-initiated radical reactions induce high stereoselectivity.

10.1 REDUCTION

Since Et_3B-initiated radical reactions with Bu_3SnH under aerobic conditions, can be carried out at 0 °C or lower temperature, stereoselective reduction of reactive α-haloesters or α-haloketones can be performed. Eq. 10.1 shows Et_3B-mediated Bu_3SnH reduction of β-methoxy-α-bromo ester in the presence and absence of $MgBr_2$ (Lewis acid) at 0 °C. *Anti/syn* diastereoselectivity depends on the absence or presence of a Lewis

OCH3 t-Bu CO2CH3 CH3 Br **1** —(Bu_3SnH, Et_3B / CH_2Cl_2, 0 °C)→ OCH3 t-Bu CO2CH3 CH3 **2a** *syn* + OCH3 t-Bu CO2CH3 CH3 **2b** *anti*

	Yield	*syn*		*anti*
$MgBr_2$(5 eq.)	91%	(33	:	1)
—	75%	(1	:	21)

(10.1)

[t-Bu, CH3, CO2CH3, CH3O, H, Bu3Sn–H] ⇒ *anti* (without $MgBr_2$)
[I]

[Bu3Sn–H, H, CH3, O····Mg2+, CH3, C=O, OCH3, t-Bu] ⇒ *syn* (with $MgBr_2$)
[II]

acid, and can be explained as follows. Both transition states **[I]** and **[II]** in the hydrogen atom abstraction from Bu_3SnH by α-methoxycarbonyl radical are shown below. Bu_3SnH donates a hydrogen atom to the α-methoxycarbonyl radical from the less-hindered side to provide *anti*-isomer in the absence of a Lewis acid as shown in [**I**], while the opposite *syn*-isomer is formed by the chelation between Mg^{2+} and two oxygen atoms of the methoxy and carbonyl groups in the presence of $MgBr_2$, as shown in [**II**]. Thus, Bu_3SnH donates a hydrogen atom from the less-hindered side of the formed α-methoxycarbonyl radical in both cases [1–4].

Experimental procedure 1 (eq. 10.1).

To a solution of $MgBr_2 \cdot Et_2O$ in dichloromethane was added α-bromo ester (0.1 M) of dichloromethane. The mixture was stirred for 5 min at 25 °C, and then cooled at 0 °C. After addition of Bu_3SnH (2 eq.), Et_3B (1 M, hexane solution), divided to 3 parts (0.2 eq. × 3 = 0.6 eq.), was added to the mixture within 15 min. After 2 h at 0 °C, *m*-dinitrobenzene (0.5 eq.) was added to the reaction mixture, which was further washed with aq. $NaHCO_3$ solution, extracted with dichloromethane, and dried over $MgSO_4$. After filtration and removal of the solvent, the residue was poured into hexane and $Bu_4N^+F^-$ (2.5 eq.) was added to the hexane solution and stirred for 5 min at 25 °C. After column chromatography (eluent: ethyl acetate/hexane = 15/85), a reduction product was obtained in 91% yield with *syn*:*anti* = 33:1 [1].

The following boronate derivative, formed by the reaction of 1,3-diol (**3**) with Et_3B in the presence of air, reacts with Bu_3SnH to form *anti* reduction product (**4**) with high stereocontrol (eq. 10.2) [1–4].

OH OH O, OtBu, Me SePh, **3** —(Et_3B, air)→ [boronate: Et, B, O, O, O, OtBu, Me] —(Bu_3SnH)→ OH OH O, OtBu, Me, **4** (*anti* : *syn* = 20 : 1) (10.2)

5 (R, Ph, O, CH_3) —($C_{12}H_{25}SH$, ${}^tBuO{-}N{=}N{-}OBu^t$ (initiator), $(Me_3Si)_3SiH$, 30 °C)→ [R, Ph, $OSi(SiMe_3)_3$, CH_3] → **6a** (R, Ph, OH, CH_3) + **6b** (R, Ph, OH, CH_3) (10.3)

R = Me (2.9 : 1)
= Pr^i (12.6 : 1)
= Bu^t (5.3 : 1)

Reduction of chiral ketones (**5**) with the $(Me_3Si)_3SiH/C_{12}H_{25}SH$/initiator system generates alcohols with moderate *anti*/*syn* diastereoselectivity as shown in eq. 10.3 [5]. Chiral reduction of α-haloesters (**7**), (**11**) and α-haloketone (**9**) with chiral trialkyltin hydrides (**A**), (**B**), (**C**) is known, as shown in eqs. 10.4–10.6 [6–8]. All these reactions were carried out at −78 °C initiated by Et_3B. Chiral binaphthyl-based organotin reagents (**A**) and (**B**) generate the corresponding reduced ester (**8**) and ketone (**10**) with moderate %*e.e.*, respectively, while *bis*[(1*S*, 2*S*, 5*R*)-menthyl]phenyltin hydride (**C**) provides the corresponding reduced ester (**12**) with 86%ee, in the presence of a Lewis acid.

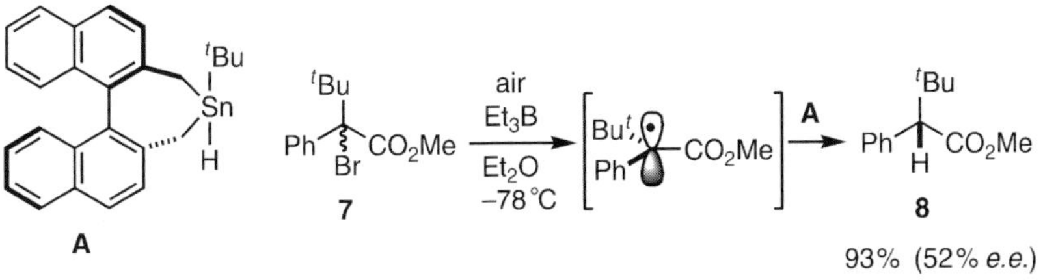

(10.4)

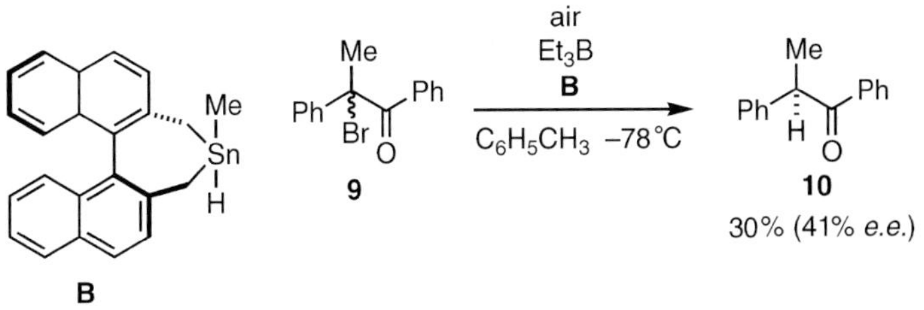

(10.5)

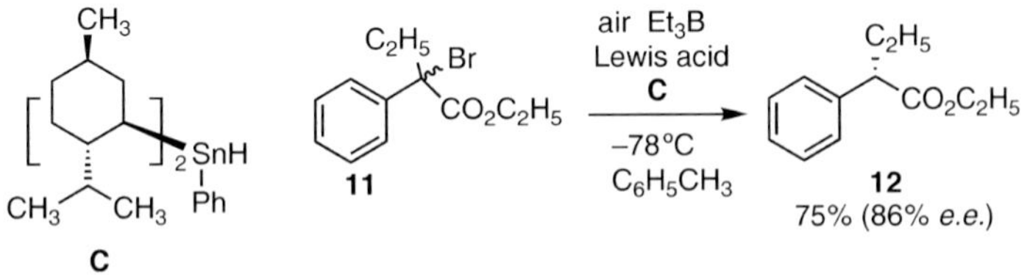

(10.6)

When the formed α-ester radicals (π radicals) abstract a hydrogen atom from the tin reagents, one of the two sides is preferential in the transition states.

10.2 CARBON–CARBON BOND FORMATION

Since Barton decarboxylation can be performed under mild conditions, thermal or photolytic treatment of the Barton ester (**13**) of propionic acid with *N*-hydroxy-2-thiopyridone in the presence of chiral menthyl acrylates generates addition products (**14**). However, diastereoselectivity is rather poor, since the chiral menthyl center is too far away from the C–C bond-forming position, as shown in eq. 10.7. When the chiral center is adjacent to the reaction position, stereocontrol is significantly affected, as shown in eq. 10.8 [9–12].

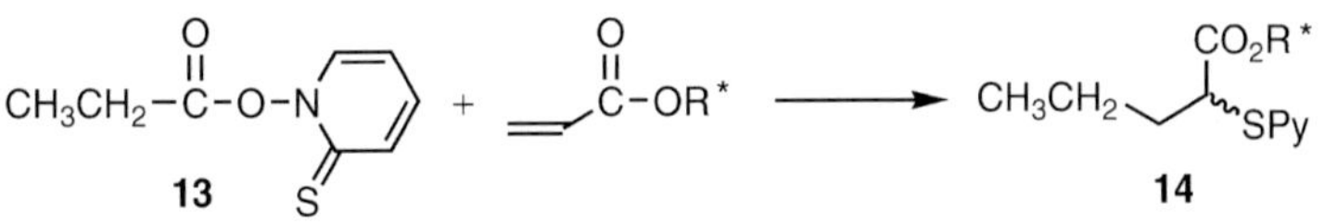

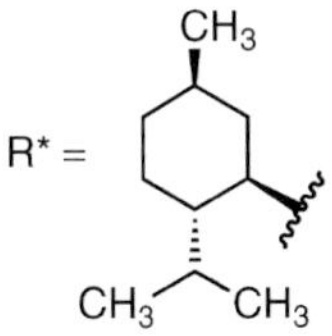

(10.7)

	Conditions	Yield	*d.e.*
Et_2O	W-$h\nu$, 15°C	45%	8%
$C_6H_5CH_3$	Δ	44%	4%

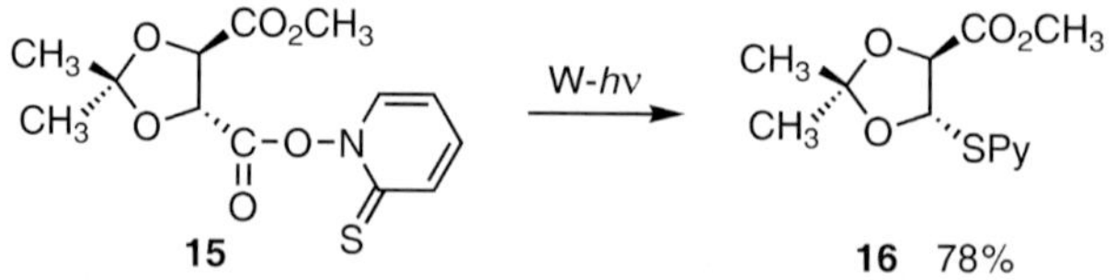

(10.8)

Stereoselective addition of an alkyl radical to the C–C double bond bearing a chiral sulfoxide has been reported. Thus, eq. 10.9 shows the addition of a *tert*-butyl radical formed from *tert*-butyl iodide with Et_3B via S_Hi reaction, to electrophilic vinyl sulfoxide (**17**), which takes a 6-membered chair conformation through intramolecular hydrogen bonding between the sulfoxide and OH groups. Since the carbon-centered radical formed by the addition of a *tert*-butyl radical abstracts a hydrogen atom of Bu_3SnH from the less-hindered side (opposite side to the β-methyl group, [**III**]), the *syn* form becomes the major product [13–17].

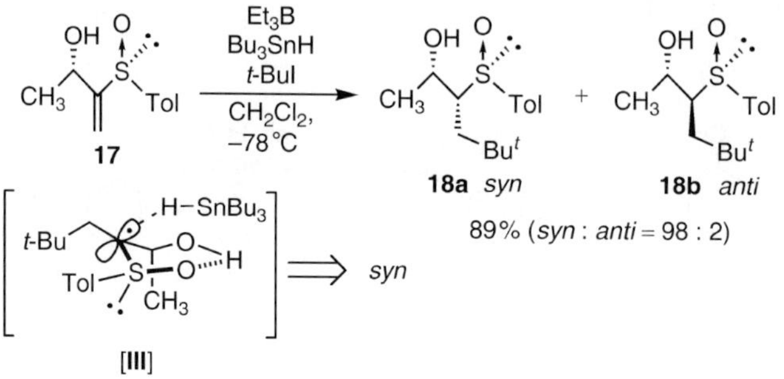

(10.9)

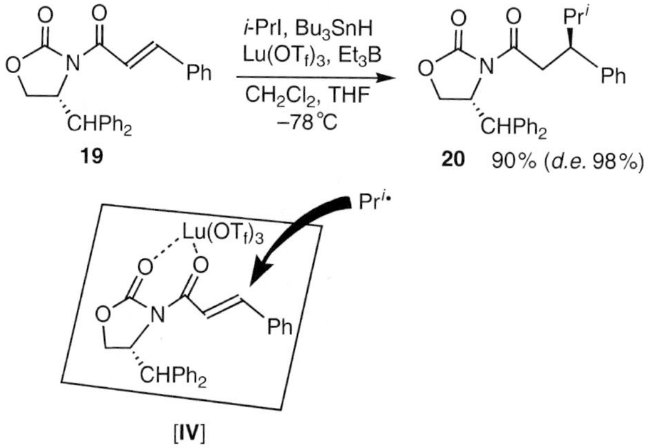

(10.10)

When the same kind of reaction is carried out in the presence of a Lewis acid, instead of intramolecular hydrogen bonding, the formation of addition product with high diastereoselectivity can be achieved [18–23]. Eq. 10.10 shows the chiral addition of an isopropyl radical to Lewis acid-chelated chiral α,β-unsaturated-*N*-enoyloxazolidinone; 6-membered rigid ring [**IV**] in the presence of $Lu(OTf)_3$ ($T_f = CF_3SO_2$).

Radical-mediated reaction of γ-hydroxy or γ-alkoxy-α-methylene ester (**21**) with *iso*-PrI in the presence of $MgBr_2$ gives the *syn*-adduct (**22a**) predominantly via the intermediate [**V**] as shown in eq. 10.11. The same chelation-controlled radical 1,3-stereo induction with γ-alkyl-α-methylene ester (**23**) in the presence of $MgBr_2$ gives predominantly the *syn* adduct (**24b**) via the intermediate [**VI**] (eq. 10.12).

$MgBr_2 \cdot Et_2O$, Et_3B, O_2, ^iPrI, Bu_3SnH, CH_2Cl_2, 0 °C

21 → **22a** *syn* + **22b** *anti* (4.5 : 1)

syn [**V**] *anti* [**V′**]

(10.11)

$MgBr_2 \cdot Et_2O$, ^tBuBr, Bu_3SnH, Et_3B, O_2, −78 °C

23 → **24a** *anti* + **24b** *syn* 83% (2 : 98)

(10.12)

[**VI**] ⟹ *syn*

Based on a chiral template or chiral catalyst, radical allylation via S_H2' can be carried out with high diastereoselectivity [24–31]. Eq. 10.13 shows the allylation of α-phenylseleno-acetylacetamide (**25**) bearing a chiral oxazolidine group with allyltributyltin at −78 °C under irradiation with a mercury lamp. Et_3B-initiated 1,3-diastereoselective radical allylation of α-hydroxy-α′-bromoketone (**27**) with allyltributyltin in the presence of $ZnCl_2$ or $MgBr_2$ provides superior diastereoselectivity, as shown in eq. 10.14. Radical reaction of ethyl γ-methoxy-α-methylenealkanoate (**29**) with EtI and allyltributyltin in the presence of $MgBr_2$ initiated by Et_3B and air at 0 °C generates exclusively ethyl (2*R*,4*R*)-2-allyl-4-methoxy-2-propylalkanoate (**30a**) in good yield, through the addition of $Et^{\cdot}$ to the reactive methylene group and subsequent S_H2'

allylation, based on the chelation of methoxy and ester carbonyl groups on Mg^{2+} (eq. 10.15a and 10.15b). The same type of addition-allylation of fumarate (**31**) with RI and allyltributyltin in the presence of $Sm(OTf)_3$ at − 78 °C initiated by Et_3B-air generates 1,2-disubstituted succinate (**32**) with high diastereoselectivity (eq. 10.15b).

Hg-*hν*, allyl-$SnBu_3$, CH_2Cl_2, −78 °C: **25** → **26** 100%; 0 °C *d.e.* = 32 : 1; −78 °C *d.e.* = 100 : 1 (10.13)

27 → $MgBr_2$, −78 °C, CH_2Cl_2-Et_2O → [Mg^{2+} chelate] → allyl-$SnBu_3$, Et_3B, O_2, −78 °C → **28a** + **28b** 81%(100 : 1) (10.14)

29 → EtI, $MgBr_2 \cdot Et_2O$, Bu_3Sn-allyl, Et_3B, O_2, CH_2Cl_2, 0 °C → **30a** + **30b** 53%(12 : 1) (10.15a)

31 → allyl-$SnBu_3$, RI, Et_3B, O_2, Sm $(OTf)_3$, −78 °C → **32** (10.15b)

R = Et 87%
= ^{i}Bu 85%
= ^{i}Pr 82%
= ^{t}Bu 64%

33 → ^{i}PrI, Bu_3SnH, Et_3B, O_2, CH_2Cl_2, −78 °C → **34** 73% (94 : 6) ---→ Amino acids (10.16)

Bu_3Sn (allyl)
Lewis acid, ^iPrI,
Et_3B, O_2, CH_2Cl_2

35 → 36

(*bis*oxazoline)

(10.17)

Lewis acid (eq)	Temp °C	Yield (%)	*d.e.*	*e.e.* (%)
MgI_2 (1)	−78	82	19:1	86
MgI_2 (0.3)	−78	93	37:1	93
MgI_2 (0.3)	−40	70	36:1	65

Highly diastereoselective alkyl radical addition to Oppolzer's camphorsultam derivative (**33**) of oxime provides enantiomerically pure α-alkyl-α-amino acid derivative (**34**) at −78 °C by the same method as shown in eq. 10.16. Moreover, enantioselective tandem radical 1,2-difunctionalization of cinnamamide (**35**) can be carried out with high stereoselectivity, using the chelation manner of the cinnamamide and a chiral bisoxazoline ligand on MgI_2, as shown in eq. 10.17.

In stereoselective cyclization reactions [32–39], treatment of 2-bromoacetal (**37**) bearing an allene group and a chiral template with Bu_3SnH/Et_3B generates the corresponding chiral β-vinyl-γ-lactone (**38**) via 5-*exo-trig* manner, as shown in eq. 10.18. Here, [**VII**] is the transition state for the cyclization. With the same method, but using chiral perhydro-1,3-benzoxazine, preparation of 3-alkylpyrrolidines and lactams with high %*e.e.* can be carried out.

1) Bu_3SnH, Et_3B 88%
2) HCl, THF 63%
3) PCC, Al_2O_3 90%

37 → 38 (99% *e.e.*)

[VII]

(10.18)

Bu_3SnH / cat. Et_3B

39 → 40a + 40b

(10.19)

Lewis acid		Yield
—	THF	92% (90 : 10)
Et_2AlCl	$C_6H_5CH_3$	86% (10 : 90)

Lewis acid	Yield	*cis / trans*
—	77%	1 / 4.5
Me_3Al	74%	5.8 / 1
$(i\text{-}Bu)_3Al$	69%	6.8 / 1

(10.20)

Treatment of α-alkylidene-γ-lactones (**39**) with the Bu_3SnH/Et_3B system produces the corresponding 5-*exo-trig* cyclization products with high stereoselectivity (eq. 10.19). The effect of Lewis acids in the radical cyclization of β-allyloxyalkyl phenyl selenide (**41**) was investigated, as shown in eq. 10.20. Normal cyclization without a Lewis acid occurs to give mainly *trans*-2,4-disubstituted tetrahydrofuran (**42**), while cyclization in the presence of a Lewis acid such as trialkylaluminium predominantly provides the corresponding *cis* isomer (**42**). Both transition states [**VIII**] and [**IX**] are shown. Inexpensive carbohydrates can be used as removable chiral scaffolds for 5-*exo-trig* cyclization. Eq. 10.21 shows the preparation of ester (**44**) with high diastereoselectivity from compound (**43**).

(10.21)

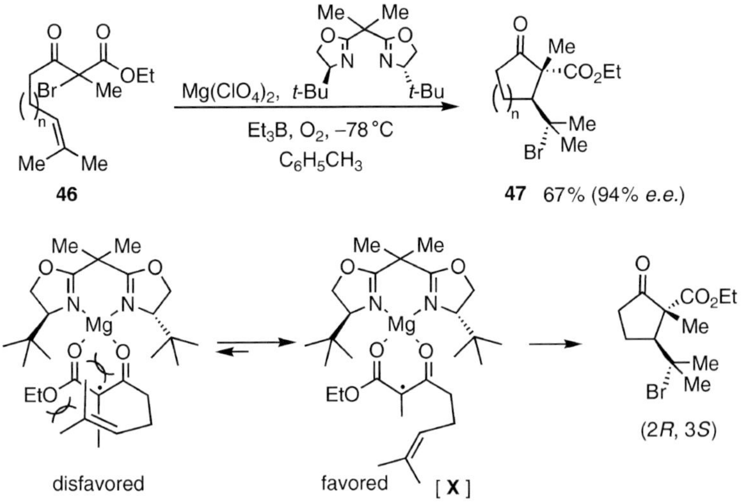

(10.22)

Chiral Lewis acid-catalyzed atom-transfer cyclization is performed by the reaction of α-bromo-α-methyl-β-keto esters (**46**) in the presence of $Mg(ClO_4)_2$ and a chiral ligand such as *bis*oxazoline, initiated by Et_3B, as shown in eq. 10.22. The Lewis acid promotes not only the atom-transfer radical cyclization, but also high stereoselectivity through the chelation onto Mg^{2+}. Intermediate [**X**] is a plausible chelated radical intermediate.

As a further stereoselective organic synthesis [40–47] using reactive sp^2 carbon-centered radicals, eq. 10.23 shows the preparation of chiral 4-*tert*-butylcyclohexene (**49**) from the optically pure *o*-bromophenyl sulfoxide (**48**) through 1,5-H shift by sp^2 carbon-centered radical, followed by β-elimination. This reaction looks like a thermal concerted intramolecular elimination reaction (Ei).

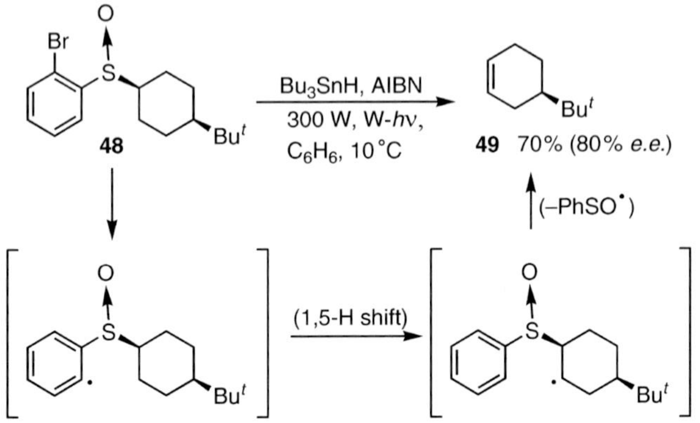

(10.23)

Since the amide group and the aromatic ring are perpendicular (atropisomer) in *N*-methyl-*N*-(2-iodo-4,6-dimethyl)phenyl acrylamide, free rotation between the C–N bond does not occur, so each enantiomer can be separated. Once an sp^2 carbon-centered radical is formed from each enantiomer, chiral oxindole (**51a** or **51b**) with high *%e.e.* can be obtained via 5-*exo-trig* cyclization as shown in eq. 10.24a and 10.24b.

50a — Et_3B, Bu_3SnH, C_6H_6, 25 °C → **51a** (84% *e.e.*) (10.24a)

50b — Et_3B, Bu_3SnH, C_6H_6, 25 °C → **51b** (82% *e.e.*) (10.24b)

52 — $Eu(OTf)_3$, Et_3B, O_2, CH_2Cl_2, −78 °C → **53** 71% (12.5 : 1) (10.25)

→ K_2CO_3, THF, H_2O → **54** (74% *e.e.*)

Diastereoselective atom-transfer addition reaction of bromoacetate (**52**) derived from D-xylose as a chiral auxiliary, with 1-hexene initiated by Et_3B at − 78 °C proceeds effectively, as shown in eq. 10.25. Treatment of the adduct (**53**) under basic conditions generates (*S*)-γ-butyl-γ-lactone with 74*%e.e.*

Under oxidative conditions, treatment of chiral menthyl β-ketoesters (**55**) with $Mn(OAc)_3/Yb(OT_f)_3$ (Lewis acid) in CF_3CH_2OH generates the corresponding cyclized products with high diastereoselectivity. Eq. 10.26 shows the treatment of chiral menthyl β-keto esters (**55**) with $Mn(OAc)_3$-mediated oxidative free radical cyclization to form a single isomer (**56a**) predominantly. This method can be applied to the enantioselective synthesis of (+)-triptophenolide with 90*%e.e.*, which is a biologically active natural product.

$Mn(OAc)_3$

55 56a 56b (10.26)

R	*d.e.*	Yield
Ph	38:1	77%
3,5-$(MeO)_2C_6H_3$	86:1	75%
3,5-$(i\text{-}PrO)_2C_6H_3$	>99:1	53%
2-naphthyl	>99:1	67%

(+)-Triptophenolide

Recently, as mentioned above, stereoselective organic synthesis with high *d.e.* (diastereomer excess) and high *e.e.* (enantiomer excess) by means of the radical reactions, has been partly established under mild conditions.

REFERENCES

1. Y Guindon and J Rancourt, *J. Org. Chem.*, 1998, **63**, 6554.
2. B Giese, W Damm, F Wetterich, HG Zeitz, J Rancourt and Y Guindon, *Tetrahedron Lett.*, 1993, **34**, 5885.
3. Y Guindon, G Jung, B Guerin and WW Ogilvie, *Synlett*, 1998, 213.
4. J-P Bouvier, G Jung, Z Liu, B Guerin and Y Quindon, *Org. Lett.*, 2001, **3**, 1391.
5. B Giese, W Damm, J Dickhaut, F Wetterich, S Sum and DP Curran, *Tetrahedron Lett.*, 1991, **32**, 6097.
6. D Nanni and DP Curran, *Tetrahedron Asymmetry*, 1996, **7**, 2417.
7. M Blumenstein, K Schwarzkopf and JO Metzger, *Angew. Chem. Int. Ed.*, 1997, **36**, 235.
8. D Dakternieks, K Dunn, VT Perchyonok and CH Schiesser, *Chem. Commun.*, 1999, 1665.
9. D Crich and JW Davies, *Tetrahedron Lett.*, 1987, **28**, 4205.
10. DHR Barton, AG Olesker, SD Gero, B Lacher, C Tachdjian and SZ Zard, *Chem. Commun.*, 1987, 1790.
11. P Garner and JT Anderson, *Tetrahedron Lett.*, 1997, **38**, 6647.
12. P Garner and JT Anderson, *Org. Lett.*, 1999, **1**, 1057.
13. NA Porter, DM Scott, B Lacher, B Giese, HG Zeitz and HJ Lindner, *J. Am. Chem. Soc.*, 1989, **111**, 8311.
14. G Kneer and J Mattay, *Tetrahedron Lett.*, 1992, **33**, 8051.
15. T Morikawa, Y Washio, M Shiro and T Taguchi, *Chem. Lett.*, 1993, 249.
16. N Mase, S Wake, Y Watanabe and T Toru, *Tetrahedron Lett.*, 1998, **39**, 5553.
17. N Mase, Y Watanabe and T Toru, *Tetrahedron Lett.*, 1999, **40**, 2797.
18. M Nishida, E Ueyama, H Hayashi, Y Ohtake, Y Yamaura, E Yanaginuma, O Yonemitsu, A Nishida and N Kawahara, *J. Am. Chem. Soc.*, 1994, **116**, 6455.
19. M Nishida, H Hayashi, Y Yamaura, E Yanaginuma and O Yonemitsu, *Tetrahedron Lett.*, 1995, **36**, 269.
20. H Nagano, S Toi and T Yajima, *Synlett*, 1999, 53.
21. MP Sibi, J Ji, JB Sausker and CP Jasperse, *J. Am. Chem. Soc.*, 1999, **121**, 7517.
22. H Miyabe, K Fujii and T Naito, *Org. Lett.*, 1999, **1**, 569.
23. A Hayen, R Koch and JO Metzger, *Angew. Chem. Int. Ed.*, 2000, **39**, 2758.
24. NA Porter, JH Wu, G Zhang and AD Reed, *J. Org. Chem.*, 1997, **62**, 6702.
25. IJ Rosenstein and TA Tynan, *Tetrahedron Lett.*, 1998, **39**, 8429.
26. H Miyabe, C Ushiro, M Ueda, K Yamakawa and T Naito, *J. Org. Chem.*, 2000, **65**, 176.
27. H Nagano, T Hirasawa and T Yajima, *Synlett*, 2000, 1073.

28. Y Watanabe, N Mase, R Furue and T Toru, *Tetrahedron Lett.*, 2001, **42**, 2981.
29. MP Sibi and J Chen, *J. Am. Chem. Soc.*, 2001, **123**, 9472.
30. MP Sibi and H Hasegawa, *Org. Lett.*, 2002, **4**, 3347.
31. EJ Enholm, S Lavieri, T Cordova and I Ghiviriga, *Tetrahedron Lett.*, 2003, **44**, 531.
32. F Villar and P Renaud, *Tetrahedron Lett.*, 1998, **39**, 8655.
33. C Andres, JP Duque-Soladana and R Pedrosa, *J. Org. Chem.*, 1999, **64**, 4273 see also 4282.
34. M Matsugi, K Gotanda, C Ohira, M Suemura, A Sano and Y Kita, *J. Org. Chem.*, 1999, **64**, 6928.
35. W Wang, M-H Xu, X-S Lei and G-Q Lin, *Org. Lett.*, 2000, **2**, 3773.
36. K Kim, S Okamoto and F Sato, *Org. Lett.*, 2001, **3**, 67.
37. C Ericsson and L Engman, *Org. Lett.*, 2001, **3**, 3459.
38. D Yang, S Gu, Y-L Yan, N-Y Zhu and K-K Cheung, *J. Am. Chem. Soc.*, 2001, **123**, 8612.
39. EJ Enholm, JS Cottone and F Allais, *Org. Lett.*, 2001, **3**, 145.
40. L Giraud and P Renaud, *J. Org. Chem.*, 1998, **63**, 9162.
41. C Imboden, T Bourquard, O Corminboeuf, P Renaud, K Schenk and M Zahouily, *Tetrahedron Lett.*, 1999, **40**, 495.
42. C Imboden, F Villar and P Renaud, *Org. Lett.*, 1999, **1**, 873.
43. DP Curran, W Liu and CH Chen, *J. Am. Chem. Soc.*, 1999, **121**, 11012.
44. R Gosain, AM Norrish and ME Wood, *Tetrahedron Lett.*, 1999, **40**, 6673.
45. D Yang, X-Y Ye and M Xu, *J. Org. Chem.*, 2000, **65**, 2208.
46. D Yang, M Xu and M-Y Bian, *Org. Lett.*, 2001, **3**, 111.
47. EJ Enholm and A Bhardawaj, *Tetrahedron Lett.*, 2003, **44**, 3763.

– 11 –

Free Radicals Related to Biology

Molecular oxygen and related oxygen-centered radicals were mentioned in Chapter 1 as typical free radicals in living bodies. However, other radicals such as carbon- and nitrogen-centered radicals also exist in living bodies. Vitamin B_{12} is the most typical.

11.1 VITAMIN B_{12}

Vitamin B_{12}, the coenzyme, is sometimes called cyanocobalamine, and is contained in the liver in ppm level. It is comprised of Co^{2+}, porphyrine, cyanide, and porphyrine-side chained nucleoside, and is a complicated complex, as can be seen in Figure 11.1. It is well known that the first total synthesis of vitamin B_{12} was carried out by Woodward. Vitamin B_{12} plays a variety of important roles such as rearrangement of C–C bonds (1,2-acyl transfer) and rearrangement of C–N bonds, through radical species. Eq. 11.1 shows a typical acyl-transfer between glutamic acid and

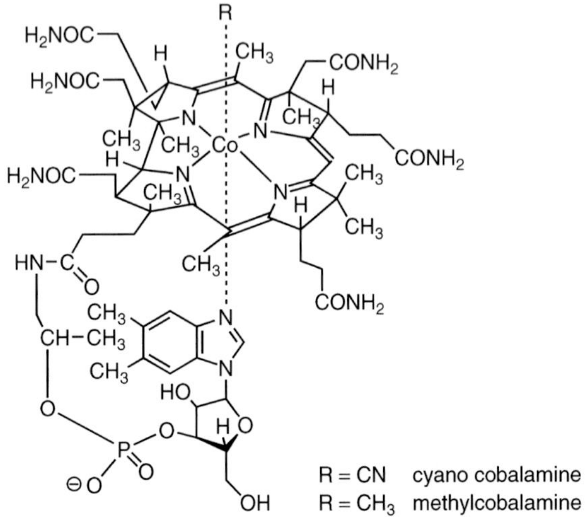

R = CN cyano cobalamine
R = CH_3 methylcobalamine

Figure 11.1 Structure of vitamin B_{12}.

β-methylaspartic acid.

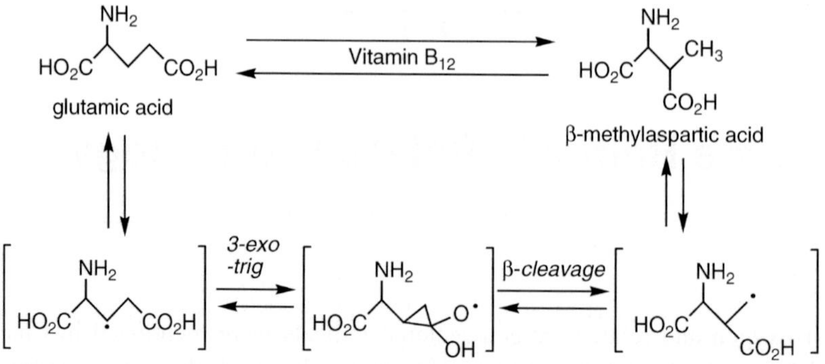

(11.1)

This reaction mechanism is similar to the chain-extension reaction of β-ketoester to γ-ketoester, and ring-expansion reaction of cyclic β-ketoester to cyclic γ-ketoester, through the formation of the corresponding cyclopropanoxyl radicals (3-*exo-trig*), followed by the β-cleavage reactions, as mentioned in Chapter 3. It is the formation of cyclopropanoxyl radicals that is the key point. Eq. 11.2 shows a model reaction of vitamin B_{12} with Co^{3+} complex (**A**). Reduction of the Co^{3+} complex (**A**) with $NaBH_4$ or an electrochemical method generates a Co^{1+} complex that has high nucleophilicity (super nucleophile). This super-nucleophilic Co^{1+} complex reacts with alkyl bromide (RBr) (**1**) bearing a cyclohexenone group, to form R–Co complex. Once it is formed, R˙ is generated under thermal or irradiation conditions, followed by thermodynamically stable 6-*endo-trig* cyclization to give bicyclic ketone (**2**) [1].

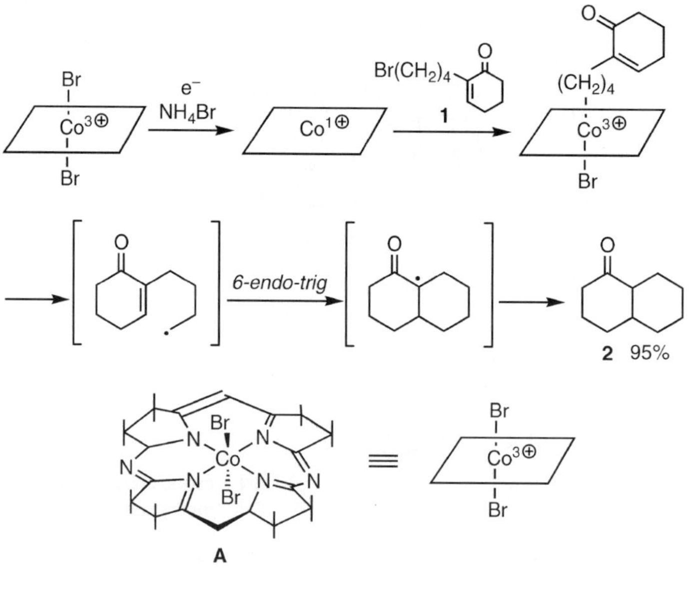

(11.2)

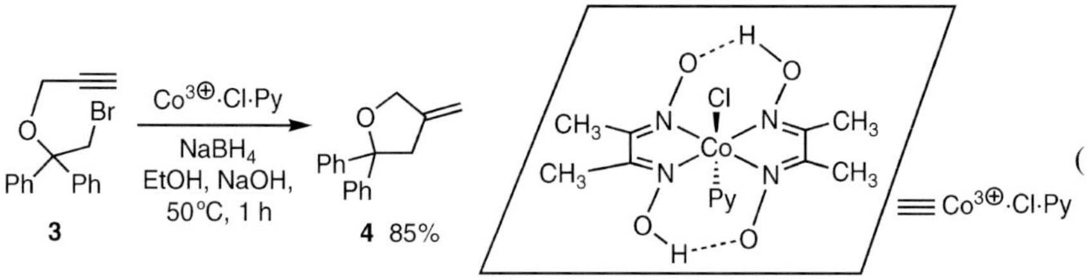

(11.3)

B

The same radical cyclization can be performed using a more simplified vitamin B_{12} model, such as *bis*(dimethylglyoximato)(pyridine)Cobalt chloride: $Co^{3+}\cdot Cl^{-}\cdot Py$ complex (**B**), as shown in eq. 11.3. Treatment of propargyl β-bromoethyl ether (**3**) with $NaBH_4$ and a catalytic amount of the cobalt complex (**B**) provides β-*exo*-methylene tetrahydrofuran via 5-*exo-dig* manner [2–7].

As a model study of methyl cobalamine (methyl transfer) in living bodies, a methyl radical, generated by the reduction of the *bis*(dimethylglyoximato)(pyridine)Co^{3+} complex to its Co^{1+} complex, reacts on the sulfur atom of thiolester via S_H2 to generate an acyl radical and methyl sulfide. The formed methyl radical can be trapped by TEMPO or activated olefins [8–13]. As a radical character of real vitamin B_{12}, the addition of zinc to a mixture of alkyl bromide (**5**) and dimethyl fumarate in the presence of real vitamin B_{12} at room temperature provides a C–C bonded product (**6**), through the initial reduction of Co^{3+} to Co^{1+} by zinc, reaction of Co^{1+} with alkyl bromide to form R–Co bond, its homolytic bond cleavage to form an alkyl radical, and finally the addition of the alkyl radical to diethyl fumarate, as shown in eq. 11.4 [14].

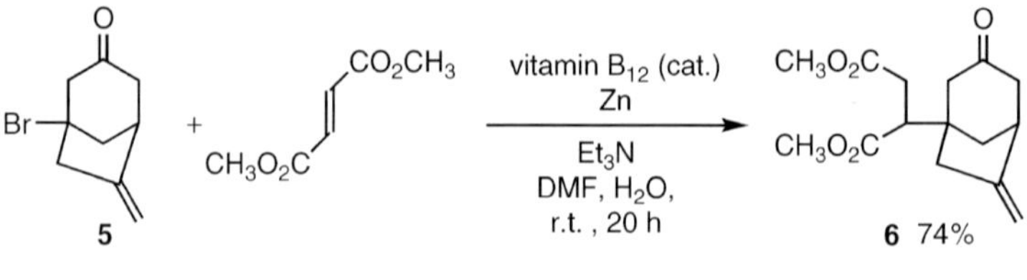

(11.4)

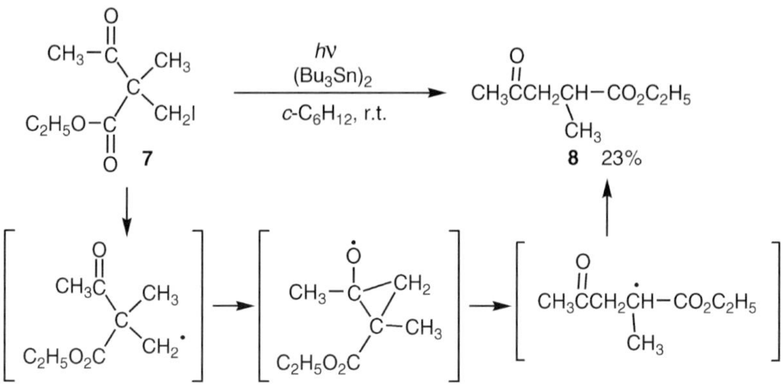

(11.5)

Vitamin B_{12} also plays a role in 1,2-acyl transfer. A related model reaction is shown in eq. 11.5. Photolytic treatment of methyl α-methyl-α-iodomethylacetoacetate (**7**) with $(Bu_3Sn)_2$ generates the corresponding α-substituted methyl radical. The formed α-substituted methyl radical cyclizes onto the carbonyl group to form a cyclopropoxyl radical derivative via 3-*exo-trig* manner, then its β-cleavage occurs to form the stable α-ester radical. This is the 1,2-acetyl transfer from the quaternary carbon to a primary carbon. This then, is the model experiment in the 1,2-acyl transfer with vitamin B_{12} [15].

11.2 ENE–DIYNE REACTIONS: BERGMAN CYCLIZATION

The ene–diyne reaction (Bergman cyclization) chemistry described below is not part of the biologically related radical reactions. However, after the successive discovery of

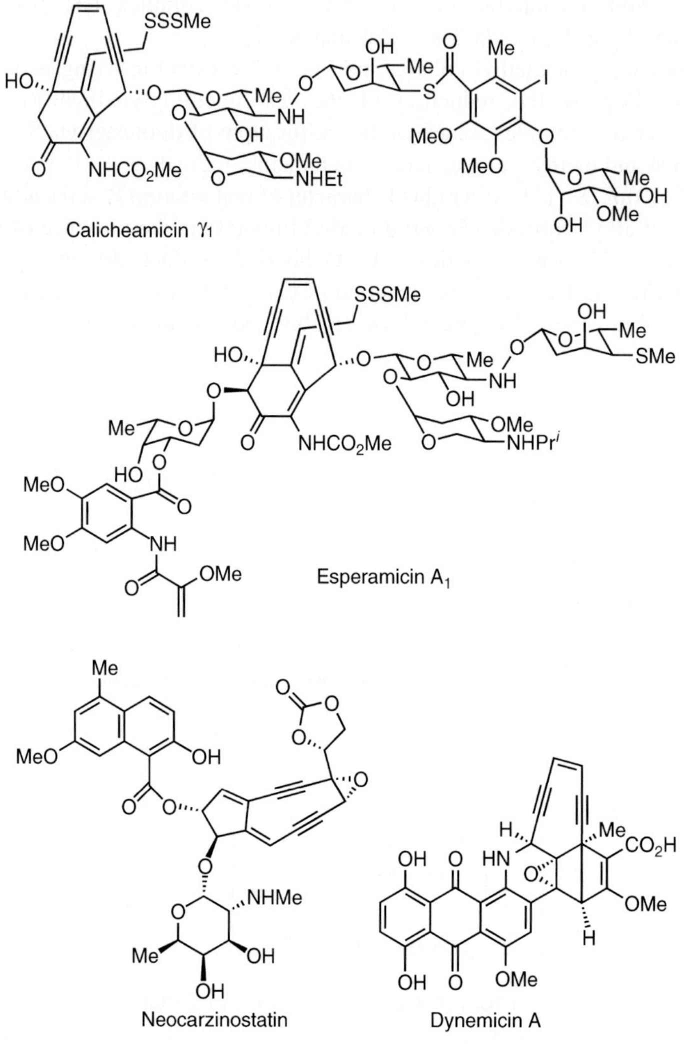

Calicheamicin γ_1

Esperamicin A_1

Neocarzinostatin

Dynemicin A

Figure 11.2 Natural ene–diyne products.

natural ene–diyne products in 1985, many natural and unnatural ene–diynes have been prepared, because most natural ene–diyne products do have potent antitumor and cytotoxicity activities (antibiotics), and their structures are really beautiful, systematic, and artistic. The structures of esperamicins, dynemicins, calicheamicins, and neocarzinostatins are shown in Figure 11.2. These compounds bear an aromatic group and a sugar group for DNA recognition site. The aromatic group acts as an intercalation and the sugar group acts as molecular recognition via hydrogen bonding. The essential role of these ene–diynes is the formation of two reactive sp^2 carbon-centered radicals and their hydrogen atom abstraction from substrates. Their potent biological activity has been ascribed to DNA damage resulting from hydrogen atom abstraction from the sugar backbone by the sp^2 carbon-centered biradical species formed through Bergman cyclization of the ene–diyne upon triggering [16–23]. The mechanism of the Bergman cyclization and hydrogen atom abstraction from the sugar backbone is shown in eq. 11.6. The *p*-phenylene biradical abstracts hydrogen atoms from the 4- and/or 5-position of the sugar moiety in nucleosides to generate an sp^3 carbon-centered radical, which further reacts with molecular oxygen. The formed peroxide reacts with glutathione RSH (tripeptide bearing L-cysteine), and induces O–C bond cleavage via hydrolysis to form the aldehyde and the phosphate groups. These ene–diynes induce DNA cleavage. The driving force of these reactions is the conversion of rather reactive sp^2 carbon-centered radicals to sp^3 carbon-centered radicals. This is a reflection of the difference of bond dissociation energies between $C(sp^2)$–H (approximately 112 kcal/mol) and $C(sp^3)$–H (91 ~ 98 kcal/mol). The hydrogen atom abstraction by the sp^2 carbon-centered radical to form the sp^3 carbon-centered radical is an exothermic reaction, and ΔH is approximately −15 kcal/mol.

(11.6)

Figure 11.3 shows the result on the direct observation by NMR of the interaction between duplex DNA and esperamicin. Thus, the planer aglycone group **E** in esperamicin intercalates between C7′···G2 and C6′···G3 of the duplex DNA, and the

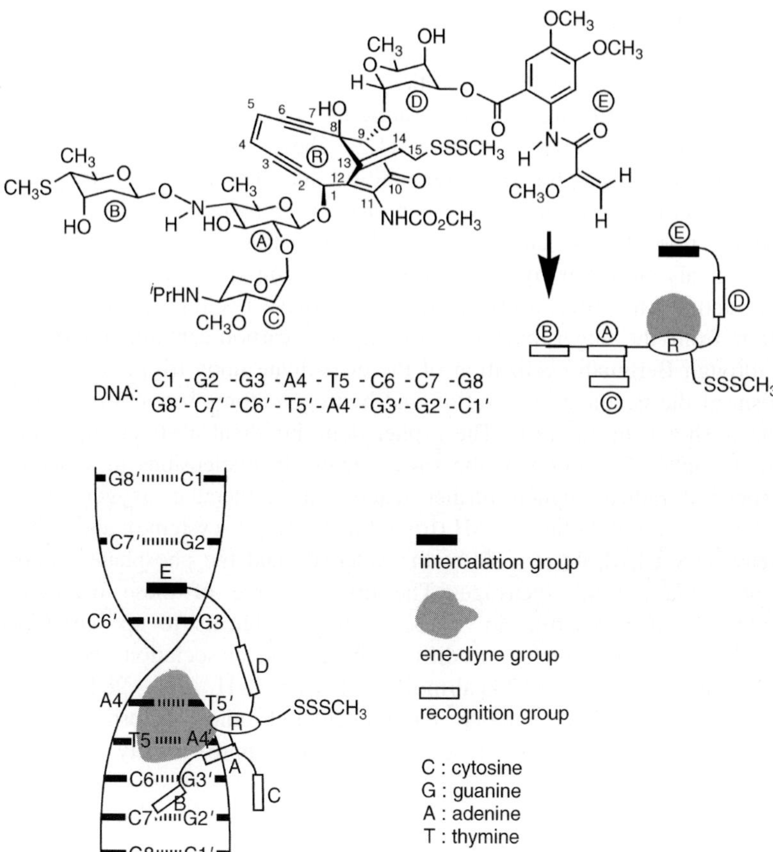

Figure 11.3 Interaction between DNA and esperamicin.

C_3 position of the ene–diyne group **R** in esperamicin is close to the sugar group $C_{1'}$–H in cytosine of the duplex DNA, about 2.1 Å distance. Once the Bergman cyclization occurs, both hydrogen atoms of $C_{5'}$–H in cytosine C6 and sugar group $C_{6'}$–H in cytosine C6′ are abstracted [24]. It is a really artistic structure and an intelligent reaction mechanism, and it is also surprising that microorganisms produce these beautiful compounds in their living bodies.

unstable —(1,4-cyclohexadiene)→ [ ] →

$E_a = 32$ kcal/mol

1,4-cyclohexadiene

E_a = 23.8 kcal/mol

(11.7)

1,4-cyclohexadiene

E_a = 25.1 kcal/mol

Now we turn to the Bergman cyclization with simple ene–diynes. Eq. 11.7 shows the thermal Bergman cyclizations and their activation energies E_a in the presence of 1,4-cyclohexdiene, as a hydrogen atom donor, to generate benzene derivatives [25–34]. Thus, E_a in the most simple *cis*-hex-3-ene-1,5-diyne is 32 kcal/mol, and the rate constant at 156 °C is approximately $10^{-4}\ s^{-1}$. As the alkyl group or benzo group is introduced into the basic structure, E_a in the Bergman cyclization is decreased. Generally, the rate-determining step in the Bergman cyclization is the cyclization step via 6-membered transition state, to generate a *p*-phenylene biradical. Once this *p*-phenylene biradical is formed, it rapidly abstracts hydrogen atoms from other molecules, since the sp^2 carbon-centered radical is extremely reactive. The formed *p*-benzophenylene biradical can be trapped by TEMPO (2,2,6,6-tetramethylpiperidine-1-oxyl radical) to generate a radical coupling product, followed by O–N homolytic bond cleavage to provide 1,4-naphthoquinone as shown eq. 11.8 [35].

•O–N

PhCl
150 °C

9

O–N

O–N

–2 •N

10 65%

(11.8)

An ene–diyne compound (**C**) bearing a pyrene group, a large π planar group, can intercalate into the space between the base pairs of the DNA (Figure 11.4). When this ene–diyne is irradiated in the presence of DNA plasmid, this large π planar group intercalates into the space between the base pairs of DNA, and the Bergman cyclization occurs to destroy the DNA plasmid through hydrogen atom abstraction [36].

The following are examples of other generation methods of the same kind of reactive sp^2 carbon-centered radicals. Treatment of aromatic diazocarboxylate ester (**11**) at pH 7.2 forms the phenyl radical, through hydrolysis of the ester, decarboxylation to the phenyldiimide, and finally, reaction with molecular oxygen (eq. 11.9a). Electron transfer reduction of 1,4-diazonium (**12**) with Cu^+ generates the corresponding *p*-phenylene biradical (probably step-by-step formation) (eq. 11.9b). These simple sp^2 carbon-centered radicals also destroy DNA plasmid at pH 7.6, under living-body conditions, like esperamicin [37–39].

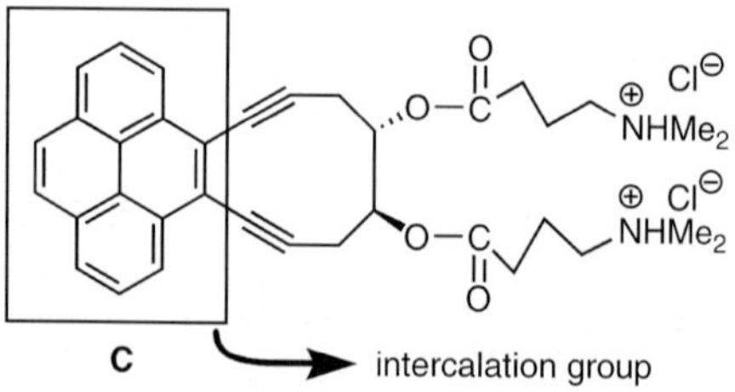

C → intercalation group

Figure 11.4

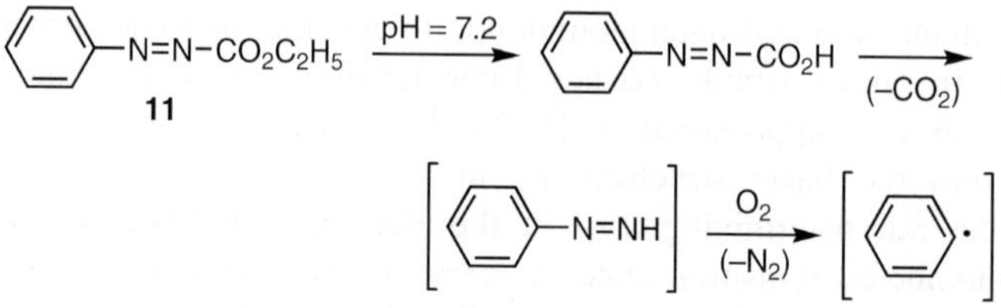

(11.9a)

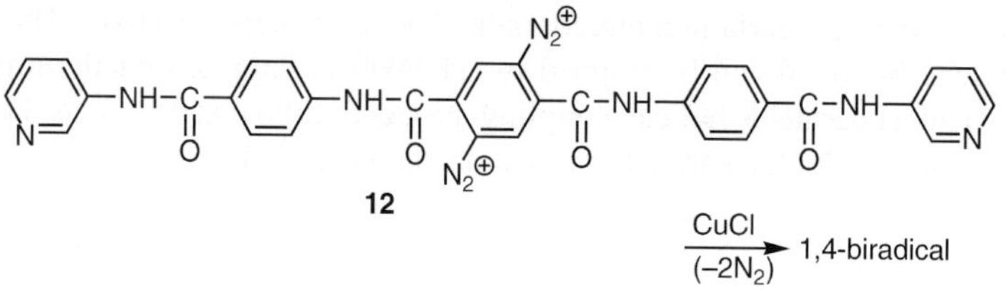

(11.9b)

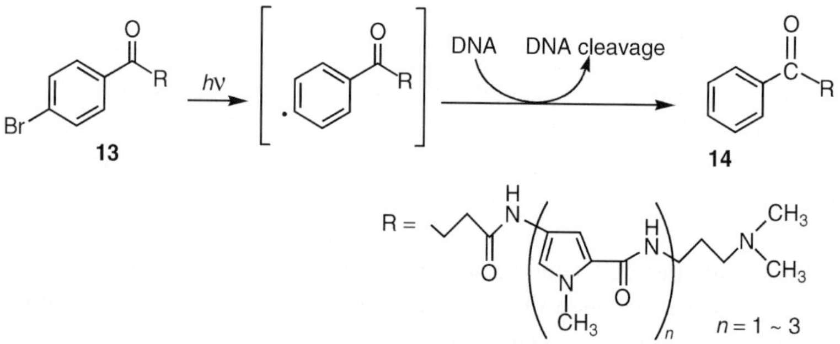

(11.10)

By the photolytic treatment of *p*-bromoacetophenone derivative (**13**) in the presence of supercoiled DNA, the formed phenyl radical derivative abstracts a hydrogen atom from the sugar moieties of DNA to induce the DNA destruction [40]. Another reactive hydrogen atom-abstraction species is superoxide anion radical, $O_2^{-\cdot}$ which is formed by the single electron reduction of molecular oxygen. The superoxide anion radical also abstracts a hydrogen atom from DNA sugar moieties. Figure 11.5 shows the iron–molecular oxygen complex of bleomycin (**D**), and fullerenes (**E**), (**F**) bearing side-chain carboxylic acid. The fullerene carboxylic acid displays inhibitory activity toward various

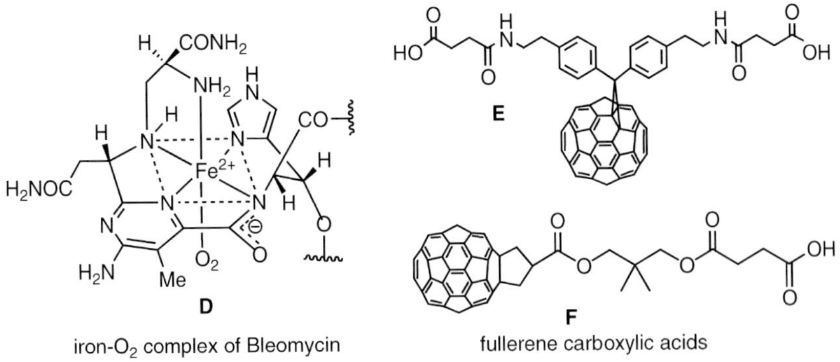

Figure 11.5

enzymes, cytotoxicity against tumor cells, and DNA cleaving activity, under irradiation with visible light, through the formation of a superoxide anion radical [41–43].

11.3 RADICAL 1,2-ACYLOXY TRANSFER

Alkyl or aryl 1,2-transfer reactions like the Wagner–Meerwein rearrangement and the Stevens rearrangement are well known in polar reactions. 1,2-Acyloxy rearrangement, on the other hand, is recognized in radical reactions. This radical 1,2-acyloxy rearrangement is interesting and important; probably, it is related to the evolutionary progress from D-ribose to 2-deoxy-D-ribose and RNA to DNA. The most popular method for generation of the initial β-acyloxyalkyl radical is the reaction of β-acyloxyalkyl halide (**15**) with $Bu_3Sn^{\cdot}$ initiated by AIBN, as shown in eq. 11.11a, where the rearranged radical then abstracts a hydrogen atom from Bu_3SnH to complete the rearrangement. The driving force of radical 1,2-acyloxy rearrangement is the formation of a more stable β-acyloxyalkyl radical [44–56]. That is, therefore, rearrangement from a primary alkyl radical to a tertiary alkyl radical, and from an alkyl radical to a benzylic radical. This 1,2-acyloxy rearrangement of β-acyloxyethyl radical can be applied to the β-phosphate ethyl radical, β-nitrate ethyl radical, and β-sulfonate ethyl radical. Thus, treatment of α-phenyl-β-bromoethyl phosphate (**16**), β-bromobenzocyclopentyl nitrate (**18**), and α-phenyl-β-bromoethyl sulfonate (**20**) with Bu_3SnH in the presence of AIBN under benzene refluxing conditions generates the corresponding radicals. These radicals rearrange to the corresponding benzylic radicals via radical-1,2-phosphate, -1,2-nitrate, and -1,2-sulfonate rearrangement, like 1,2-acyloxy rearrangement to provide the corresponding rearranged-reduction products (**17**), (**19**), and (**21**) in good yields, respectively (eqs. 11.12a–11.12c). Here again, the driving force of the 1,2-rearrangement is the formation of more stable benzylic radicals, through a thermodynamically controlled pathway.

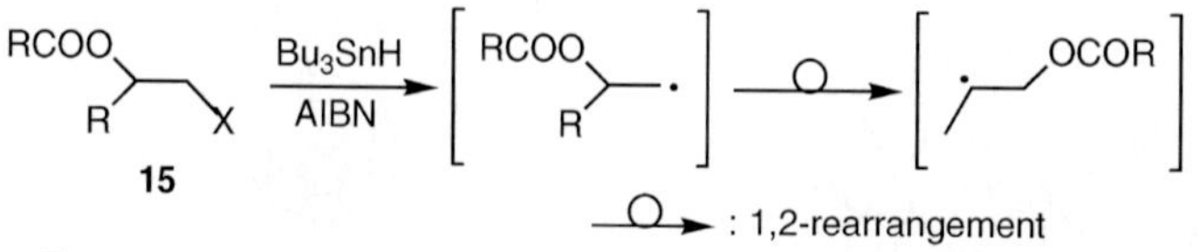

(11.11a)

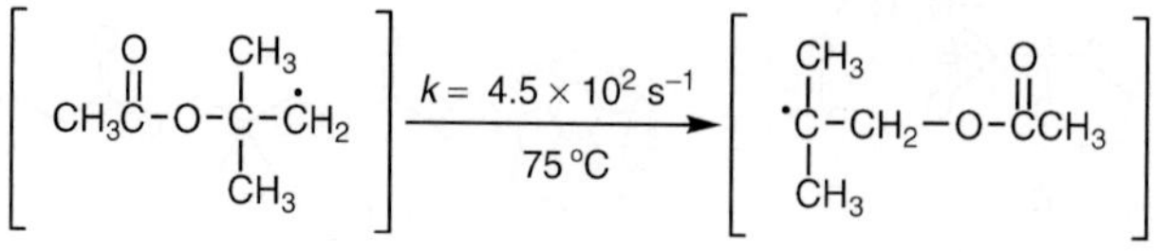

(11.11b)

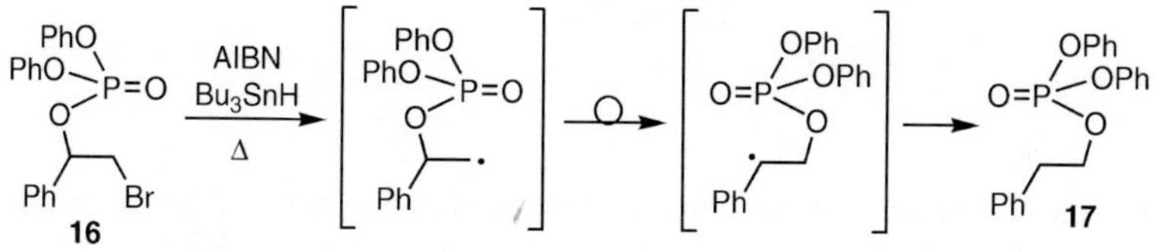

(11.12a)

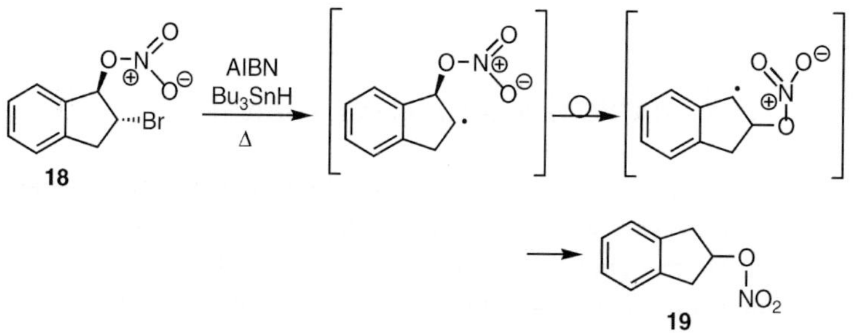

(11.12b)

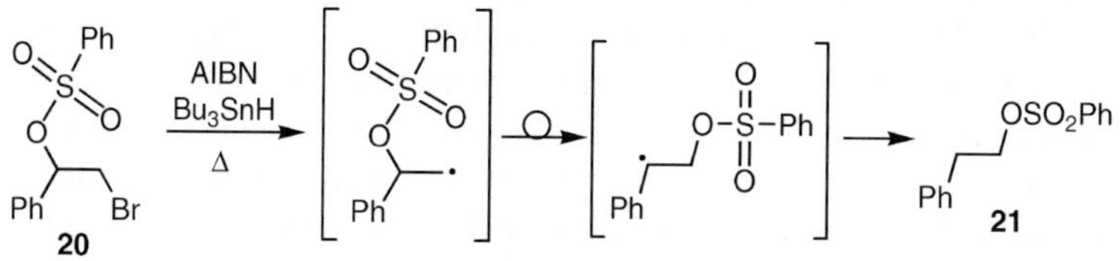

(11.12c)

Generally, radical 1,2-acyloxy rearrangement proceeds through the 5-membered transition state as shown in eq. 11.13. The rate constant for 1,2-acyloxy transfer is about $10^2\ s^{-1}$ at 75 °C (eq. 11.11b). Treatment of 1-bromo-3,4,6-tri-*O*-acetyl-2-*O*-benzoyl-[^{18}O]-D-glucose (**22**) with Bu_3SnH in the presence of AIBN under benzene refluxing conditions generates 1,2-benzoate rearranged product (**23**), in which the ^{18}O atom is directly bonded to the glucosyl anomer carbon. This indicates that the 1,2-rearrangement proceeds via 5-membered transition state [**I**]. What is its driving force? The initially formed radical, 3,4,6-tri-*O*-acetyl-2-*O*-benzoyl-D-1-glucosyl radical, adopts $B_{2,5}$ conformation (boat form), and the rearranged 3,4,6-tri-*O*-acetyl-1-*O*-benzoyl-D-2-glycosyl radical adopts a more stable $^4C_1$ conformation (chair form), so the driving force comes from the conformational stabilization effect [57–64].

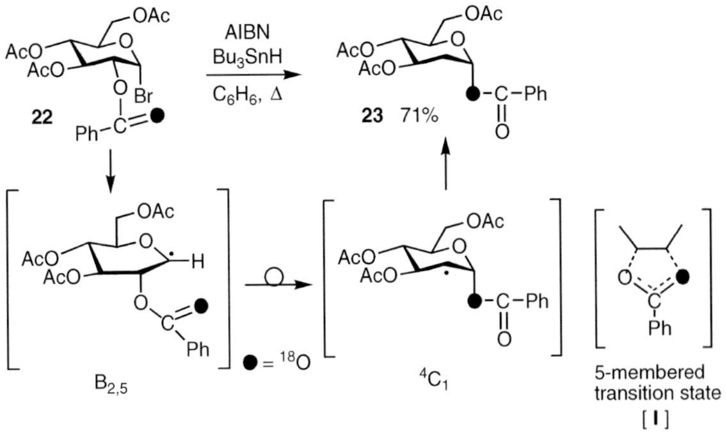

(11.13)

Experimental Procedure 1 (eq. 11.13).

To a refluxing benzene (80 ml) solution of 3,4,6-tri-*O*-acetyl-2-*O*-benzoyl-[^{18}O]-α-D-glucopyranosyl-1-bromide (1.8 g, 3.8 mmol) under an argon atmosphere was added dropwise a benzene (20 ml) solution of Bu_3SnH (4.8 mmol) and AIBN (80 mg) over 8 h. After the reaction, the solvent was removed and the residue was recrystallized from a mixture of *t*-butyl methyl ether and hexane (1:1) to provide 3,4,6-tri-*O*-acetyl-1-*O*-benzoyl-[^{18}O]-α-D-glucose in 71% yield [58].

Experimental Procedure 2.

To a benzene (80 ml) solution of 2,3,4,6-*tetra-O*-acetyl-α-D-glucopyranoyl-1-bromide (8.22 g, 20 mmol) was added dropwise a benzene (14 ml) solution of Bu_3SnH (7.0 g, 24 mmol) and AIBN (0.41 g, 2.5 mmol) over 10 h under an argon atmosphere. After the reaction, the solvent was removed and the residue was chromatographed on silica gel to give 1,3,4,6-*tetra-O*-acetyl-2-deoxy-α-D-arabinohexapyranose in 80% yield [59].

Instead of Bu_3SnH, treatment of 2,2,2-trichloroethyl carboxylate (**24**) with CuCl (SET agent) in the presence of bipyridyl (1:1) generates α-chlorovinyl carboxylate (**25**) through the same radical 1,2-acyloxy transfer, as shown in eq. 11.14.

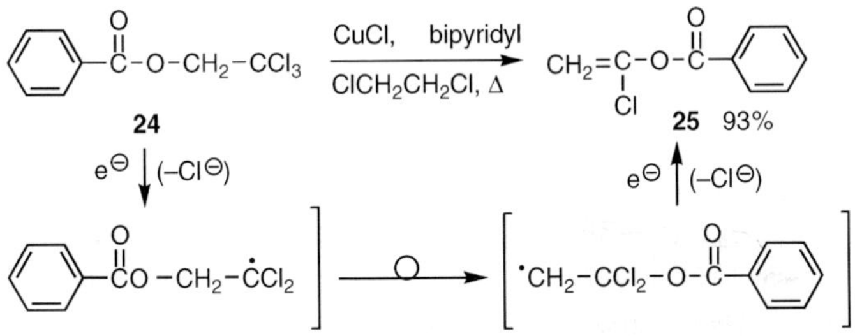

(11.14)

Related reactions are outlined in eqs. 11.15 and 11.16. Eq. 11.15 shows the reductive conversion of bicyclic 6-membered carbamate (**26**) to 5-membered bicyclic carbamate (**27**) through radical 1,2-acyloxy transfer. Eq. 11.16 shows the radical conversion of 1,3,2-dioxaphosphepane (**28**) to 1,3,2-dioxaphosphorinane (**29**) with retention of configuration at the phosphorus atom. These reactions proceed through the generation of carbon-centered radicals, heterolytic (polar) bond cleavage of β-acyloxy and β-phosphate groups, polar cyclization to form stable benzylic radicals, and finally, hydrogen atom abstraction from Bu_3SnH.

Et_3B, Bu_3SnH, $C_6H_5CH_3$ Δ; **26** → **27** 70% (11.15)

Bu_3SnH, AIBN, cat. PhSeH, C_6H_6, 80°C; **28** → **29** (11.16)

An interesting feature of radical 1,2-acycloxy rearrangement is probably reflected in the conversion of D-ribose to 2-deoxy-D-ribose, and D-glucose to 2-deoxy-D-glucose [65]. Thus, it seems that DNA derived from RNA.

The 4′-nucleotide radical is a very important species in biology. $O_2^{\cdot-}$ (or HOO˙) and $^1O_2$ play important roles in human immunity for defense against viruses. However, these reactive radicals also induce cancer, since they rapidly react with DNA, RNA, and peptides. So, these oxygen-centered radical species not only kill viruses, but also induce cancer, depending on the environment or field of these radicals generated. An important intermediate in this process is the formation of a 4′-deoxyribonucleotide radical, as shown in eq. 11.17.

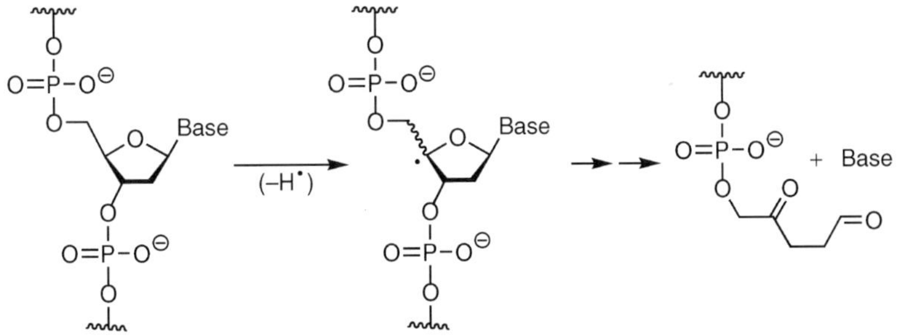

(11.17)

Once a 4′-nucleotide radical is formed, it induces direct fragmentation via β-cleavage as mentioned above, or reacts with molecular oxygen. The following eq. 11.18, is a model reaction, and the rate constant for β-cleavage of the 4′-nucleotide radical [**II**] is about $10^8\ s^{-1}$, which is rather rapid. The 4′-nucleotide radical [**II**] is formed by the addition of a thiyl radical to 4′-*exo*-methylene nucleotide (**30**). β-Cleavage of the 4′-nucleotide radical is the key step in DNA cleavage. Therefore, when an aqueous solution of oligonucleotide (**31**) is irradiated with a mercury lamp for 1 h under aerobic conditions, cleavage product (**32**) is predominantly obtained. This model study indicates that the 4′-nucleotide radical [**III**] cleaves off the oligomeric phosphates at the 3′- or 5′-positions. In the RNA model, the same cleavage was also observed [66–75].

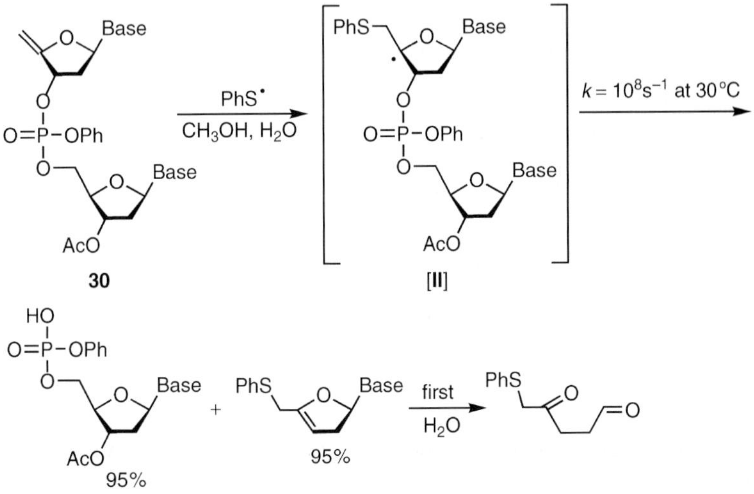

(11.18)

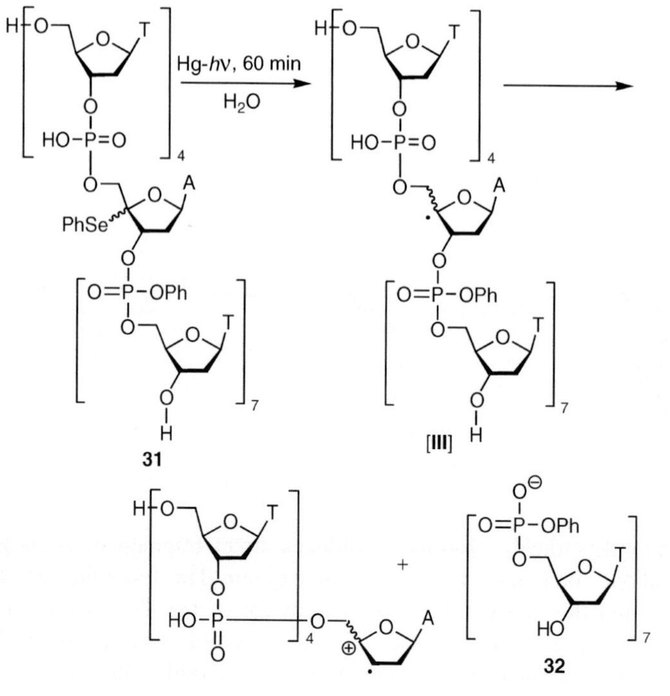

(11.19)

REFERENCES

1. R Scheffold, M Dike, S Dike, T Herold and L Walder, *J. Am. Chem. Soc.*, 1980, **102**, 3642.
2. M Okabe and M Tada, *Chem. Lett.*, 1980, 831.
3. M Okabe, M Abe and M Tada, *J. Org. Chem.*, 1982, **47**, 1775.
4. VF Patel, G Pattenden and JJ Russell, *Tetrahedron Lett.*, 1986, **27**, 2833.
5. JE Baldwin and CS Li, *Chem. Commun.*, 1987, 166.
6. B Giese, P Erdmann, T Gobel and R Springer, *Tetrahedron Lett.*, 1992, **33**, 4545.
7. M Tada, *Rev. Heteroatom Chem.*, 1999, **20**, 97.
8. VF Patel and G Pattenden, *Chem. Commun.*, 1987, 871.
9. VF Patel and G Pattenden, *Tetrahedron Lett.*, 1988, **29**, 707.
10. BP Branchaud, MS Meier and Y Choi, *Tetrahedron Lett.*, 1988, **29**, 167.
11. M Tada, T Hirokawa and T Tohma, *Chem. Lett.*, 1991, 857.
12. TM Brown, CJ Cooksey, D Crich, AT Dronsfield and R Ellis, *J. Chem. Soc., Perkin Trans. 1*, 1993, 2131.
13. H Huang and CJ Forsyth, *Tetrahedron Lett.*, 1993, **34**, 7889.
14. GA Kraus, TM Siclovan and B Watson, *Synlett*, 1995, 201.
15. WM Best, APF Cook, JJ Russell and DA Widdowson, *J. Chem. Soc. Perkin Trans. 1*, 1986, 1139.
16. K Edo, M Mizuyaki, Y Koide, H Seto, K Furihata, N Otake and N Ishida, *Tetrahedron Lett.*, 1985, **26**, 331.
17. MD Lee, TS Dunne, MM Siegel, CC Chang, GO Morton and DB Borders, *J. Am. Chem. Soc.*, 1987, **109**, 3464.
18. MD Lee, TS Dunne, GA Ellestad, MM Siegel, CC Chang, GO Morton, WJ McGahren and DB Borders, *J. Am. Chem. Soc.*, 1987, **109**, 3466.

19. N Zein, AM Sinha, WJ McGahren and GA Ellestad, *Science*, 1988, **240**, 1198.
20. N Zein, M Poncin, R Nilakantan and GA Ellestad, *Science*, 1989, **244**, 697.
21. M Konishi, H Ohkuma, K Saitoh, H Kawaguchi, J Golik, G Dubay, G Groenewold, B Krishnan and TWJ Doyle, *J. Antibiot.*, 1985, **38**, 1605.
22. M Konishi, H Ohkuma, K Matsumoto, T Tsuno, H Kamei, T Miyaki, T Oki, H Kawaguchi, GD Duyne and J Clardy, *J. Antibiot.*, 1989, **42**, 1449.
23. H Strittmatter, A Dussy, U Schwitter and B Giese, *Angew. Chem. Int. Ed.*, 1999, **38**, 135.
24. N Ikemoto, RA Kumar, PC Dedon, SJ Danishefsky and DJ Patel, *J. Am. Chem. Soc.*, 1994, **116**, 9387.
25. RR Jones and RG Bergman, *J. Am. Chem. Soc.*, 1972, **94**, 660.
26. TP Lockhart, CB Mallon and RG Bergman, *J. Am. Chem. Soc.*, 1980, **102**, 5976.
27. KC Nicolaou, Y Ogawa, G Zuccarello and H Kataoka, *J. Am. Chem. Soc.*, 1988, **110**, 7247.
28. KC Nicolaou, G Zuccarello, Y Ogawa, EJ Schweiger and T Kumazawa, *J. Am. Chem. Soc.*, 1988, **110**, 4866.
29. JW Grissom, TL Calkins, HA McMillen and Y Jiang, *J. Org. Chem.*, 1994, **59**, 5833.
30. DL Boger and J Zhou, *J. Org. Chem.*, 1993, **58**, 3018.
31. CS Kim and KC Russell, *Tetrahedron Lett.*, 1999, **40**, 3835.
32. JW Grissom and TL Calkins, *J. Org. Chem.*, 1993, **58**, 5422.
33. MF Semmelhack, T Neu and F Foubelo, *J. Org. Chem.*, 1994, **59**, 5038.
34. T Kaneko, M Takahashi and M Hirama, *Tetraheron Lett.*, 1999, **40**, 2015.
35. JW Grissom and GU Gunawardena, *Tetrahedron Lett.*, 1995, **36**, 4951.
36. RL Funk, ERR Young, RM Williams, MF Flanagan and TL Cecil, *J. Am. Chem. Soc.*, 1996, **118**, 3291.
37. DJ Jebaratnam, S Kugabalasooriar, H Chen and DP Arya, *Tetrahedron Lett.*, 1995, **36**, 3123.
38. DP Arya and DJ Jebaratnam, *Tetrahedron Lett.*, 1995, **36**, 4369.
39. R Braslau and MO Anderson, *Tetrahedron Lett.*, 1998, **39**, 4227.
40. PA Wender and R Jeon, *Org. Lett.*, 1999, **1**, 2117.
41. M Otsuka, H Satake and Y Sugiura, *Tetrahedron Lett.*, 1993, **34**, 8497.
42. SH Friedman, DL Decamp, RP Sijbesma, G Srdanov, F Wudl and GL Kenyon, *J. Am. Chem. Soc.*, 1992, **115**, 6506.
43. E Nakamura, H Tokuyama, S Yamago, T Shiraki and Y Sugiura, *Bull. Chem. Soc. Jpn.*, 1996, **69**, 2143.
44. LRC Barclay, J Lusztyk and KU Ingold, *J. Am. Chem. Soc.*, 1984, **106**, 1793.
45. S Saebo, ALJ Beckwith and L Radom, *J. Am. Chem. Soc.*, 1984, **106**, 5119.
46. ALJ Beckwith and PJ Duggan, *Chem. Commun.*, 1988, 1000.
47. ALJ Beckwith, L Radom and S Saebo, *J. Am. Chem. Soc.*, 1984, **106**, 5119.
48. LRC Barclay, J Lusztyk and KU Ingold, *J. Am. Chem. Soc.*, 1984, **106**, 1793.
49. D Crich and GF Filzen, *Tetrahedron Lett.*, 1993, **34**, 3225.
50. D Crich and Q Yao, *J. Am. Chem. Soc.*, 1993, **115**, 1165.
51. D Crich and Q Yao, *Tetrahedron Lett.*, 1993, **34**, 5677.
52. D Crich and GF Filzen, *J. Org. Chem.*, 1995, **60**, 4834.
53. ALJ Beckwith, D Crich, PJ Duggan and Q Yao, *Chem. Rev.*, 1997, **97**, 3273.
54. D Crich, X Huang and M Newcomb, *Org. Lett.*, 1999, **1**, 225.
55. D Crich, K Ranganathan and X Huang, *Org. Lett.*, 2001, **3**, 1917.
56. AL Williams, TA Grillo and DL Comins, *J. Org. Chem.*, 2002, **67**, 1972.
57. HG Korth, R Sustmann, J Dupris and B Giese, *J. Chem. Soc., Perkin Trans. II*, 1986, 1453.
58. HG Korth, R Sustmann, KS Groninger, M Leisung and B Giese, *J. Org. Chem.*, 1988, **53**, 4364.
59. B Giese, KS Groninger, T Witzel, HG Korth and R Sustmann, *Angew. Chem. Int. Ed. Engl.*, 1987, **26**, 233.
60. B Giese, S Gilges, KS Groninger, C Lamberth and T Witzel, *Liebigs Ann. Chem.*, 1988, 615.
61. T Gimisis, G Ialongo, M Zamboni and C Chatgilialoglu, *Tetrahedron Lett.*, 1995, **36**, 6781.
62. D Crich, F Sartillo-Piscil, L Quintero-Cortes and DJ Wink, *J. Org. Chem.*, 2001, **67**, 3360.
63. D Crich, F Sartillo-Piscil, L Quintero-Cortes and DJ Wink, *J. Org. Chem.*, 2002, **67**, 3360.
64. RN Ram and NK Meher, *Org. Lett.*, 2003, **5**, 145.
65. A Koch, C Lamberth, F Wetterich and B Giese, *J. Org. Chem.*, 1993, **58**, 1083.

66. D Schulter-Frohlinde and C Sonntag, *Isr. J. Chem.*, 1972, **10**, 1139.
67. B Giese, J Burger, TW Kang, C Kesselheim and T Wittmer, *J. Am. Chem. Soc.*, 1992, **114**, 7322.
68. B Giese, X Beyrich-Graf and J Burger, *Angew. Chem. Int. Ed. Engl.*, 1993, **32**, 1472.
69. B Giese, P Erdmann, L Giraud, T Göbel, M Petretta, T Schäfer and M von Raumer, *Tetrahedron Lett.*, 1994, **35**, 2683.
70. S Peukert, R Batra and B Giese, *Tetrahedron Lett.*, 1997, **38**, 3507.
71. F Wackernagel, U Schwitter and B Giese, *Tetrahedron Lett.*, 1997, **38**, 2657.
72. B Giese, A Dussy, C Elie, P Erdmann and U Schwitter, *Angew. Chem. In. Ed. Engl.*, 1994, **33**, 1861.
73. D Crich and X-C Mo, *J. Am. Chem. Soc.*, 1997, **119**, 249.
74. H Togo, W He and Y Waki, *Synlett*, 1998, 700.
75. H Strittmatter, A Dussy, U Schwitter and B Giese, *Angew. Chem. Int. Ed.*, 1998, **38**, 135.

– 12 –

Free Radicals for Green Chemistry

Green chemistry means environmentally friendly organic synthesis. The essential aims are to reduce the amounts of dangerous, toxic starting materials and byproducts (waste disposal), and to reduce damage to the natural environment. Thus, green chemistry is the study to remove these risks fundamentally during the preparation and isolation of chemical materials, based on molecular chemistry; it is not, therefore, the treatment of symptoms [1]. What it does do is substitute solvents and reagents with safer ones.

In the previous chapters, Bu_3SnH has been used as a typical and useful radical reagent in a benzene solvent. Generally, radical reactions with Bu_3SnH initiated by AIBN, proceed effectively in benzene, which bears a conjugated π-system. Probably, the formed radicals are somewhat stabilized through the SOMO–LUMO or SOMO–HOMO interaction between the radicals and benzene.

Occasionally, it may be required to study the fundamental radical reactions with organotins in benzene. However, the use of radical reactions with such toxic reagents and solvents cannot be considered in the chemical and pharmaceutical industries, even if the results in terms of organic synthesis are excellent and effective. Even in a fundamental study, it might not be wise to use 1 g of Bu_3SnH and 5 ml of benzene. Hence, radical chemists should develop new and less toxic radical reagents and reaction media in order to reduce the damage to nature.

Fortunately, radicals are a neutral species in general. Thus, they are not affected by the various kinds of solvents (reaction media), i.e. protic polar solvents such as ethanol and water, aprotic polar solvents such as acetonitrile, dimethyl sulfoxide, and non-polar solvents such as hexane and benzene. Moreover, radicals are not affected fundamentally by basic species or acidic species. Radical reactions should take place not only in benzene, but also in water, and proceed not only in 1 N aqueous HCl solution, but also in 1 N aqueous NaOH solution. This is the fundamental character of radicals and radical reactions, and is a great advantage; an advantage that should be reflected in green chemistry.

12.1 DESIGN OF FREE RADICAL PRECURSORS

Bu_3SnH is rather expensive, highly toxic, and has an unpleasant odor; furthermore, it is rather difficult to remove organotin species completely from the reaction mixture. These factors mean that reactions with organotin species cannot be used in pharmaceutical companies. Therefore at this juncture, we introduce the study on the improvement of radical reactions with organotin reagents.

Bis(*tri-n*-butylstannyl)benzopinacolate (**A**) generates $Bu_3Sn^{\cdot}$ under heating conditions through the homolytic C–C bond cleavage of compound (**A**), followed by β-cleavage of the formed radical as shown in eq. 12.1. $Bu_3Sn^{\cdot}$ is formed without a radical initiator such as AIBN.

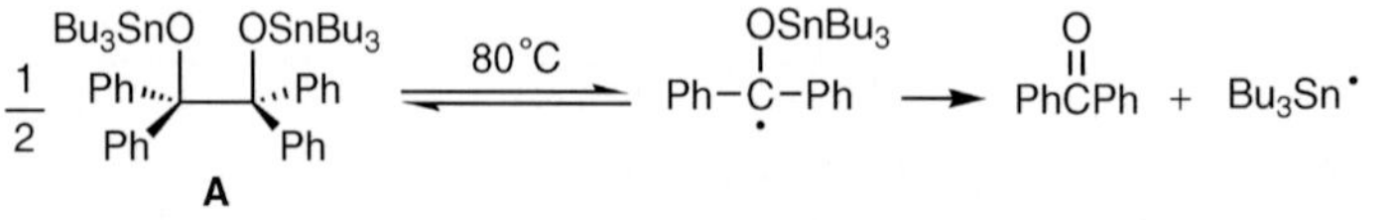

$$\frac{1}{2}\ \mathbf{A} \xrightleftharpoons{80^{\circ}C} Ph{-}\dot{C}(OSnBu_3){-}Ph \longrightarrow PhC(=O)Ph + Bu_3Sn^{\bullet} \tag{12.1}$$

$$\mathbf{1} \xrightarrow[\text{AIBN, }\Delta]{\text{10 mol\% }Bu_3SnCl\text{, PMHS, aq. KF}} \mathbf{2}\ \ 55\% \tag{12.2}$$

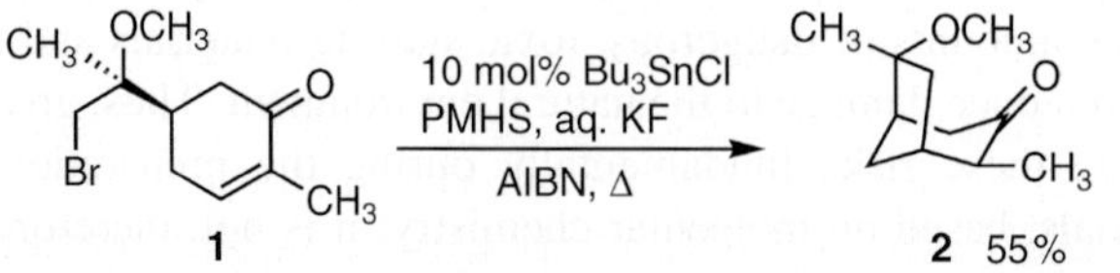

Bu_3SnX can be reduced by $NaBH_4$ or hydrosilane. Thus, radical reactions with a catalytic amount of Bu_3SnCl in the presence of $NaBH_4$ or PMHS (polymethylhydrosiloxane) can be used for the same reactions as use $Bu_3Sn\underline{H}$ (eq. 12.2). The Bu_3SnCl/PMHS system does not reduce any carbonyl groups [2–4] (Figure 12.1).

In order to simplify the removal operation of organotin species from the reaction mixture, organotin reagent (**B**) bearing a pyridyl group was prepared. After the radical reaction with reagent (**B**), the organotin species can be removed by treatment with 6 M hydrochloric acid solution. However, reagent (**B**) is not particularly stable and has to be

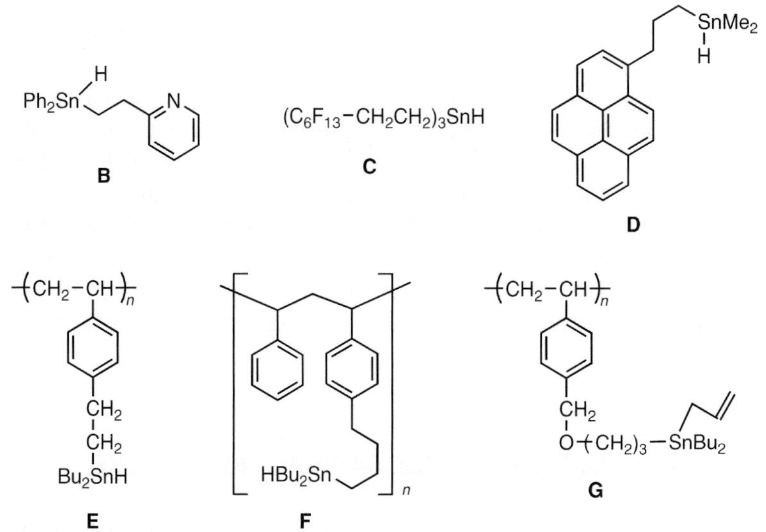

Figure 12.1

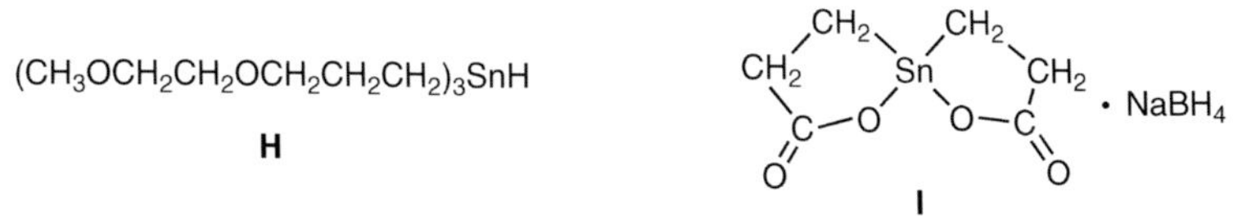

$(CH_3OCH_2CH_2OCH_2CH_2CH_2)_3SnH$

H

$\cdot\ NaBH_4$

I

Figure 12.2

used within two days [5]. *Tris*(2-perfluorohexylethyl)tin hydride (**C**) is soluble in $C_6H_5CF_3$ and has no unpleasant odor like Bu_3SnH, while $C_6H_5CF_3$ is insoluble in general organic solvents. Thus, after the radical reactions with reagent (**C**) in $C_6H_5CF_3$, the organotin species derived from reagent (**C**) can be removed from the product by ether extraction of the reaction mixture [6–8]. An advantage of the method is the easy separation of the tin species from the reaction mixture. Pyrene-supported dimethyltin hydride (**D**) can be used for easy isolation of the product by the adsorption of the pyrene-supported tin species with activated carbon [6–8]. Moreover, polystyrene-supported di-*n*-butyltin hydrides (**E**) and (**F**), polystyrene-supported allyldibutyltin (**G**) are regenerative and reusable organotin reagents, and can be used for the reduction and allylation of halides and xanthates, respectively [9–14]. However, these reactions still use toxic organotin species.

As water soluble organotin reagents, *tris*[3-(2-methoxyethoxy)propyl]tin hydride (**H**) and bicyclic tin reagent (**I**) can reduce alkyl halides in water [15, 16]. The solubility of the former reagent in water is about 30 mM, and the latter reagent requires $NaBH_4$ to generate a tin hydride species bearing two carboxylates in water (Figure 12.2).

This reaction in water appears to be attractive and environmentally friendly, since we do not need to use toxic solvents. However, if the water-soluble organotin species can not be recovered completely from the water used as a reaction medium, it becomes more toxic and dangerous than Bu_3SnH.

12.2 APPLICATION OF FREE RADICALS TO ENVIRONMENTALLY BENIGN SYNTHESIS

Against this background, *tris*(trimethylsilyl)silane, $(Me_3Si)_3SiH$ (**J**) was developed as an alternative to Bu_3SnH. Bond dissociation energy of Si–H in $(Me_3Si)_3SiH$ is about 79 kcal/mol, while that of Sn–H in Bu_3SnH is about 74 kcal/mol. This indicates that $(Me_3Si)_3SiH$ is less reactive than Bu_3SnH. However in practice, the former reagent shows the same reactivity as the latter, and can be used for the reduction of halides, chalcogenides, xanthates, ketones, and C–C bond formations [17]. It is commercially available (Figure 12.3).

Similarly, heptamethyltrisilane (**K**), heptamethyltrisilanethiol (**L**), 5,10-dihydrodisilanthracene (**M**) also reduce organic halides and xanthates, though the reactivities are decreased and they cannot be used for C–C bond formation [18, 19]. Reagent (**N**) is water-soluble monosilane and it can reduce sugar halides and halo-nucleosides in water [20]. However, the reactivity is low and it can only be used for the reduction of organic

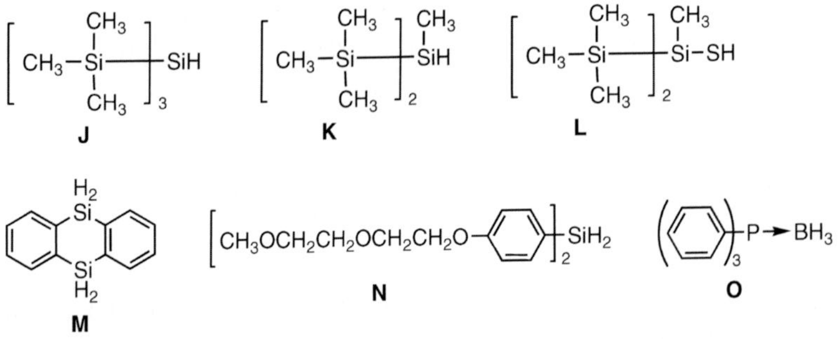

Figure 12.3

halides. Treatment of organic halides and xanthates with a triphenylphosphine–borane complex (**O**) in the presence of peroxide generates the corresponding reduction products respectively, and here the reactive species is a triphenylphosphine–boranyl radical, $Ph_3P{-}BH_2^{\cdot}$ [21–27]. The amine–borane complex/peroxide system can also be used for the same radical reduction of organic halides.

Generally, potassium persulfate in the presence of Ag^+ is used for the Hunsdiecker type radical decarboxylation of carboxylic acids in water. $(Bu_4N^+)_2S_2O_8^{2-}$ (**P**) is soluble in THF, and a sulfate anion radical [**I**] is formed under refluxing conditions. Thus, refluxing treatment of *bis*(tetrabutylammonium) persulfate (**P**) in the presence of alcohol in THF provides tetrahydrofuryl-protected alcohol (**4**), through the abstraction of α-H from THF by sulfate anion radical [**I**], followed by oxidation to a tetrahydrofuryl cation, as shown in eq.12.3 [28].

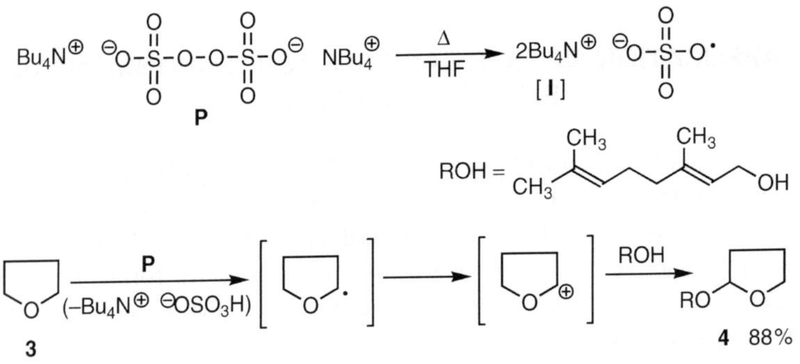

(12.3)

...$(Me_3Si)_3SiH$ is a less stable oil and is oxidized under aerobic conditions, while 1,1,2,2-...henyldisilane (TPDS) (**Q**) is stable crystal under aerobic conditions, and remains ...even when left on a table for 6 months. TPDS can be used for the reduction of ...halides, chalcogenides, and xanthates, and can be used for C–C bond formations in ethanol or ethyl acetate [29–34]. Now it is commercially available and three examples of reactions in ethanol are shown in eq. 12.4.

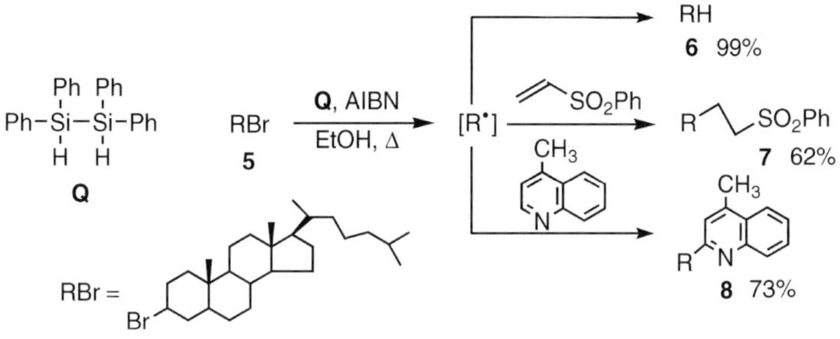

(12.4)

Experimental Procedure 1 (eq. 12.4).

A mixture of 3-bromocholestane (0.3 mmol), phenyl vinyl sulfone (0.9 mmol), TPDS (0.75 mmol), and AIBN (0.15 mmol) in ethanol (3.6 ml) was refluxed for 14 h under an argon atmosphere. After the reaction, the solvent was removed and the residue was purified by column chromatography on silica gel to give the reductive addition product in 62% yield [30].

Experimental Procedure 2 (eq. 12.4).

To a mixture of 1-bromoadamantane (0.3 mmol), TPDS (0.45 mmol), and caffeine salt (1.5 mmol) with camphorsulfonic acid was added AIBN (0.45 mmol) over 8 h (5 times in 2 h intervals). After 4 h, TPDS (0.45 mmol) was added again. After a total of 22 h, sat. aq. $NaHCO_3$ solution was added and the mixture was extracted with ethyl acetate. The organic layer was dried over Na_2SO_4. After removal of the solvent, the residue was chromatographed on silica gel to provide 2-(1-adamantyl)caffeine in 55% yield [30].

TPDS can be also used for the preparation of biaryl through the intramolecular free radical *ipso*-substitution of *N*-(2-bromoaryl)arenesulfonamide (**9**) and radical ring expansion of α-haloalkyl cyclic β-keto ester (**11**) (eqs. 12.5a and 12.5b).

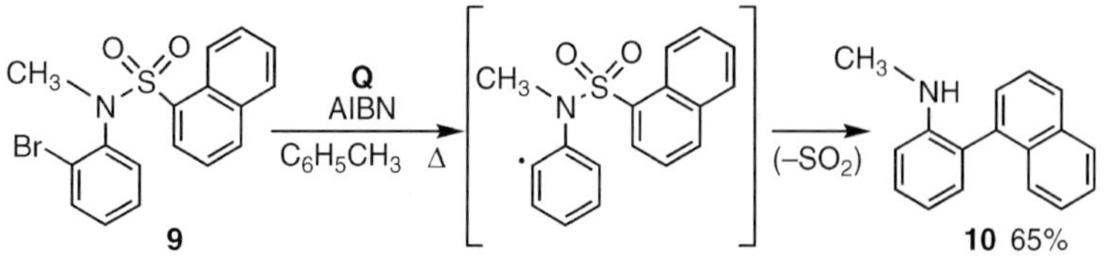

(12.5a)

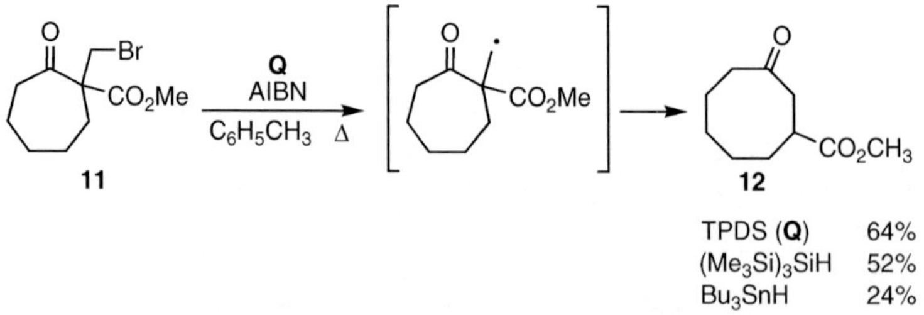

(12.5b)

TPDS (**Q**)	64%
$(Me_3Si)_3SiH$	52%
Bu_3SnH	24%

Hypophosphorus acid, H_3PO_2 in the presence of AIBN and a base (amine or $NaHCO_3$) can reduce organic halides and xanthates in dioxane or aqueous alcohols. Eq. 12.6a to 12.6d shows the reduction of xanthate, thionocarbonate, and iodoarene, and 5-*exo-trig* cyclization respectively [35–43].

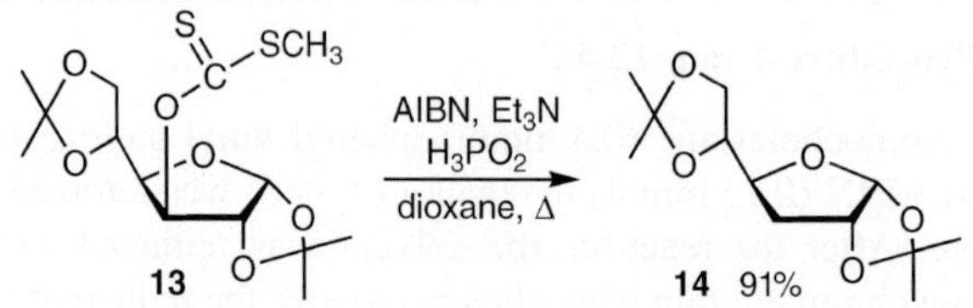

(12.6a)

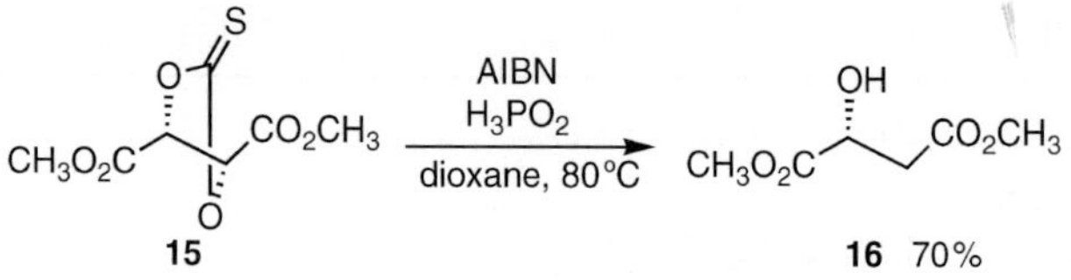

(12.6b)

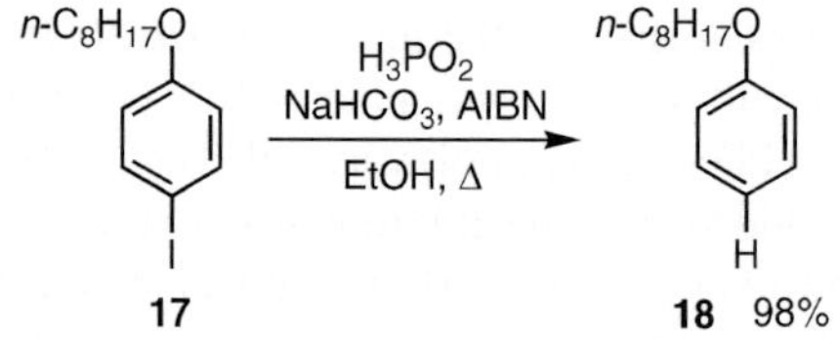

(12.6c)

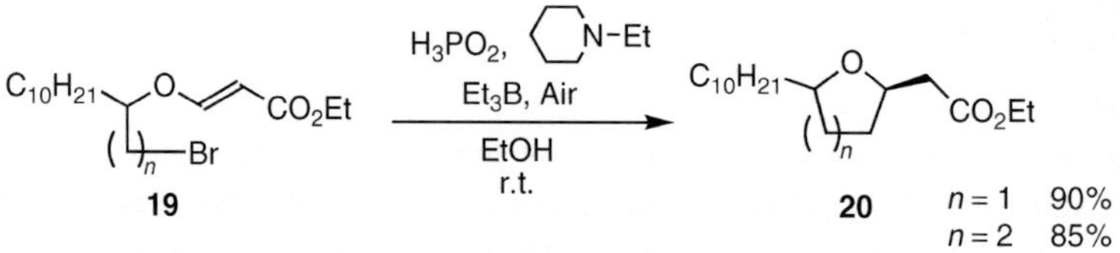

(12.6d)

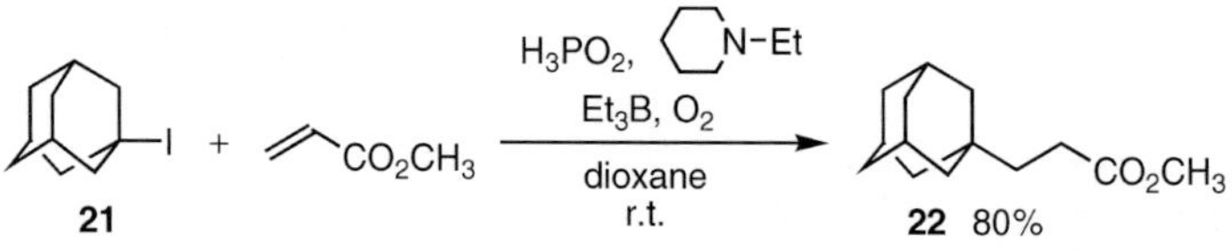

(12.6e)

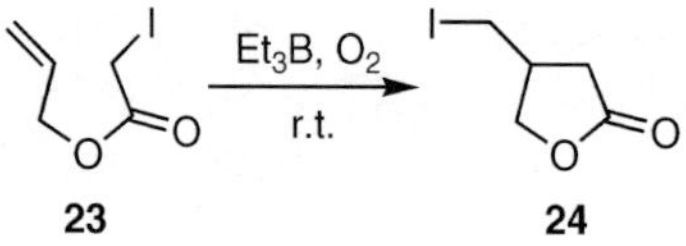

solvent	Yield
$CH_3(CH_2)_4CH_3$	0%
C_6H_6	0%
H_2O 30 ml	67%
H_2O 100 ml	78%

(12.6f)

$$\underset{\mathbf{25}}{\mathrm{MeO_2C\diagup\!\!=NOBn}} \xrightarrow[\substack{H_2O\,:\,MeOH = 1:1 \\ 20\,^\circ C}]{i\text{-PrI, } Et_3B} \underset{\mathbf{26}\ 99\%}{\mathrm{MeO_2C\!-\!CH(}i\text{-Pr})\mathrm{\!-\!NHOBn}} \tag{12.6g}$$

Experimental Procedure 3 (eq. 12.6b).

A mixture of cyclic thionocarbonate (0.36 mmol), H_3PO_2 (50% aqueous solution, 0.13 ml, 1.26 mmol), and AIBN (12 mg) in dioxane (3 ml) was heated at 80 °C under an argon atmosphere for 30 min. After the reaction, dichloromethane was added to the reaction mixture. The obtained mixture was then washed with sat. aq. $NaHCO_3$ solution. The organic layer was dried over $MgSO_4$. After removal of the solvent, the residue was chromatographed on silica gel (eluent: hexane/ethyl acetate = 8/2) to give the reduction product in 70% yield [38].

Experimental Procedure 4 (eq. 12.6g).

To a mixture of glyoxylic oxime ether (0.26 mmol) in H_2O/MeOH (1/1, 10 ml) were added isopropyl iodide (7.8 mmol) and Et_3B (1 M hexane solution, 1.3 ml, 1.3 mmol) at 20 °C. After 2 h, the solvent was removed, and the residue was diluted with aqueous $NaHCO_3$ solution and then extracted with dichloromethane. The organic layer was dried over $MgSO_4$ and the residue was chromatographed on silica gel (eluent: hexane/ethyl acetate = 12/1) to give the adduct in 99% yield [40].

Treatment of adamantyl iodide (**21**) and methyl acrylate with H_3PO_2/*N*-ethylpiperidine in the presence of Et_3B induces a radical chain reaction to form the addition product (**22**) (eq. 12.6e).

Et_3B (cat.)-induced radical cyclization of allyl α-iodoacetate (**23**) readily happens in water to give the corresponding γ-lactone (**24**) in good yield (eq. 12.6f). Interestingly, γ-lactone is not formed in organic solvents such as hexane or benzene. This is an atom-transfer reaction, because there is no hydrogen atom donor. Under the same conditions, Et_3B-initiated reaction of glyoxylic oxime ether (**25**) and isopropyl iodide provides α-(*N*-benzyloxy)amino ester (**26**) (eq. 12.6g).

Sunlight irradiation (solar photochemical synthesis) of 1,4-naphthoquinone (**27**) in the presence of aldehyde (**28**) in a mixture of *t*-butanol and acetone gives a good yield of the corresponding acyl hydroquinone (**29**) through the abstraction of the formyl hydrogen atom of aldehyde by the excited triplet biradical derived from 1,4-naphthoquinone, followed by the reaction of the acyl radical with 1,4-naphthoquinone (eq. 12.7). Here, kg-scale preparation is possible with sunshine [44].

$$\mathbf{27} + \underset{\mathbf{28}}{CH_3CH_2CH_2CH{=}O} \xrightarrow[\text{t-BuOH, acetone}]{\text{Sunlight}} \mathbf{29}\ 90\% \tag{12.7}$$

$$\mathbf{30} \xrightarrow[\text{Solvent free}]{\text{Conditions A or B}} \mathbf{31} \quad \text{A } 68\%,\ \text{B } 73\% \tag{12.8a}$$

Condition **A** : NBS(1.05 eq.), $(PhCO_2)_2$(0.04 eq.), 90 ~ 95 °C
Condition **B** : NBS(1.2 eq.), AIBN(0.1 eq.), 60 ~ 65 °C

$$\mathbf{30} \xrightarrow[\text{[bmim]}PF_6]{\text{Conditions}} \mathbf{31} \tag{12.8b}$$

$[bmim]PF_6$: CH_3-N(imidazolium)$^{\oplus}$N-$C_4H_9{}^n$ $PF_6^{\ominus}$

Condition **A**:

Regeneration	0	1	2	3	4
Yield (%)	76	74	72	69	73

Condition **B**:

Regeneration	0	1	2	3	4
Yield (%)	79	79	78	78	78

Wohl–Ziegler bromination generally requires carbon tetrachloride as a solvent. However, Wohl–Ziegler bromination of a benzylic methyl group bearing various other functional groups on the aromatics, with *N*-bromosuccinimide (NBS) can be effectively carried out in a solvent-free system (eq. 12.8a) and in an ionic liquid (1-butyl-3-methylimidazolium hexafluorophosphate) system (eq. 12.8b) respectively [45]. The ionic liquid can be used repeatedly, maintaining high yields. Particularly in the solvent-free system, the crude product can be easily obtained from the filtrate without further purification, by the addition of ether to the reaction mixture for the removal of succinimide.

$Mn_2(CO)_{10}$ **R**; $Fe^{\oplus}$ $PF_6^{\ominus}$ (ferrocenium) **S**; $[CpFe(CO)_2]_2$ **T**

Figure 12.4

Experimental Procedure 5 (eq. 12.8a).

To a flask were added ethyl *p*-toluate (6 mmol), NBS (7.2 mmol), and AIBN (0.6 mmol) under an argon atmosphere. The mixture was well mixed and stirred for 2 min. Then, the mixture was heated at 60 °C for 1 h. After the reaction, ether was added to the mixture, and the formed succinimide was removed by filtration. The filtrate was evaporated and the residue was chromatographed on silica gel (eluent: hexane/ethyl acetate = 1/1) to provide ethyl *p*-(α-bromo)toluate in 73% yield [45].

Other organometallic radical reagents are shown in Figure 12.4, though environmental requirements are not satisfied. Dimanganese decacarbonyl (**R**), ferronium cation (**S**) containing Fe^{3+}, and an iron–carbonyl complex (**T**) reacts with alkyl halides to generate the corresponding alkyl radicals [46–48].

Finally, atom (group)-transfer reaction is also important for green chemistry, since the atom-transfer reaction does not lose any functional groups. Thus, those functional groups can be used for the next functional group conversion. One example is the thermal reaction of propargyl cyclohexenyl ether (**32**) with *Se*-phenyl selenosulfonate in the presence of AIBN (cat.). This is just a coupling reaction to form the addition-cyclization product (**33**) bearing phenylselenyl and *p*-toluenesulfonyl groups. By treatment of the addition-cyclization product (**33**) with *m*CPBA, the corresponding olefin (**34**) is obtained, as shown in eq. 12.9 [49–51].

AIBN, $TolSO_2SePh$, Δ; *m*CPBA; SO_2Tol; H; SePh

32 → **33** 54% → **34** 74% (12.9)

The study of green chemistry with free radical reactions has just begun; work on reagents, procedures, and reaction systems is in progress and not yet established. Therefore, to move toward green chemistry based on the advantages of free radical reactions, much hard work will need to be done in the future.

REFERENCES

1. PT Anastas and JC Warner, *Green Chemistry*, 1999.
2. DJ Hart and FL Seely, *J. Am. Chem. Soc.*, 1988, **110**, 1631.
3. DJ Hart, R Krishnamurthy, LM Pook and FL Seely, *Tetrahedron Lett.*, 1993, **34**, 7819.
4. I Terstiege and RE Maleczka, Jr., *J. Org. Chem.*, 1999, **64**, 342.
5. DLJ Clive and W Yang, *J. Org. Chem.*, 1995, **60**, 2607.
6. K Olofsson, SY Kim, M Larhed, DP Curran and A Hallberg, *J. Org. Chem.*, 1999, **64**, 4539.
7. DP Curran, S Hadida, SY Kim and Z Luo, *J. Am. Chem. Soc.*, 1999, **121**, 6607.
8. S Gastaldi and D Stien, *Tetrahedron Lett.*, 2002, **43**, 4309.
9. M Gerlach, F Jordens, H Kuhn, WP Neumann and M Peterseim, *J. Org. Chem.*, 1991, **56**, 5972.
10. WP Neumann and M Peterseim, *Synlett*, 1992, 801.
11. EJ Enholm and JP Schulte, II, *Org. Lett.*, 1999, **1**, 1275.

12. EJ Enholm, ME Gallagher, KM Moran, JS Lombardi and JP Schulte, II, *Org. Lett.*, 1999, **1**, 689.
13. AM Yim, Y Vidal, P Viallefont and J Martinez, *Tetrahedron Lett.*, 1999, **40**, 4535.
14. P Boussaguet, B Delmond, G Dumartin and M Pereyre, *Tetrahedron Lett.*, 2000, **41**, 3377.
15. J Light and R Breslow, *Tetrahedron Lett.*, 1990, **31**, 2957.
16. R Rai and DB Collum, *Tetrahedron Lett.*, 1994, **35**, 6221.
17. C Chatgilialoglu, *Acc. Chem. Res.*, 1992, **25**, 188.
18. T Gimisis, M Ballestri, C Ferreri and C Chatgilialoglu, *Tetrahedron Lett.*, 1995, **36**, 3897.
19. J Daroszewski, J Lusztyk, M Degneil, C Navarro and B Maillard, *Chem. Commun.*, 1991, 356.
20. O Yamazaki, H Togo, G Nogami and M Yokoyama, *Bull. Chem. Soc. Jpn*, 1997, **70**, 2519.
21. JA Baban and BP Roberts, *J. Chem. Soc. Perkin Trans 1*, 1984, 1717.
22. JA Baban and BP Roberts, *J. Chem. Soc. Perkin Trans. II*, 1986, 1607.
23. VPJ Marti and BP Roberts, *J. Chem. Soc. Perkin Trans. II*, 1986, 1613.
24. JA Baban and BP Roberts, *J. Chem. Soc. Perkin Trans. II*, 1988, 1195.
25. V Paul and BP Roberts, *J. Chem. Soc. Perkin Trans. II*, 1988, 1895.
26. PLH Mok and BP Roberts, *Tetrahedron Lett.*, 1992, **33**, 7249.
27. DHR Barton and M Jacob, *Tetrahedron Lett.*, 1998, **39**, 1331.
28. JC Jung, HC Choi and YH Kim, *Tetrahedron Lett.*, 1993, **34**, 3581.
29. O Yamazaki, H Togo and S Matsubayashi, *Tetrahedron Lett.*, 1998, **39**, 1921.
30. O Yamazaki, H Togo and S Matsubayashi, *Tetrahedron*, 1999, **55**, 3735.
31. O Yamazaki, H Togo and M Yokoyama, *J. Chem. Soc. Perkin Trans. 1*, 1999, 2891.
32. H Togo, S Matsubayashi, O Yamazaki and M Yokoyama, *J. Org. Chem.*, 2000, **65**, 2816.
33. A Ryokawa and H Togo, *Tetrahedron*, 2001, **57**, 5915.
34. M Sugi and H Togo, *Tetrahedron*, 2002, **58**, 3171.
35. DHR Barton, DO Jang and JC Jaszberenyi, *J. Org. Chem.*, 1993, **58**, 6838.
36. H Yorimitsu, K Wakabayashi, H Shinokubo and K Oshima, *Tetrahedron Lett.*, 1999, **40**, 519.
37. H Yorimitsu, H Shinokubo and K Oshima, *Chem. Lett.*, 2000, 104.
38. DO Jang and SH Song, *Tetrahedron Lett.*, 2000, **41**, 247.
39. K Wakabayashi, H Yorimitsu, H Shinokubo and K Oshima, *Bull. Chem. Soc. Jpn*, 2000, **73**, 2377.
40. H Miyabe, M Ueda and T Naito, *J. Org. Chem.*, 2000, **65**, 5043.
41. H Yorimitsu, H Shinokubo and K Oshima, *Bull. Chem. Soc. Jpn*, 2001, **74**, 225.
42. DO Jang, DH Cho and CM Chung, *Synlett*, 2001, 1923.
43. E Lee and HO Han, *Tetrahedron Lett.*, 2002, **43**, 7295.
44. C Schiel, M Oelgemöller and J Mattay, *Synthesis*, 2001, 1275.
45. H Togo and T Hirai, *Synlett*, 2003, 702.
46. BC Gilbert, W Kalz, CI Lindsay, PT McGrail, AF Parsons and DTE Whittaker, *Tetrahedron Lett.*, 1999, **40**, 6095.
47. U Jahn, *J. Org. Chem.*, 1998, **63**, 7130.
48. G Thoma and B Giese, *Tetrahedron Lett.*, 1989, **30**, 2907.
49. AC Serra and CMMS Correa, *Tetrahedron Lett.*, 1991, **32**, 6653.
50. JE Brumwell, NS Simpkins and NK Terrett, *Tetrahedron Lett.*, 1993, **34**, 1219.
51. JE Brumwell, NS Simpkins and NK Terrett, *Tetrahedron Lett.*, 1993, **34**, 1215.

Index